LEMENTS°

	IB	IIB	IIIA	IVA	VA	VIA	VIIA	0
								2 **He** 4.0026
			5 **B** 10.811	6 **C** 12.01115	7 **N** 14.0067	8 **O** 15.9994	9 **F** 18.9984	10 **Ne** 20.183
			13 **Al** 26.9815	14 **Si** 28.086	15 **P** 30.9738	16 **S** 32.064	17 **Cl** 35.453	18 **Ar** 39.948
8 **Ni** 71	29 **Cu** 63.54	30 **Zn** 65.37	31 **Ga** 69.72	32 **Ge** 72.59	33 **As** 74.9216	34 **Se** 78.96	35 **Br** 79.909	36 **Kr** 83.80
6 **d** .4	47 **Ag** 107.870	48 **Cd** 112.40	49 **In** 114.82	50 **Sn** 118.69	51 **Sb** 121.75	52 **Te** 127.60	53 **I** 126.9044	54 **Xe** 131.30
09	79 **Au** 196.967	80 **Hg** 200.59	81 **Tl** 204.37	82 **Pb** 207.19	83 **Bi** 208.980	84 **Po** (209)	85 **At** (210)	86 **Rn** (222)

6	64 **Gd** 157.25	65 **Tb** 158.924	66 **Dy** 162.50	67 **Ho** 164.930	68 **Er** 167.26	69 **Tm** 168.934	70 **Yb** 173.04	71 **Lu** 174.97
	96 **Cm** (247)	97 **Bk** (247)	98 **Cf** (251)	99 **Es** (254)	100 **Fm** (253)	101 **Md** (256)	102 **No** (253)	103 **Lw** (257)

heses.

DATE DUE

Aug 7 '70	Dec 15 78		
May 1 '71	Nov 29 79		
Oct 11 '71	Dec 7 '81		
Apr 24 '72			
Nov 15 '72			
Sep 3 '73			
Mar 11 '77			
May 16 '77			
Jul 6 '77			
Dec 13 '77			
Nov 1 78			
GAYLORD M-2			PRINTED IN U.S.A.

Chemical Separation Methods

John A. Dean
University of Tennessee

VAN NOSTRAND REINHOLD COMPANY

New York Cincinnati Toronto London Melbourne

Van Nostrand Reinhold Company Regional Offices:
Cincinnati, New York, Chicago, Millbrae, Dallas

Van Nostrand Reinhold Company Foreign Offices:
London, Toronto, Melbourne

Library of Congress Catalog Card Number 74-90773

Published by Van Nostrand Reinhold Company
450 West 33rd Street, New York, N. Y. 10001

Published simultaneously in Canada by
D. Van Nostrand Company (Canada), Ltd.

15 14 13 12 11 10 9 8 7 6 5 4 3 2 1

Preface

This text is an outgrowth of lectures which have been given by the author to members of his graduate analytical course since 1950 and, more recently, to senior chemistry majors. Classical gravimetric separation methods are not included; they have been admirably described over the decades. Nor are separation methods that are clearly physical in nature. Admittedly the exact classification as "chemical" or "physical" leads to arguments as a rather wide grey territory exists between clearly physical and clearly chemical methods. A short introductory chapter outlines the organization of the book and the interrelationship among the topics.

Individual chapters are designed, in general, to stand alone. Consequently, the order of presentation is not critical. It seems desirable, however, to introduce early the complex equilibria involving metals, masking and demasking reactions (Chapter 2). Liquid–liquid extractions logically precede the chromatographic methods. To emphasize the underlying unity among the several divisions of chromatographic methods, the general principles and theory common to all are treated first in Chapter 4. To aid further in emphasizing the similarity among the methods, a consistent set of symbols are employed. The unfortunate lack of uniformity among authors in regard to use of symbols certainly confuses the student and individual only just beginning to probe into chemical separation methods.

The great emphasis on gas chromatographic methods seems justified by the great importance of these methods and the almost universal application of these particular separation methods. Extensive coverage is also given to liquid–liquid extraction, ion exchange methods, ion exclusion methods, and electrophoresis.

The text is designed for a one-semester course consisting of three lectures per week and a single laboratory period. For a shorter course, the instructor could select Chapters 2 to 5, 10, and 11, plus possibly 12 and/or 14.

In style the book follows closely its companion volume on *Instrumental Methods of Analysis,* the fourth edition of which has been widely accepted by students and instructors throughout the world. Sample calculations show how the formulas given can be applied to practical cases. There are 222 numerical problems and questions, many with multiple parts. Many of the problems reflect data that would be accumulated in the course of laboratory exercises. Answers to all the problems are collected together and appear at the end of the book. Literature and collateral reading references are included in each chapter.

In the laboratory sections, the experiments are basic ones to apply the methods studied in the corresponding chapters. Only general and brief directions or suggestions are provided, rather than finished exercises for unvaried repetition, in the belief that the student should be encouraged to use his imagination and to vary operating conditions and experimental parameters.

The author is greatly indebted to the instrument and supply houses who have so generously furnished schematic diagrams, photographs and technical information of their products. Thanks are expressed also to many students who have struggled with the problems in their initial format and pointed out errors of omission and commission.

John A. Dean
Knoxville, Tennessee

Contents

APPENDICES

List of Experiments

Abbreviations

alternating current (adj.)	a-c
ampere	A
ampere-hour	A-hr
angstrom	Å
boiling point	b.p.
butyl	Bu
cubic centimeter	cm^3
curie	Ci
degrees Celsius (centigrade)	°C
degrees Kelvin (absolute)	°K
direct current (adj.)	d-c
electron volt	eV
ethylenediamine tetraacetic acid	EDTA
hour	hr
inch	in.
inside diameter	i.d.
kilocalorie	kcal
liter	liter
melting point	m.p.
meter	meter
microampere	μA
micron	μ
microgram	μg
milliampere	mA
milliequivalents	mEq
milliliter	ml
milligram	mg

nanogram	ng
ohm	Ω
percent extracted	%*E*
radio frequency	r-f
second	sec
versus	vs
volt	V

Symbols

A	area
A_M	cross-sectional area of mobile phase
A_S	cross-sectional area of solid or liquid surface
B	adduct species
C	concentration
C_M	concentration of solute in mobile phase
C_m	concentration of central metal ion
C_{max}	solute concentration in effluent at peak maximum
C_o	solute concentration in organic phase
C_S	solute concentration in stationary phase
D	distribution ratio
D_f	diffusion coefficient in liquid film
D_M	diffusion coefficient in mobile phase
D_r	diffusion coefficient within resin
D_S	diffusion coefficient in stationary phase
D_v	volume distribution ratio
d	distance traveled
d_p	particle diameter
E	potential difference
$E_H^{M/n}$	selectivity coefficient
e	electron (symbol); Naperian base (logarithms)
$\mathscr{F}$	faraday
F_c	flow rate of mobile phase (carrier gas)
f	solute–eluent parameter (adsorption chromatography)
G	Gibb's free energy
H	plate height
ΔH	enthalpy change
ΔH_s	heat of vaporization

$\bar{h}$	reduced plate height
i	current
j	pressure-gradient correction or compressibility factor
K	general constant
K'	conditional constant
K_a	acid dissociation constant
K_{av}	partition coefficient (exclusion chromatography)
K_d	partition coefficient; molar distribution coefficient
K_f	formation constant
K_2	dimerization constant
k	partition or capacity ratio (chromatography); stepwise stability constant
L	column length; ligand (complexation)
M	molar (concentration unit); central metal ion (complexation)
N	plate number
N_s	mole fraction
P_i	inlet column pressure
P_o	outlet column pressure
p	fraction of solute in mobile phase
p-	negative logarithm of (prefix)
$pH_{1/2}$	extraction pH when $D = 1$
Q	quantity; relative peak sharpness
Q_v	volume capacity
Q_w	weight capacity
q	fraction of solute in stationary phase
R	retardation value
R_f	retardation factor based on relative distances traveled
Rs	resolution
$\mathscr{R}$	gas constant
r	(subscript) denotes resin phase; radius
r°	heating rate (degrees Celsius)
r_{max}	tube containing maximum amount of solute
S	solvent (symbol); water content of resin in weight percent
S^o	adsorbate adsorption energy
S_r	solvent regain or uptake
T	temperature
T_b	temperature at boiling point
T_c	column temperature
T_R	retention temperature
t	time
t_M	time in mobile phase

t_R	retention time
t_S	time in stationary phase
u	chemical potential
V	volume
V_a	adsorbent surface volume
V_b	bed volume
V_e	elution volume (exclusion chromatography)
V_g	specific retention volume
V_i	inner volume of liquid in gel phase
V_M	volume of mobile phase
V_{max}	volume of effluent when peak emerges
V_N	net retention volume
V_o	void outer or interstitial volume
V_r	resin or gel volume
V_R	retention volume
V'_R	adjusted retention volume
V_S	volume of stationary phase
V_T	retention volume at temperature T
w	weight
w_L	weight of liquid phase
W	zone width at baseline
W_e	zone width at C_{max}/e or at 0.368 C_{max}
$W_{1/2}$	zone width at height/2 or at $0.368C_{max}$
X	field strength

1 *Introduction*

In all analytical methods one is interested in measuring a signal which can be related to the concentration of a particular species in the original sample. An ideal analytical method would enable a species to be determined directly on diverse matrices and with a limited quantity of sample. Unfortunately, few analytical methods are sufficiently selective, and the analyte is not always present in adequate concentration. Hence, a major problem in analysis is the isolation or preconcentration of a constituent to remove or minimize the ever-present possibility of interferences or to provide an adequate amount of test substance for the measurement to be employed. The specific requirements for separation and enrichment are intimately related to the sample under consideration and the technique of measurement employed.

In this introduction to chemical separations, the most essential and fundamental theoretical concepts and equations will be introduced, but in an elementary manner. References cited and the general bibliography appended to each chapter will enable anyone interested in any topic to pursue it in depth. Omitted from consideration in this book are the "classical" precipitation methods and electrodeposition methods; these methods should have been introduced in the first course on analytical chemistry.

Classification of Chemical Separation Methods

For two completely miscible substances to be separable, their molecules must differ in at least one physical-chemical property. Chemical separation methods may be conveniently grouped as shown in Table 1-1. The two major approaches employ either methods that create a second phase with a different concentration of the desired component or else methods that exploit differences within the single phase.

TABLE 1-1. Chemical Separation Methods

	Methods that create a second phase with a different concentration of desired component		*Methods that exploit differences within a single phase*	
Basis of separation	*By means of heat*	*By means of extraneous chemical reagents*	*By using barriers*	*By using non-uniformities of concentration*
Volatility	Distillation			
Partition coefficient		Gas-liquid chromatography Solvent extraction Partition chromatography		
Exchange equilibrium		Ion exchange		
Surface activity		Adsorption chromatography Gas-solid chromatography		Foam fractionation
Molecular geometry			Molecular sieves Gel filtration Gel permeation Gas diffusion Inclusion complexes Ultrafiltration Dialysis Electrodialysis	
Electromigration				Electrophoresis
Solubility	Zone refining			

Adapted from T. Melnechuk, *International Science and Technology* (February 1963).

TABLE 1-2. Chromatography

Adsorption chromatography		*Partition chromatography*		*Exclusion chromatography*		*Ion-exchange chromatography*	
Solution-solid adsorption (Chap. 7)	Gas-solid adsorption (Chap. 11)	Solution-liquid partition (Chap. 8)	Gas-liquid partition (Chap. 11)	Solution-solid exclusion (Chap. 12)	Gas-solid exclusion (Chap. 12)	Resinous materials (Chap. 5)	Cellulosic and inorganic (Chap. 6)

Techniques
Columnar methods (Chapters 4–8, 11, 12)
Paper chromatography (Chapter 9)
Thin-layer chromatography (Chapter 10)

The basis of separation depends upon the existence of a difference in one physical-chemical property between two compounds. Oftentimes the difference is very subtle, but exists nevertheless. Physical-chemical properties invoked in chemical separation methods include volatility, solubility, partition coefficient, exchange equilibrium, surface activity, molecular geometry, electromigration, and molecular kinetic energy.

In the first group, chemical separation is achieved by adding or removing heat or by deliberate addition of extraneous chemical agents. The desired material partitions between the two phases created. This category embraces liquid–liquid extraction (Chapter 3) and the several chromatographic columnar and sheet methods (Table 1-2). Within the second general grouping of chemical separation methods, the desired result is achieved through the use of barriers or membrane systems which will pass some components of a mixture and reject others, or will pass all components at different rates (Chapters 12 and 13). Another approach in the second category involves the use of non-uniformities of concentration, either natural or contrived, as in electrophoresis (Chapter 14).

Selection and Evaluation of Separation Methods

The form of separation should be chosen so that a good separation from other components is assured. As is often the case in analytical chemistry, practice has consistently outstripped theory in many of the fields to be discussed. As a consequence, a certain amount of trial-and-error will be inevitable, unless a procedure can be patterned after an analogous procedure found in the literature. Owing to distribution or solubility equilibria, separation is never theoretically complete. For the determination of major constituents, methods of practical value are those in which the separation is complete for both components within the limits of error of the subsequent determinations. When interest centers around purity of a desired component, the separation factor for an undesired constituent with respect to the desired component is of more concern than recovery or yield. On the other hand, the inverse situation pertains when the goal is the enrichment of some component that is present in a particular matrix. Partial separation also may be combined with the use of specific properties to facilitate analysis in cases where neither complete separation nor specificity can be obtained for the entire mixture. Sometimes the interfering elements or functional groups can be made "unreactive" through formation of masked species (Chapter 2).

BIBLIOGRAPHY

Berg, E. W., *Physical and Chemical Methods of Separation,* McGraw-Hill, New York, 1963.

Heftmann, E. (Ed.), *Chromatography,* 2nd ed., Reinhold, New York, 1967.

Kolthoff, I. M., and P. J. Elving (Eds.), *Treatise on Analytical Chemistry,* Part 1, Vol. 3, Interscience, New York, 1961.

Melnechuk, T., "Chemical Separations," *International Science and Technology* (February 1963), p. 26.

Morrison, G. H. (Ed.), *Trace Analysis, Physical Methods,* Interscience, New York, 1965.

Walton, H. F., *Principles and Methods of Chemical Analysis,* 2nd ed., Prentice-Hall, Englewood Cliffs, New Jersey, 1964.

2 *Complexation Reactions*

Complex formation is important in two ways. It may produce a species that has more useful characteristics for a particular chemical separation method and, alternatively, the concentrations of particular species can be diminished to levels below those at which they interfere in reactions designed to separate other molecules or ions.

Complex Equilibria Involving Metals

Practically every metal forms complex ions of some kind. Some metals form more numerous and more stable complexes than others, but in the reactions of analytical chemistry, the possibility of complex formation must always be considered. Metal ions in aqueous solution possess solvent molecules in their primary solvation shell. Attraction between them is weak usually and the number of solvent molecules immediately surrounding each metal ion is variable. However, in the transition-metal ions and higher-valent metal ions, definite complexes such as $Cu(H_2O)_4^{+2}$ and $Al(H_2O)_6^{+3}$ exist in aqueous solutions. For this reason complex formation in aqueous solutions is really a replacement process in which solvent molecules in the coordination sheath surrounding a metal ion are replaced stepwise by other ligands.

Every anion is a potential electron donor and therefore can be considered as a potential ligand. The stability of the complexes formed is related to the effective charge of the metal ion, the availability of orbitals suitable for the formation of covalent bonds, and the electronegativity of the bonding atoms in the ligand group.

At least two sets of equilibria are involved in the process of complex formation. There is competition between solvent molecules and the ligand for the metal ion, and simultaneous competition between protons and the metal ion for the ligand. Reaction rates must also be considered.

Inert and Labile Complexes The formation and decomposition of complex ions on a time scale is important. The addition of aqueous ammonia to a solution of a silver salt will convert the hydrated silver ion into an ammonia complex in no more time than it takes to mix the solutions together. If acid is added, the ammonia complex decomposes to give back hydrated silver ions. On the other hand, the direct reaction between Fe(III) and CN^- is too slow to be useful, although the stability of $Fe(CN)_6^{-3}$, once formed, is well known. Obviously, the reaction equation represents only the initial and final states and tells nothing about what goes on inbetween. A thorough understanding of ligand substitution dynamics is needed.[1]

Taube[2] has classified a large number of complexes on the basis of the rates of formation of complexes and substitution reactions. Complexes which come to equilibrium rapidly with their dissociation products are called *labile* complexes. Those that decompose very slowly and never come to equilibrium with their components are called *inert* complexes. The inert complexes are characterized by a structure in which all the inner d orbitals are occupied by at least one electron. These include complexes of V(II), Cr(III), Mo(III), Mn(III), Co(III), and some of the complexes of Fe(II) and Fe(III). The relatively few inert complexes involving outer orbital configurations are characterized by a central ion of high charge, such as SiF_6^{-2}.

Mononuclear Complexes Turning now to stoichiometry, only reactions leading to mononuclear compounds, ML_n, where M denotes the central metal ion and L the complexing ligand, will be considered in this brief treatment. The treatise by Ringbom treats the problem of mononuclear and polynuclear compounds in considerable detail.

The simplest case is represented by the reaction, for which $n = 1$,

$$M + L \rightleftharpoons ML \tag{2-1}$$

For sake of simplicity here and in the following discussion, charges will be omitted as will solvated species. The stability or formation constant of the reaction is

$$k_1 = [ML]/[M][L] \tag{2-2}$$

Instead of stability constants, some authors use the inverse values, i.e., the dissociation or instability constants.

[1] See, for example, F. Basolo and R. G. Pearson, *Mechanism of Inorganic Reactions*, 2nd ed., Wiley, New York, 1967; and C. H. Langford and H. B. Gray, *Ligand Substitution Processes*, W. A. Benjamin, New York, 1965.

[2] H. Taube, *Chem. Rev.* **50,** 69 (1952).

If more than one ligand is bound to the central ion, the complex formation will occur stepwise. There will be as many stability constants as there are complexes. They can be represented as follows:

Reaction	Constant
$M + L \rightleftharpoons ML$	$k_1 = [ML]/[M][L]$
$ML + L \rightleftharpoons ML_2$	$k_2 = [ML_2]/[ML][L]$
. . .	. . .
$ML_{n-1} + L \rightleftharpoons ML_n$	$k_n = [ML_n]/[ML_{n-1}][L]$

The overall stability product of the complex ML_n is the product of the consecutive stepwise stability constants; $K_n = k_1k_2 \ldots k_n$. The stability products are denoted by capital K, and are given by

$$K_n = [ML_n]/[M][L]^n \tag{2-3}$$

When ambiguity seems likely, the released species will be indicated by a superscript to K, viz, $K^L_{ML} = K_1$, $K^{2L}_{ML_2} = K_2$, whereas $K^L_{ML_2} = k_2$. Methods for the determination of stability constants and stability products are exhaustively described by Rossotti and Rossotti.[3]

The distribution of the various complexes with a given ligand can be obtained as a set of fractions Φ_0 to Φ_n. Each represents the ratio of the concentration of metal-ligand species to the total concentration of the central metal ion, C_m. Thus

$$\Phi_0 = [M]/C_m \tag{2-4}$$

$$\Phi_1 = [ML]/C_m \tag{2-5}$$

. . .

$$\Phi_n = [ML_n]/C_m \tag{2-6}$$

and

$$\Phi_0 + \Phi_1 + \cdots + \Phi_n = 1 \tag{2-7}$$

If the stability constants are known, it is possible to calculate the distribution of the various complexes solely from the concentration of ligand, since these fractions are a function of the concentration of "free" ligand:

$$\Phi_0 = 1/(1 + k_1[L] + k_1k_2[L]^2 + \cdots + k_n[L]^n) \tag{2-8}$$

$$\Phi_1 = k_1[L]/(1 + k_1[L] + k_1k_2[L]^2 + \cdots + k_n[L]^n) \tag{2-9}$$

. . .

$$\Phi_n = k_n[L]^n/(1 + k_1[L] + k_1k_2[L]^2 + \cdots + k_n[L]^n) \tag{2-10}$$

[3] F. J. C. Rossotti and H. Rossotti, *The Determination of Stability Constants,* McGraw-Hill, New York, 1961.

In this series of expressions, each term in the denominator becomes in turn the numerator. If the ligand concentration is small, the terms with powers of $[L]$ lower than the coordination number of the metal ion are predominant. On the other hand, if the ligand concentration is large, the terms with power of $[L]$ higher than the coordination number predominate. Thus, for any complex, except for saturated ones, there is a certain anion concentration at which the fraction of the complex attains a maximum value. Usually only a few of the potential complexes coexist in significant concentrations. The overall cation concentration has relatively little effect on the fractions.

Example 2-1

The use of Eqs. (2-4) through (2-10) will be illustrated for the various species in mixtures of silver ion and aqueous ammonia containing these equilibrium concentrations of NH_3: 1.0×10^{-5}, 1.0×10^{-4}, 1.0×10^{-3}, and $1.0 \times 10^{-2} M$. The total silver ion concentration is taken as $1.0 \times 10^{-2} M$. The stepwise stability constants for the silver ammine complexes are: $\log k_1 = 3.20$ and $\log k_2 = 3.83$. The solution also contains $2M$ NH_4NO_3 to maintain the ionic strength constant and to keep the pH sufficiently low to prevent the formation of silver-hydroxy complexes in the range of low $[NH_3]$.

Substitution of the appropriate values of NH_3 and the stability constants is done successively in Eqs. (2-8) to (2-10). A sample calculation will be done for Φ_1 when $[NH_3] = 1.0 \times 10^{-3} M$:

$$\Phi_1 = 1.6 \times 10^3(1.0 \times 10^{-3})/[1 + (1.6 \times 10^3)(1.0 \times 10^{-3}) + (1.6 \times 10^3)(6.8 \times 10^3)(1.0 \times 10^{-3})^2]$$

$$= 1.6/(1 + 1.6 + 10.9) = 0.12$$

Therefore, from Eq. (2-5),

$$[Ag(NH_3)^+] = \Phi_1(1.0 \times 10^{-2}) = 1.2 \times 10^{-3} M$$

The remainder of the fractional ratios of each complex species and the free silver ion concentration can be obtained in a similar manner. The values are tabulated below and the plot is shown in Fig. 2-1.

$[NH_3]$, M	Φ_0	Φ_1	Φ_2
1.0×10^{-5}	0.983	0.0157	1.07×10^{-3}
1.0×10^{-4}	0.78	0.13	0.086
1.0×10^{-3}	0.074	0.12	0.81
1.0×10^{-2}	9×10^{-4}	1.4×10^{-3}	0.998

Although excellent compilations of successive stability constants of metal-ion complexes are available, tedious and troublesome calculations (without the aid of computers) are required to obtain specific information from these data. For practical purposes diagrammatic representation, as in

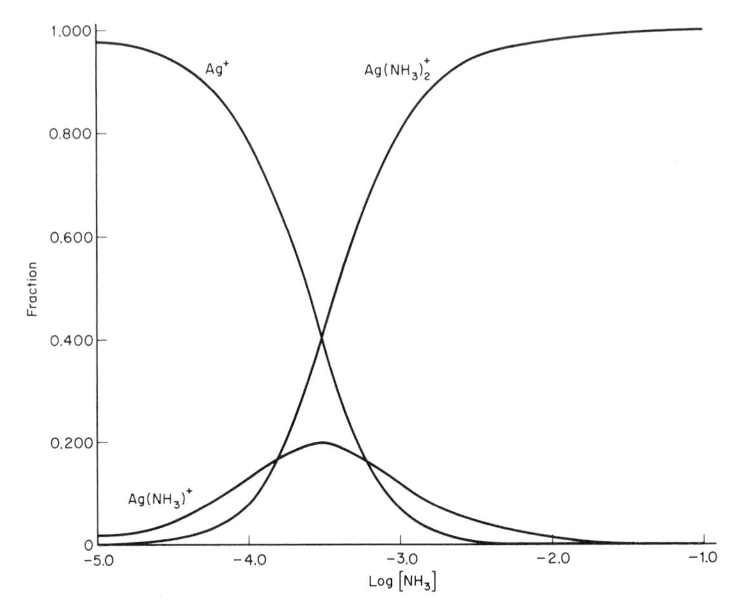

FIG. 2-1. Silver ammines.

Fig. 2-1, is usually more informative than compilations. Diagrams [4] show at a glance the ligand concentration that can be tolerated before complexing of a metal ion begins, as well as the species present in a solution at a given ligand concentration. The diagrams serve as guides for the preparation of solutions that contain a desired species.

Conditional Constants A ligand is usually an anion or a neutral molecule with basic properties so that, at sufficiently low pH values, it becomes extensively protonated, with consequent reduction in its complex-forming ability. Other side reactions may involve the central metal ion, perhaps as metal hydroxo complexes. In order to keep the equilibrium equation in a simple form, it is convenient to define a new constant

$$K' = K_{M'L'} = [ML]/[M'][L'] \qquad (2\text{-}11)$$

which will be called the *conditional constant*. K' is not constant but depends on the experimental conditions. In this expression $[M']$ denotes the concentration of all the metal in solution that has not reacted with some complexing agent. Rearranging Eq. (2-4), the variation of $[M']$ can be expressed as

$$[M'] = \Phi_0 C_m \qquad (2\text{-}12)$$

[4] See G. Goldstein, *Equilibrium Distribution of Metal-Ion Complexes,* U.S. Atomic Energy Commission, ORNL 3620, Nov. 1964.

or, if necessary, by a series of correction factors. In a corresponding manner, $[L']$ represents the concentration of the free ligand under the specified operating value of pH, where

$$[L'] = \alpha C_L \qquad (2\text{-}13)$$

The fractions α_0, α_1, α_2 of ligand in the forms H_2L, HL^-, and L^{-2}, respectively, are given by expressions as follows:

$$\alpha_0 = [H_2L]/C_L = [H^+]^2/([H^+]^2 + K_1[H^+] + K_1K_2) \qquad (2\text{-}14)$$

$$\alpha_1 = [HL^-]/C_L = K_1[H^+]/([H^+]^2 + K_1[H^+] + K_1K_2) \qquad (2\text{-}15)$$

$$\alpha_2 = [L^{-2}]/C_L = K_1K_2/([H^+]^2 + K_1[H^+] + K_1K_2) \qquad (2\text{-}16)$$

The conditional constant thus gives the relationship between the quantities in which the analyst is actually interested, namely, the concentration of the product formed $[ML]$, the total concentration of uncomplexed (in terms of the principal reaction) metal $[M']$, and the total concentration of the uncomplexed reagent $[L']$.

Example 2-2

Calculate $[Al^{+3}]$ in a solution obtained by mixing 1.0×10^{-3} moles of $Al(NO_3)_3$ and 0.1 mole of acetylacetone, adjusting the pH to 2.5, then diluting to 1 liter. The pK_a of acetylacetone is 8.8; the successive stability products: AlL, AlL_2 and AlL_3 are 8.1, 15.7, and 21.2, respectively. First, it is necessary to calculate $[L']$ of acetylacetone at pH 2.5.

$$[L'] = 1.6 \times 10^{-9}(0.1)/(7.0 \times 10^{-3})$$
$$= 2.3 \times 10^{-8}$$

Next, assuming no other competing aluminum complexes, Φ_0 is calculated

$$\Phi_0 = 1/[1 + (1.3 \times 10^8)(2.3 \times 10^{-8}) + (5 \times 10^{15})(2.3 \times 10^{-8})^2 + (1.6 \times 10^{21})(2.3 \times 10^{-8})^3$$
$$= 1/(1 + 3 + 2.65 + 0.02) = 0.15$$

From this value,

$$[Al^{+3}] = \Phi_0 C_{Al} = 0.15\ (1.0 \times 10^{-3}) = 1.5 \times 10^{-4} M$$

Only 85% of the aluminum exists in some acetylacetone complex. If the pH were raised to 3.5, $\Phi_0 = 1/[1 + 30 + 265 + 20] = 0.0032$, and an almost negligible quantity remains uncomplexed. At pH 3.5, the dominant species is AlL_2^+; the conditional stability product for AlL_3 becomes

$$K' = K_{M'L_3'} = [ML_3]/[M'][L']^3$$
$$= (1 \times 10^{-3})/(0.0032)(1 \times 10^{-3})(2.3 \times 10^{-8})^3$$
$$= [ML_3]/(\Phi_0 C_m (L')^3$$
$$= 2.6 \times 10^{25}$$

Complexation Ability of Ligands The most important points for the characterization of a ligand are the nature and basicity of its ligand atom. Oxygen donors and fluoride are general complexing agents, combining with any metal ion with a charge more than one. Acetates, citrates, tartrates, and β-diketones sequester all metals in general. Purely electrostatic phenomena predominate. The strength of the coordinating bond formed with the central metal ion increases enormously with the charge of the metal ion and decreases with its radius. Upon comparing various oxygen donors, bond stability is found to increase regularly with the basicity, as measured by proton addition, of the ligand atom.

Cyanide, heavy halides, sulfur donors, and to a smaller extent the nitrogen donors, are more selective complexing agents than are oxygen donors.[5] These ligands do not combine with the cations of the A-metals of the periodic table. Only the cations of B-metals and transition-metal cations are coordinated to carbon, sulfur, nitrogen, chlorine, bromine, and iodine. The bonds in the complexes are predominately covalent. Bond strength increases with the ease by which the metal ion accepts electrons and the ease by which the ligand atom donates electrons. A decisive factor is the difference in the electronegativity of the metal ion and the donor atom. Highly polarizable ligands are favored, especially if the latter have suitable vacant orbitals into which some of the *d* electrons from the metal ion can be "back-bonded." This condition favors sulfur-containing ligands.

Chelates A large and important group of metal complexes contain a number of ligands which is half the usual coordination number of the metal ion involved. These organic ligands have a dual character and are of sufficient importance to be discussed separately. The ligand is usually a weak, polyfunctional organic acid. It must also possess a pair of unshared electrons on an oxygen, nitrogen, or sulfur atom which is available for coordination (I).

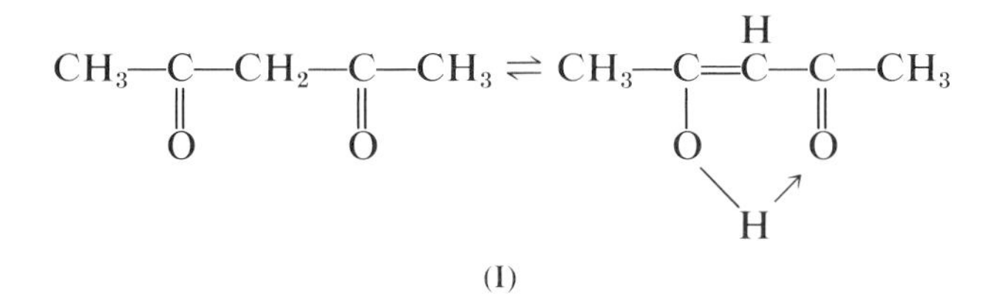

(I)

Furthermore, the acidic and basic groups in the ligand must be so situated that a ring formation, involving the metal ion, can proceed relatively free from strain (II).

[5] G. Schwarzenbach, *Anal. Chem.* **32,** 6 (1960).

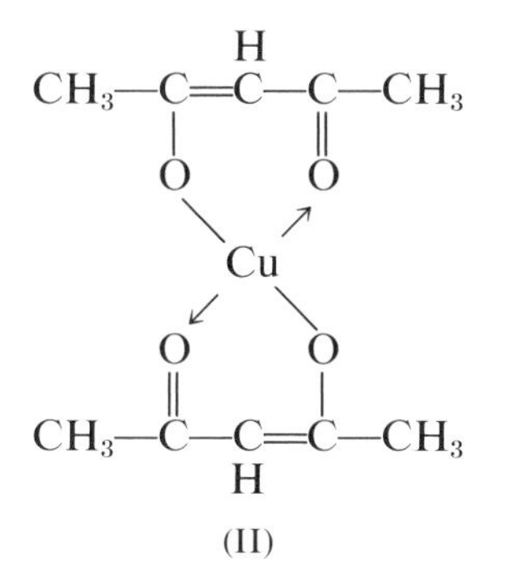

(II)

The resulting compound is called a chelate from reference to the pincer-like action of the ligand on the metal ion. A neutral chelate is often called an inner complex salt; "inner" because of the formation of a ring structure, "complex" because of the coordination of the ligands with the metal ion, and "salt" because of the cancellation of charges. The metal chelate presents essentially a hydrocarbon-like surface when the coordination number of the metal ion is exactly twice the charge of the ion. However, if not all coordination sites around the central metal ion are occupied by the complexing ligand, some secondary solvation remains (III). On the other hand, a charged

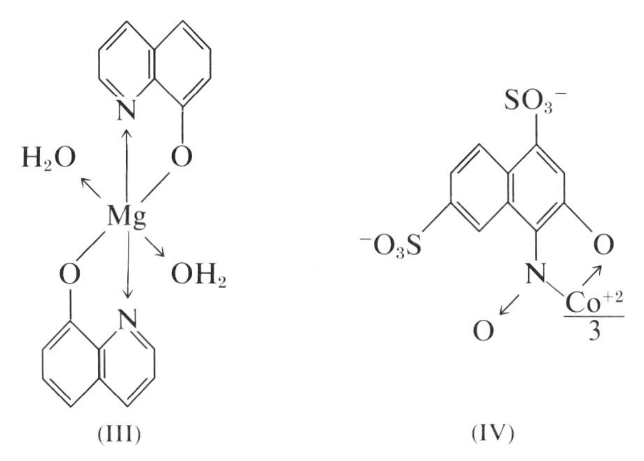

(III) (IV)

anionic chelate results when the coordination number is greater than the valence of the metal ion and all the coordination sites are occupied by the complexing ligand (e.g., MgY^{-2} where Y denotes the anion of EDTA). Metal chelate ions (inner complex anions) also result when the complexing ligand contains a hydrophilic group that is not involved in chelation (IV).

Many properties of chelates are determined by the nature of the organic ligand which combines with the metal ion. Practically all chelates have five- or six-membered rings. In general, the five-membered ring is more stable when the ring is entirely saturated, but six-membered rings are favored when one or more double bonds are present. Multiple ring systems formed with a given metal ion markedly improve stability. When two or more donor groups are tied together to form an additional chelate ring without materially altering the donor groups, the increased stability of the chelate is due almost entirely to an increase of entropy. This is due to the increasing number of positions occupied by the chelating ligand in the metal coordination sphere and the improvement in isolation of the central metal atom from the influence of solvent hydration.

Stability of chelates is influenced by stereochemical effects and the resonance effect. The latter refers to the stability increase that results from the contribution of resonance structures of the chelate ring to the structure of the metal chelate. Stability parallels the number of resonating structures that can be written for the chelate species. Spatial considerations, size of the metal ion and space available between the coordination and ligand sites determine whether a given metal chelate can form.

Reagents giving uncharged complexes are precipitating and extracting agents. The pure oxygen donors among them, like the β-diketones and cupferron, are general reagents. Little selectivity is found in 8-quinolinol (8-hydroxyquinoline), which donates one oxygen and one nitrogen atom to each metal. On the other hand, the nitrogen and sulfur donors, such as dithizone and diethyldithiocarbamate, are highly selective for the B-metals and noble transition metals.

Reagents giving charged chelates are sequestering (masking) agents. The pure oxygen donors—oxalates, tartrates, and citrates—are non-selective. Higher complex stabilities are reached with aminopolycarboxylic acid anions, which are also general masking agents because of the greater number of oxygens in comparison to nitrogens among their ligand atoms. Polyamines are rather selective agents, combining only with B-metals and transition metals, but having hardly any affinity for the A-metal cations. Still greater selectivity is found with sulfur donors like thiourea, dithiocarbamate, and dithiophosphate.

Masking

Masking involves the addition of a chemical that reacts with some or all of a particular species to form a stable complex. After the transformation, certain reactions are prevented. Masking has proven to be one of the most effective means of achieving selectivity in chemical separation methods. Oftentimes a non-specific separation method can be rendered highly selective through elimination of interferences by masking.

Demasking is the process in which a masked substance is released from its masked form and regains its ability to enter into certain reactions.

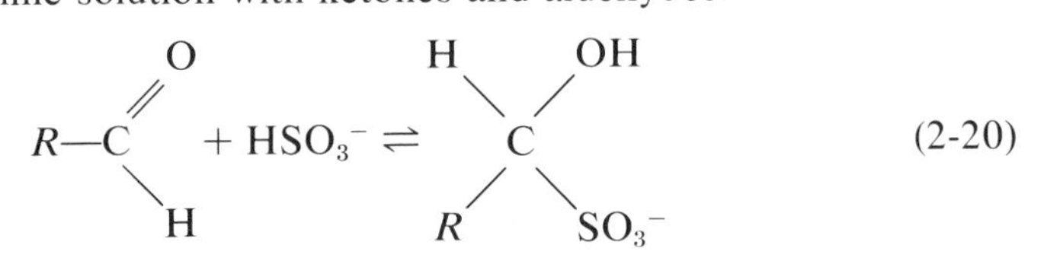

Principles Masking never completely removes certain ionic or molecular species, but only lowers their concentrations. The extent of this lowering determines which reactions can be prevented. Of particular concern is the amount of principal reagent and masking agent to be used to control the equilibrium concentration of the species of interest. Theoretically, any reaction equilibrium can be shifted completely in one direction or the other if enough reactant or product is added or removed. Practically, the shift of equilibrium is limited because the concentration of reactant in a solvent is limited.

Equilibrium data for many reactions likely to be encountered in inorganic work are contained in the publication by Ringbom, cited in the Bibliography, and the publication by Cheng.[6]

For the major part, masking reactions which occur in solution and lead to soluble compounds are equilibrium reactions. They usually require the use of an excess of the masking agent and can be reversed again by removal of the masking agent. If a metal ion is in solution with two competing complexing agents, L and A, then the metal ion concentration must satisfy both equilibria:

$$[M] = \frac{[ML_n]}{[L]^n K_{ML_n}} \tag{2-17}$$

and

$$[M] = \frac{[MA_n]}{[A]^n K_{MA_n}} \tag{2-18}$$

Assuming both L and A are singly charged anions,

$$\frac{[MA_n]}{[ML_n]} = \frac{[A]K_{MA_n}}{[L]K_{ML_n}} \tag{2-19}$$

The principal reaction (assumed to involve L) is favored if pM_L [the negative logarithm of M in Eq. (2-17)] is larger than pM_A [where A is assumed to be the masking agent, Eq. (2-18)]. The appearance of the principal reaction depends also on the magnitude of pM_L. Regardless of the presence of a competing masking agent, the higher the value of pM_L, the easier is the formation of the principal complex or product.

Through deliberate selection of suitable masking agents it is possible to secure a precise, fractional masking action. For example, copper is extracted with 8-hydroxyquinoline in chloroform at a much lower pH than is uranium(VI). If EDTA is added, however, the uranium extraction proceeds

[6] K. L. Cheng, *Anal. Chem.* **33,** 783 (1961); see also A. Hulicka, *Talanta* **9,** 549 (1962) who disagrees with the approach used by Cheng.

as before but the copper is not extracted until the solution pH is raised by more than five units.

Certain organic substances have no charge at any pH, but form complexes with substances that do have a charge. The sugars and polyalcohols form such complexes in the pH range between 9 and 10 with a number of anions, including borate, molybdate, and arsenite. Such complexes are usually unstable and can be maintained only if the complexing agent is present in the solution in large excess so that mass action maintains the compounds in the complexed condition. Another example is the reaction of bisulfite ion in alkaline solution with ketones and aldehydes:

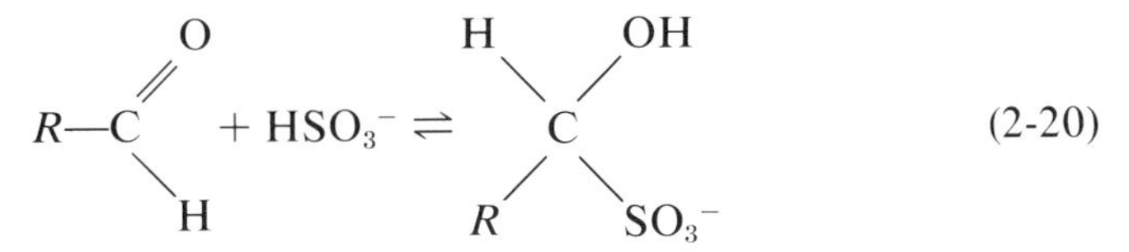

$$R\text{—}C(=O)H + HSO_3^- \rightleftharpoons R\text{—}C(H)(OH)SO_3^- \qquad (2\text{-}20)$$

The carbon–oxygen double bond of the carbonyl group is opened, and the bisulfite radical is added. An increase in temperature reverses the reaction more easily for ketones than for aldehydes.

Demasking The discussion of the masking of reactions makes it clear that a system masked against a certain reagent, despite the lowering of the concentration of the principal species, is not necessarily masked against another but more aggressive reagent. The methods used in demasking fall into two main categories. One approach is to change drastically the hydrogen-ion concentration of the solution. In most cases a strong mineral acid is added, and the ligand is removed from the coordination sphere of the complex through the formation of a weak acid. Examples include the formation of volatile HCN, or the formation of undissociated acids among the polyprotic (citric, tartaric, EDTA, and nitriloacetic) acids.

The second type of demasking involves formation of new complexes or other compounds that are more stable than the masked species. For example, the borate ion is used to demask fluoride complexes of tin(IV) and molybdenum(VI); and formaldehyde is often used to remove the masking action of cyanide ions through the reaction:

$$CN^- + HCHO \rightleftharpoons OCH_2CN \qquad (2\text{-}21)$$

which forms glycollic nitrile. Selectivity is evident in that $Zn(CN)_4^{-2}$ is demasked whereas $Cu(CN)_3^{-2}$ is not.

Problems

1. Calculate the fraction of $[CN^-]$ at pH of 8.00, 9.00, and 10.00.
2. Using the values of the stepwise dissociation constants given in the Appendix

for ethylenediaminetetraacetic acid (EDTA), compute the fractions α_0, α_1, α_2, α_3, and α_4 of the ligand. Plot the results as a function of pH.

3. Graph the fraction of each species of tartaric acid as a function of pH.
4. Zinc forms a 1:1 complex with $(H_2N—CH_2—CH_2—NCH_2—)_2$, "triene." Calculate the zinc ion concentration in the solution resulting from mixing equal volumes of 0.04M zinc nitrate and 0.04M "triene" and adjusting the pH to 10.0. The formation constant of the zinc triene complex is 1.0×10^{12}.
5. Calculate the concentrations of the various species at equilibrium in a solution prepared by diluting 1.40 moles of KCN and 0.10 mole of $Cd(NO_3)_2$ to 1 liter with water.
6. Calculate $\beta_{Cd}{}^0$ for cadmium in a solution containing 0.1M NH_4^+ and 0.1M NH_3 (pH = 9.2).
7. Calculate the concentration of uncomplexed copper(II) in the presence of 0.1M EDTA and these pH values: 4, 6, 8, and 9.
8. Calculate the conditional formation constant of the copper(II)–EDTA complex in an ammonia buffer solution containing 0.1M NH_3 and 0.019M NH_4^+ (pH = 9.0). Compare with the formation constant of calcium–EDTA.
9. Calculate the concentration of uncomplexed $[Cd^{+2}]$ in Problem 5 when the solution is adjusted to the pH values of Problem 1.
10. Why is a precipitate of aluminum hydroxide formed when cyanide, sulfide, and alkyl amines are added to a solution of aluminum, whereas aluminum forms soluble complexes with fluoride, hydroxide, citrate and tartrate?
11. Suggest a method to avoid losses of BF_3 in the determination of boron in silicon while the latter is removed as SiF_4.
12. For the analysis of trace admixtures in uranium, suggest a method for separating the traces while keeping the uranium in solution.

BIBLIOGRAPHY

Bailer, J. C., Jr. (Ed.), *Chemistry of the Coordination Compounds,* Reinhold, New York, 1956.

Feigl, F., *Specific, Selective, Sensitive Reactions,* Academic Press, New York, 1949.

Martell, A. E., and M. Calvin, *Chemistry of the Metal Chelate Compounds,* Prentice-Hall, Englewood Cliffs, New Jersey, 1952.

Perrin, D. D., *Organic Complexing Reagents,* Wiley, New York, 1964.

Ringbom, A., *Complexation in Analytical Chemistry,* Wiley, New York, 1963.

Rossotti, F. J. C., and H. Rossotti, *The Determination of Stability Constants,* McGraw-Hill, New York, 1961.

Welcher, F., *Organic Analytical Reagents,* 4 vols, Van Nostrand, Princeton, New Jersey, 1947.

3 *Liquid–Liquid Extraction*

Liquid–liquid extraction methods are based on the distribution of a solute between two essentially immiscible solvents. Although combinations of any two immiscible organic solvents can be employed, one phase is usually an aqueous solution, the other an organic liquid. A solute which is soluble in both phases will distribute between the two phases in a definite proportion. Equilibrium is attained when the free energy of the solute is the same in each phase.

Solvent extraction offers speed, simplicity, convenience, and scope. Extraction procedures are applicable to both trace and macro levels of concentrations. Extractions can generally be performed in a matter of minutes and use quite simple apparatus. Components can be selectively extracted from aqueous solutions into organic solvents or re-extracted from the organic phase into the aqueous one. The desired separation is achieved by adjustment of the chemical parameters—pH, masking agent, solvent, extraction agent. When the extraction is conducted under equilibrium conditions, it is possible to apply the principles of chemical equilibrium to predict accurately the course of the extraction.

Distribution Relations

Partition of a solute molecule is a dynamic process that involves a constant interchange of the solute molecules across the region of contact of the two immiscible phases. Two terms will be introduced at this point: the partition coefficient and the distribution ratio.

Partition Coefficient For ideal solutions, the partition coefficient of a substance A between two phases is related to the free energy required to transport one mole of A from one phase to the other.

If an ideal pair of solutions is considered, the chemical potentials of the partitioning solute in the two phases are represented by

$$\text{stationary: } u_S = u_S{}^0 + \mathscr{R}T \ln N_S \tag{3-1}$$

$$\text{mobile: } u_M = u_M{}^0 + \mathscr{R}T \ln N_M \tag{3-2}$$

Here u_S and u_M are the chemical potentials of the solute in the stationary and mobile phases (which are likened to the aqueous and organic phases in solvent extraction), the superscript zero refers to the standard state, and N_S and N_M are the mole fractions of the solute in the respective phases. At equilibrium the chemical potential of the solute is the same in both phases, and

$$u_M{}^0 + \mathscr{R}T \ln N_M = u_S{}^0 + \mathscr{R}T \ln N_S \tag{3-3}$$

Rearranging

$$\ln \frac{N_M}{N_S} = \frac{(u_M{}^0 - u_S{}^0)}{\mathscr{R}T} \tag{3-4}$$

or

$$\frac{N_M}{N_S} = K_d = \exp\left(-\frac{\Delta u^0}{\mathscr{R}T}\right) \tag{3-5}$$

The difference in standard chemical potentials expresses the work required to transport a mole of solute from one phase to the other. When working with relatively dilute solutions, the mole fractions of the solute may be replaced by the concentration in the respective phases. In Eq. (3-5), the term K_d is the partition coefficient.

Up to the point where one or both phases becomes saturated with a given solute, the partition coefficient is independent of the total actual concentrations in each phase. In fact, the partition coefficient is roughly a measure of the relative solubilities of the particular species in each phase and can often be estimated successfully from solubility data. When more than one component is involved in a distribution across the phase boundary, Nernst's law of independent distribution states that each component will distribute independently of all other compounds.

Distribution Ratio The distribution ratio D of the central group between an organic phase and an aqueous phase (for equal phase volumes) is given by

$$D = \frac{|[A]_o|}{|[A]|} \tag{3-6}$$

The numerator is the total concentration of solute A in all chemical forms in the organic phase (denoted by subscript o), and the denominator is the total concentration of solute A is all chemical forms in the aqueous phase. The total concentration in each phase is determined experimentally.[1]

As will be shown later, $D \simeq K_d$ for systems in which the distributing species has one and the same form in both phases. Strictly, $K_d = D(\gamma_o/\gamma_{aq})$ where the parentheses contain the activity coefficients of the distributing species in each phase. One of the most valuable features of solvent extraction procedures is the ability to alter the numerical value of the distribution ratio by means of dissociation or association equilibria to suit the problem in hand.

Completeness of Extraction The completeness of an extraction depends not only on the value of the distribution ratio but also on the volumes of the phases and on the number of extractions performed. Only when the distribution ratio is large is it possible to remove a solute quantitatively in a single extraction.

The concentration C of solute left in the aqueous phase after one equilibration using volume V of aqueous phase and volume V_o of organic phase is given by

$$C = C_o \left[\frac{V}{DV_o + V}\right] = C_o \left[\frac{1}{1 + D(V_o/V)}\right] \tag{3-7}$$

The amount of solute remaining after single extractions is clearly dependent upon two factors: the distribution ratio and the volume ratio of the phases. The process of extraction for removal can be carried out as completely as desired by repeated batch extractions or by continuous extraction methods (*q.v.*). After n extractions the concentration of solute remaining in the aqueous phase is given by

$$C_n = C_o[V/(DV_o + V)]^n \tag{3-8}$$

For a stated amount of organic phase, the extraction is more efficient if carried out several times with small portions of extracting phase. For a phase volume-ratio of unity, the denominator of Eq. (3-7) reduces to $D + 1$, and the expression becomes

$$C = C_o[1/(D + 1)] \tag{3-9}$$

The relationship between the percent extracted, $\%E$, and the distribution ratio is given by

$$\%E = 100D/(D + 1) \tag{3-10}$$

When the percent extracted approaches 100%, the distribution ratio tends toward infinity as a limit. For differences in extraction in the range 99 to

[1] Distribution ratios greater than about 10^3 become difficult to measure because the solute concentration in one phase is very large and in the other very small. The slightest error in the measurements of the lower concentration is magnified in determining D.

100%, the distribution ratio will vary from 99 to infinity. Meaningful values of D are usually limited to 10,000 or less.

Example 3-1

Knowing D and the phase ratio, it is simple to determine the number of extractions required to remove some desired amount of solute. Assume, for example, that 90% of a substance with a distribution coefficient of one is to be extracted into an organic phase. This can be achieved with a single extraction with an aqueous phase whose volume is one ninth the volume of the organic phase; i.e., $V_o = 9V$, *viz.:*

$$C = 0.1C_o[V/(V_o + V)]$$

$$0.1 = \frac{V}{V_o + V}$$

$$0.1V_o + 0.1V = V$$

$$0.1V_o = 0.9V$$

$$V_o = 9V$$

or with four extractions using fresh organic phase each time and equal volumes of each phase, *viz.:*

$$C/C_o = 0.1 = [V/(V_o + V)]^n$$

$$\log 0.1 = n \log 0.5$$

$$n = 3.3 \text{ (or 4 extractions)}$$

This second approach requires less solvent—4 volume units vs 9 volume units.

If a distribution ratio is unknown, it is obtained by equilibrating known volumes of the aqueous phase and extracting solvent, then determining the concentration of distributing species in both phases. This should be performed over a range of concentrations. On an unknown material, it is wise to perform successive extractions on the aqueous phase. A plot of the logarithm of the amount extracted against the step number should be a straight line with a negative slope such that the ratio between consecutive quantities is given by

$$\frac{1}{1 + D(V_o/V)}$$

Thus, it is possible to calculate the distribution ratio from the slope of the straight line obtained in this type of plot. Furthermore, if the data do not appear as a straight line, either the distribution ratio is not constant or the extracted material consists of two or more components with different distribution ratios. Sometimes the data on the straight line obtained from the final equilibration steps can be extrapolated to the first step. If the quantity of this component is subtracted from the total amount in the previous extractions, the distribution ratio can be obtained for a second component, should a linear relationship be found. Treatment of the results resembles the manner one handles decay curves in radiochemistry.

The requirements of a separation will set the limits on the degree of extraction required. If a practically complete extraction is taken to be 99.9% extracted, then D must be 10^3 or greater when $V = V_o$, but 10^4 or greater when $V = 10V_o$. These limits assume one equilibration only. If additional equilibration steps are employed, D could be proportionally smaller. Although a solute can be extracted, however unfavorable the value of D, practically, the use of liquid–liquid extraction is limited to $D \geqslant 1$ when performed manually.

Selectivity of an Extraction Usually an extraction for separation purposes is more involved than the singular examples discussed because two or more components are distributing. The degree of separation achievable will depend on the differences in the distribution ratios of the various components with respect to the two liquid phases. A possibility of separation always exists provided $D_1 \neq D_2$.

A measure of the separation of two substances is the separation factor α:

$$\alpha = D_1/D_2 \tag{3-11}$$

It is conventional in solvent extraction to place the larger value of D in the numerator. One always strives to make the separation factor as large as possible by proper choice of solvent extraction systems and adjustment of phase volumes. The best possible separation through adjustment of volume ratios is obtained according to the Bush–Densen equation

$$V_o/V = 1/\sqrt{D_1 D_2} \tag{3-12}$$

Equations (3-11) and (3-12) are useful in determining whether a given extraction is practical, with a reasonable phase ratio, or whether an extractant with a more favorable distribution ratio should be sought. Considering the requirements for the complete extraction of one component, stated earlier, and if less than 0.1% of the second component is to coextract, D_2 must be 10^{-3} or less, when $V = V_o$, and α must be at least 10^6. Thus, if $D_1 = 10$ and $D_2 = 0.1$, a single extraction will remove 90.9% of component one and 9.1% of component two. A second extraction of the same aqueous solution will bring the total amount of component one extracted up to 99.2%, but increases that of component two to 17.4%. More complete extraction of component one involves an increased contamination by component two. In practice, the goals of complete extractability and separability from other solutes are usually mutually antagonistic, particularly in mixtures of solutes of a very similar nature. Consequently, the analyst must decide between two aims. Is it desired to separate component one as completely as possible from other substances, irrespective of the recovery yield? Or is it desired to obtain as large a yield of component one as possible subject to contamination by component two not exceeding some predetermined percentage. Only when one of the distribution ratios is relatively large and the other very small, can almost complete separation be quickly and easily achieved (if favorable kinetics prevail). If the separation factor is large but the smaller distribution ratio is of sufficient magnitude that extraction of both components occurs, then it is necessary to resort to chemical parameters such as pH or masking agents to suppress the extraction of the unwanted component. When the separation factor approaches unity, it becomes necessary to employ countercurrent distribution methods (if equilibria are attained rapidly) in which distribution, transfer, and recombination of various fractions are performed a sufficient number of times to achieve separation.

Nature of Partition Forces

Broadly speaking, a phase distribution reflects the relative attraction or repulsion that molecules or ions of the competing liquid phases show for the solute and for themselves. The physical interactions that are of particular interest include dispersion interaction, dipole–dipole interaction, induction interaction, and the hydrogen bond interaction.

Dispersion Interaction In a nonpolar liquid, such as CCl_4, dispersion interaction is the only force present between two molecules. This interaction is produced by rapidly varying ("instantaneous") dipoles formed between nuclei and electrons at zero-point motion of the molecule acting on the polarizability of other molecules to produce induced dipoles in phase. For instantaneous dipoles moving in phase, there would arise an attractive interaction between them owing to the movement in phase of their electronic systems. This effect can be additive over large numbers of molecules in a system. We are familiar with the preference of benzene for aromatic solvents in comparison to aliphatic solvents, whereas the reverse is true in the case of cyclohexane. Benzene, although nonpolar, is highly polarizable.

Dispersion forces are relatively weak. When the CCl_4 molecules approach each other so closely that their electron shells begin to overlap, the weak attraction changes to repulsion. As a consequence, the CCl_4 molecules exist in a disordered array relative to each other. A second nonpolar solute mixes in all proportions commensurate with its solubility. Neither kind of molecule has much attraction for its fellows and therefore does not oppose molecules of the other kind intermingling.

Dipole–Dipole Interaction In a liquid composed of molecules with permanent dipoles, these molecules will be much more strongly attracted to one another than to those of a nonpolar molecule. A polar phase is one composed of polar molecules. A polar molecule is a molecule that possesses either an overall electrical dissymmetry, in which case it will show a dipole moment, or a localized polar bond—a bond that exhibits a finite separation between the centers of gravity of positive and negative charges. The solvent molecules will be oriented as $(-+)(-+)$ or $(\mp\pm)$. This dipole–dipole interaction leads to association of liquids such as water, alcohols, and acids.

In this type of solvent medium, only substances whose attraction for polar molecules of the solvent is about as strong as the attraction of solvent molecules for one another will be able to force the solvent molecules apart and mix with them. Orientation interaction of this type between polar molecules is preferred at ordinary temperatures but the interaction is weak and is readily destroyed under the buffeting that occurs at higher temperatures.

Induction Interaction When a molecule with a permanent dipolar bond approaches a nonpolar group in another molecule, but a molecule with a relatively mobile electron system, the former induces a temporary dipole in the latter so oriented that attraction interaction can occur. For example,

$$\overset{\delta+}{A} - \overset{\delta-}{B} + C::D \rightleftharpoons \begin{matrix} \overset{\delta-}{C}::\overset{\delta+}{D} \\ \overset{\delta+}{A} - \overset{\delta-}{B} \end{matrix}$$

Ion–dipole interaction occurs between an ion and a permanent dipole in a molecule. Familiar examples are solvation of ions by water, ammonia, alcohol, and other molecules which contain permanent dipoles. Sometimes this interaction is so strong that the products are stoichiometrically well characterized. An ion may also induce a dipole moment in a neutral molecule to form an ion-induced dipole bond. Such a bond is formed in the reaction:

$$I^- + I_2 \rightleftharpoons I_3^- \qquad (3\text{-}13)$$

Hydrogen Bond Interactions Dipoles capable of forming hydrogen bonds are exceptional in their behavior. Although hydrogen can form only one covalent bond, the very small size of the hydrogen atom and its lack of inner closed electron shells makes it possible to form an additional bond with electron-rich atoms outside (intermolecular) or inside (intramolecular) the molecule. This may occur between a molecule with hydrogen attached to fluorine, oxygen, or nitrogen or, in certain cases, carbon and a molecule with unshared electrons. Examples are

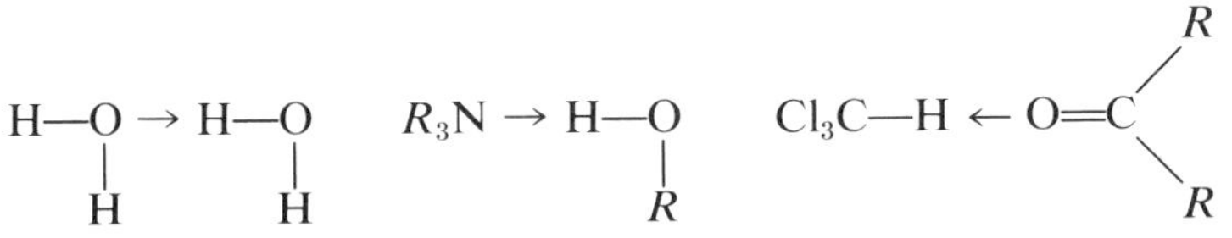

$$\begin{matrix} H\text{—}O \rightarrow H\text{—}O \\ | \qquad\quad\;\; | \\ H \qquad\quad\;\; H \end{matrix} \qquad \begin{matrix} R_3N \rightarrow H\text{—}O \\ \qquad\qquad\;\; | \\ \qquad\qquad\;\; R \end{matrix} \qquad Cl_3C\text{—}H \leftarrow O\text{=}C\begin{matrix} \diagup R \\ \diagdown R \end{matrix}$$

Fluorine forms very strong hydrogen bonds, oxygen weaker ones, and nitrogen still weaker ones. The —CH link is capable of forming hydrogen bonds to oxygen or nitrogen provided the atoms attached to the carbon are such as to increase markedly the polarity of the C—H bond (as in Cl_3CH). Among polar groups capable of forming hydrogen bonds, the hydroxyl group forms stronger bonds than the amino group. The actual strength of a hydrogen bond is dependent upon the geometry of particular combinations, upon the nature of neighboring atoms, upon resonance, and upon acid–base character. For steric reasons solvent molecules may be unable to approach closely; this will reduce the energy of association of solute with solvent.

Compounds participating in hydrogen bonding can be categorized as follows:

(1) Compounds that contain an active hydrogen but no donor atom. $CHCl_3$ is an example.
(2) Compounds which contain a donor atom but no active hydrogen. Typical classes include ethers, ketones, aldehydes, esters, tertiary amines, nitro aryls, and nitriles.
(3) Compounds containing both a donor atom and an active hydrogen atom. These include alcohols, fatty acids, phenols, primary and secondary amines, oximes, nitroalkanes with alpha-hydrogen, and the inorganic molecules NH_3, HF, N_2H_4, and HCN.
(4) Compounds capable of forming networks of multiple hydrogen bonds—the cage effect exemplified by water, glycols, amino alcohols, hydroxy acids, amides, and polyphenols.

Mixed solvents may improve the distribution ratio in favor of the organic phase through the additive effect of hydrogen bonding. Consider, for example, the hydrogen bonding between a ketone and an acid (I) and an alcohol and an acid (II)

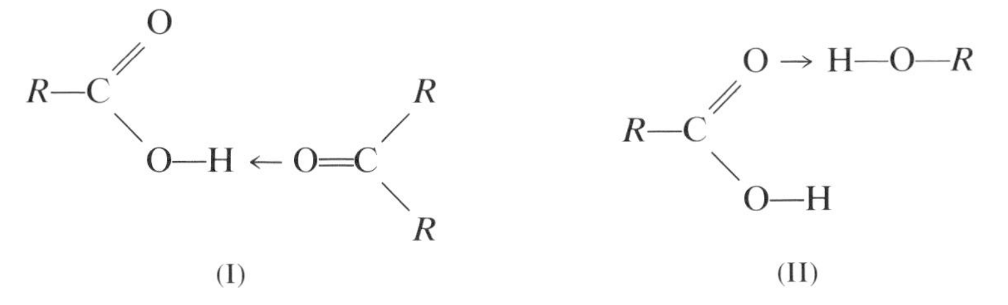

(I) (II)

with the additional effect when ketone plus alcohol is present (III).

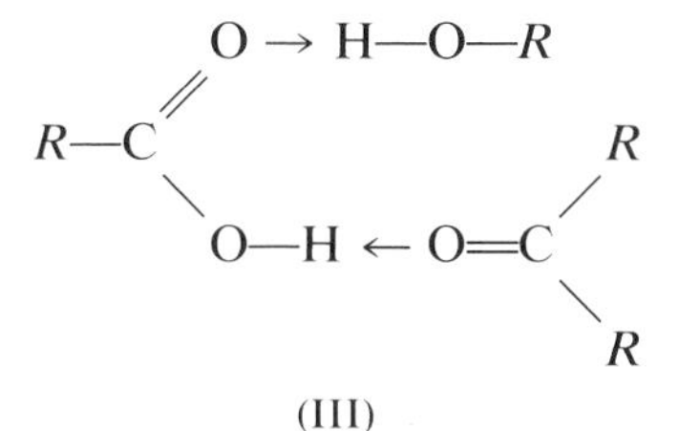

(III)

Synergistic effects may often be explained in this manner.

Water has a particularly strong tendency for formation of intermolecular hydrogen bonds and can function as an electron acceptor and an electron donor for dissolved substances. It exists in a relatively open, but highly hydrogen-bonded structure, as depicted below.

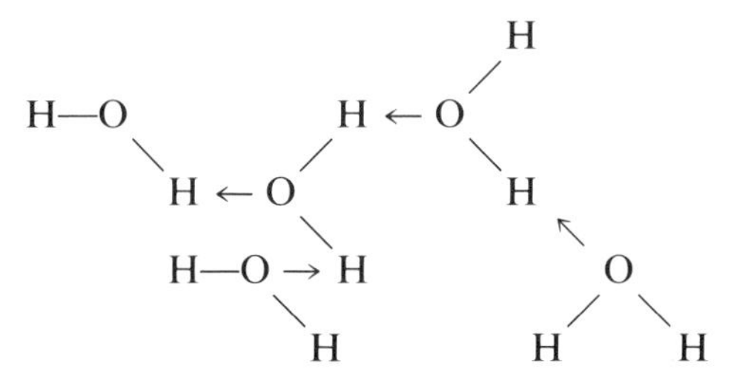

Each water molecule is bound to several others. This energy-rich, ordered array of water molecules must be broken down if solute molecules are to be accommodated. No such network exists in solvents such as ether, $CHCl_3$ or CCl_4. To break each hydrogen bond requires an expenditure of 4–6 kcal of energy. Consequently, the solubility of a solute in water is influenced strongly by its own ability to form hydrogen bonds with the water molecules or else to ionize—two means of supplying the energy required to disrupt the hydrogen bonds within water. Solvent molecules whose mutual attraction is enhanced by hydrogen bonding will resist penetration by nonpolar solute (hydrocarbon type) molecules. If these nonpolar molecules happen to be formed through a chemical reaction, they will tend to be "squeezed out" of the more polar liquid phase and into the less polar phase when the two phases are in contact at a liquid boundary.

The Dielectric Constant The force between charged particles varies directly as the product of the charges and inversely as the square of the distance between them. But the same charges, separated to the same extent, act on each other with a force which varies inversely as the dielectric constant of the medium in which they are immersed. For example, only 30% as much work is required to move the charges farther apart in water as in ethyl alcohol (dielectric constants are 80 and 24, respectively). Many organic liquids have dielectric constants between 2 and 10. Benzene, CCl_4, diethyl ether, and alkyl esters are typical of such liquids. More polar liquids, particularly the alcohols, have dielectric constants between 20 and 35. The dielectric constants of hydrogen-bonded liquids are abnormally large due to the large aggregates formed; the total dipole moment may be many times that of a simple molecule.

The dielectric constant decreases as the temperature is raised.

Methods of Conducting Extractions

The usual purpose of transferring a solute from one liquid solution to another is to effect a separation or purification of the desired component. In some cases the desired component is removed from the original solution, which then retains unwanted compounds; in other cases the desired component remains in the original solution, while the impurities are extracted. Extraction may also provide a second solution from which the desired component is more readily recoverable than from the original solution. At the same time the extraction step may effect a concentration of the desired component. The choice of procedure is sometimes a matter of convenience, but is largely dictated by the value of the distribution ratio and the particular analytical problem.

Countercurrent extraction is treated in a separate section later in this chapter.

Batch Extractions A batch extraction is employed where a large distribution ratio for the desired solute is readily obtainable. A small number of equilibration steps then suffice to remove the desired component completely. For example, if the distribution ratio is greater than 4, three extractions are adequate for removal of 99% of the substance. Equation (3-12) determines the required volume relations. The usual apparatus is a separatory funnel; the pear-shaped type is convenient. The two phases are thoroughly shaken in a separatory funnel, the phases allowed to separate, and then the lower layer is drained off through a stopcock. The length of time required to attain equilibrium must be established by preliminary experiments.

Continuous Extraction When the distribution ratio is low, but the separation factor is high, continuous methods of extraction are useful. The procedure makes use of a continuous flow of immiscible solvent phase through the solution to be extracted. If the solvent is volatile, it can be stripped and recycled by distillation and condensation. Although partition equilibrium may not be achieved during the limited time of contact of the two phases, solute is being removed continuously with the extracting phase.

Efficiency depends upon the value of the partition ratio, the viscosity of the phases, the relative phase volumes, the area of contact between the phases, and the relative velocity of the phases. As the extractant passes through the solution, baffles and stirrers may be used to bring the two phases into closer, more effective, and prolonged contact.

Many continuous extraction devices operate on the same general principle. This consists of distilling the extracting solvent from a flask that serves as a boiler, condensing the solvent, and passing the condensate through the solution to be extracted. The extracting liquid emerges as fine droplets from a dispersion disk beneath the funnel that collects the condensate. For liquids lighter than the extracting solvent, the dispersion disk is placed near the bottom of the extraction vessel. The extracting liquid, being lighter than the aqueous phase, rises to the surface of the aqueous phase and, in so doing, allows the extractable solute to partition between the droplets and the bulk solution. Eventually the extracting liquid flows back into the receiving flask (and boiler), from where it is evaporated and recycled. The extracted solute remains in the receiving flask. For liquids heavier than the phase being extracted, the condensed solvent is allowed to drop down through the phase being extracted and then it flows to the outer part of the extractor. Spent solvent rises in the outer cylinder of the extractor and exits into the boiler for stripping and recycling. A number of modifications and types of extractors are described by Morrison and Freiser.[2]

SOLVENT EXTRACTION SYSTEMS

As a resultant of the several partition forces discussed, hydrated inorganic salts tend to be more soluble in water than in organic solvents. Conversely, organic molecules tend to be more soluble in organic solvents unless they incorporate a sufficient number of hydrophilic groupings within their framework. The simple molecules constitute the first category of extraction systems. Their distribution obeys the Nernst partition law.

Ionic compounds would not be expected to extract into organic solvents from aqueous solution because of the large loss in electrostatic solvation energy which would occur. However, through addition or removal of a proton or a masking ion, an uncharged extractable species may be formed. These formally neutral or pseudomolecular species constitute the second category. The Nernst partition law is obeyed by the system as a whole when account is taken of chemical equilibria and the relative concentrations of all species can be expressed in terms of chemical equilibrium constants. Included in this category are the neutral metal chelate complexes.

In the third category, masking of ionic character takes place through association between ions. Non-specific interactions with the solvent or other solutes are often involved in ion association systems. Usually no simple mathematical treatment is possible because relevant partition coefficients are no longer constant throughout the range of experimental conditions.

[2] G. H. Morrison and H. Freiser, *Solvent Extraction in Analytical Chemistry*, Wiley, New York, 1957.

Simple Molecules

When chemical equilibria between species within a phase are not involved, the thermodynamic partition law holds. Included in this class are neutral covalent molecules which are not solvated by either of the solvents. The partition coefficient is applicable over a considerable range of solute concentrations. It corresponds in value to the ratio of the saturated solubilities of the solute in each phase.

An enormous number of organic substances distribute between aqueous and organic phases in strict accordance with the partition law. Ignoring any specific solute–solvent interactions, the solubility of such molecular species in organic solvents is generally at least an order of magnitude higher than in water; that is, $K_d \geqslant 10$. This emphasizes the preference of the covalent solute molecules for the disordered organic solvent into which they can fit with little or no energy barrier over the comparatively highly ordered aqueous phase. In each homologous series, increasing the chain length increases the partition coefficient by a factor of about 4 for each new —CH_2— group incorporated into the molecule. Branching will result in a lower K_d as compared with the *n*-isomer. When a substituent is part of an aromatic ring the effect per carbon atom is smaller. The effect of specific interactions upon the partition is shown by the introduction of hydrophilic groups into the molecule.[3] An alcoholic OH or a carbonyl group reduces the partition by a factor of 5–150; the effect is smaller if the molecule already contains other hydrophilic groups that are equally effective. Ether bridges reduce the partition by a factor of 5–20. The foregoing molecules can form hydrogen bonds with water and so are more acceptable in that phase than in the inert hydrocarbons. Substitution of a carboxyl group for a methylene group reduces the partition by a factor of 2–170, the range again reflecting the possible presence of other hydrophilic groups. Esterification has virtually the same effect as adding an equivalent number of methylene groups. Introduction of a halogen atom increases the partition by a factor of 4–40. An aliphatic amine group lowers the partition by a factor of 20–1000; conversion of a carboxyl group to an acid amide has a similar effect. For multiple hydrophilic groups in the molecule, two hydrophilic groups in the alpha (or ortho) position exert a greater effect than if their location is more remote in the molecule.

Provided their internal bonding is covalent, some neutral inorganic molecules fit into this classification but their number is quite small. Examples include iodine, the tetroxides of osmium and ruthenium, and $HgCl_2$.

Association in the Organic Phase Dimerization and higher degrees of association in the organic phase, plus any interaction between solute and

[3] R. Collander, *Acta Chem. Scand.* **3**, 717 (1949).

solvent will bring about an increase in the distribution ratio. Consider the association of fatty acid molecules to form dimers in the organic phase, expressed on the basis of the unassociated molecule,

$$2HA_{(o)} \rightleftharpoons (HA)_{2(o)} \tag{3-14}$$

The dimerization constant, K_2, is given by

$$K_2 = [(HA)_2]_o/[HA]_o^2 \tag{3-15}$$

The distribution ratio becomes

$$D = ([HA]_o + 2[(HA)_2]_o)/[HA] \tag{3-16}$$

or, incorporating the partition coefficient,

$$D = K_d(1 + 2K_2[HA]_o) \tag{3-17}$$

Dimerization results in a decrease in concentration of monomer, the species which takes part directly in the phase partition, so that the overall distribution increases. Moreover, the distribution ratio depends upon the concentration of solute distributing and upon the value of the dimerization constant. The extent of association can be ascertained by plotting the logarithm of $[HA]_o$ against the logarithm of $[HA]_{aq}$; the slope gives the n-mer value.

The tendency for a carboxylic acid to associate to a dimer depends on the tendency of the —C(→O)(—OH) groups to form hydrogen bonds. The dimerization is, however, very dependent on the solvent. Usually very large dimerization constants are obtained in hexane, benzene, and carbon tetrachloride. Smaller values are obtained with $CHCl_3$ which is a weak donor for hydrogen bonds. The dimerization of carboxylic acids is very slight in oxygen-containing solvents like ethers and ketones, which are good acceptors for hydrogen bonds, and is almost negligible in hydroxyl-containing solvents such as water.

Complexing Agent Dependence The addition of another solute to the aqueous phase may markedly affect the distribution. A classical example is the distribution of molecular iodine between water and CCl_4. Only the molecular species, I_2, exists in the organic phase. However, the addition of iodide ion to the aqueous phase results in the equilibrium.

$$I_2 + I^- \rightleftharpoons I_3^- \tag{3-18}$$

The formation constant, K_f, of the triiodide ion is given by

$$K_f = [I_3^-]/([I_2][I^-]) \tag{3-19}$$

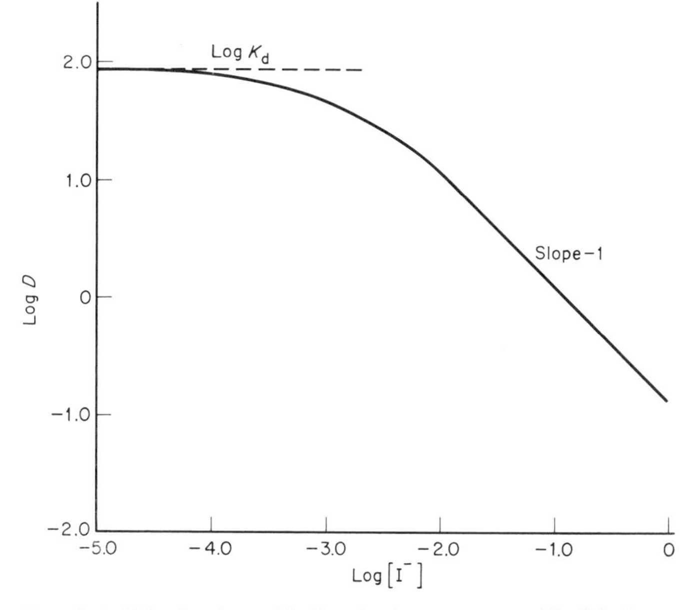

FIG. 3-1. Distribution of iodine in the presence of iodide ions.

The distribution ratio is given by

$$D = [I_2]_o/([I_2] + [I_3^-]) \tag{3-20}$$

Incorporating the expression for the formation constant into Eq. (3-20) and the partition coefficient for iodine, the distribution ratio becomes

$$D = K_d/(1 + K_f[I^-]) \tag{3-21}$$

The distribution can be followed experimentally by plotting the logarithm of the distribution ratio as a function of the logarithm of the free iodide ion concentration in the aqueous phase (Fig. 3-1). Two regions are apparent from the graph: The plateau of zero slope at low concentration of iodide ion, when $D \simeq K_d$; and the linear segment with slope of −1. When $K_f[I^-] \gg 1$, the expression for the distribution ratio reduces to

$$\log D = \log K_d - \log K_f[I^-] \tag{3-22}$$

The slope indicates the dependence of the system upon the iodide ion concentration. From a knowledge of the value of the partition coefficient, K_d, the formation constant can be evaluated.

Pseudomolecular Systems

The most obvious way to make an ionic species extractable is to destroy its charge. This can be done by combining the solute ion of interest with an ion of opposite charge to form a neutral molecular species. The larger, bulkier, and more hydrophobic the resulting molecule, the better will be its extraction.

An example is furnished by the anion of a carboxylic acid. By adjusting

$$\begin{array}{l} \qquad R\mathrm{COOH} \\ \text{org} \quad \uparrow\downarrow \\ \hline \text{aq} \\ \qquad R\mathrm{COOH} \rightleftharpoons R\mathrm{COO}^- + \mathrm{H}^+ \end{array}$$

the pH of the aqueous phase, the association of the carboxylate ion with a hydrogen ion to form the neutral acid is controlled

$$R\mathrm{COO}^- + \mathrm{H}^+ \rightleftharpoons R\mathrm{COOH} \tag{3-23}$$

The chemical equilibria can be expressed by the association constant, $K_{\mathrm{H}A}$,

$$K_{\mathrm{H}A} = [R\mathrm{COOH}]/[R\mathrm{COO}^-][\mathrm{H}^+] \tag{3-24}$$

which is recognized as the reciprocal of the acidic dissociation constant. Only the neutral molecular species partitions between the contacting phases; the partition coefficient is given by

$$K_d = [R\mathrm{COOH}]_o/[R\mathrm{COOH}] \tag{3-25}$$

The overall distribution ratio is expressed by

$$D = [R\mathrm{COOH}]_o/[R\mathrm{COOH}] + [R\mathrm{COO}^-] \tag{3-26}$$

which, by suitable substitution of Eqs. (3-24) and (3-25), becomes

$$D = \frac{K_d}{1 + 1/K_{\mathrm{H}A}[\mathrm{H}^+]} \tag{3-27}$$

or

$$\frac{1}{D} = \frac{1}{K_d} + \frac{1}{K_d K_{\mathrm{H}A}[\mathrm{H}^+]} \tag{3-27a}$$

Figure 3-2 is a plot of Eq. (3-27) for the distribution of an acid between $CHCl_3$ and aqueous solutions. Two regions are apparent. When $1/K_{\mathrm{H}A}[\mathrm{H}^+] \ll 1$, that is, $[R\mathrm{COOH}] \ll [R\mathrm{COO}^-]$, $\log D \simeq \log K_d K_{\mathrm{H}A} - \mathrm{pH}$. The plot is a straight line of unit slope. At higher concentrations of hydrogen ion, the molecular species is no longer negligible and the slope of the curve changes from unity to zero. The neutral molecular species is then the dominant species in both phases and $\log D \simeq \log K_d$. The point of intersection of the two linear segments of the graph provides the value of $\log K_{\mathrm{H}A}$. When the partition coefficient is not too large and data are limited, a plot of $1/D$ vs $1/[\mathrm{H}^+]$ is convenient. This utilizes Eq. (3-27a); the intercept gives $1/K_d$ and the slope $1/K_d K_{\mathrm{H}A}$.

An exactly analogous situation is furnished by weak bases, such as the

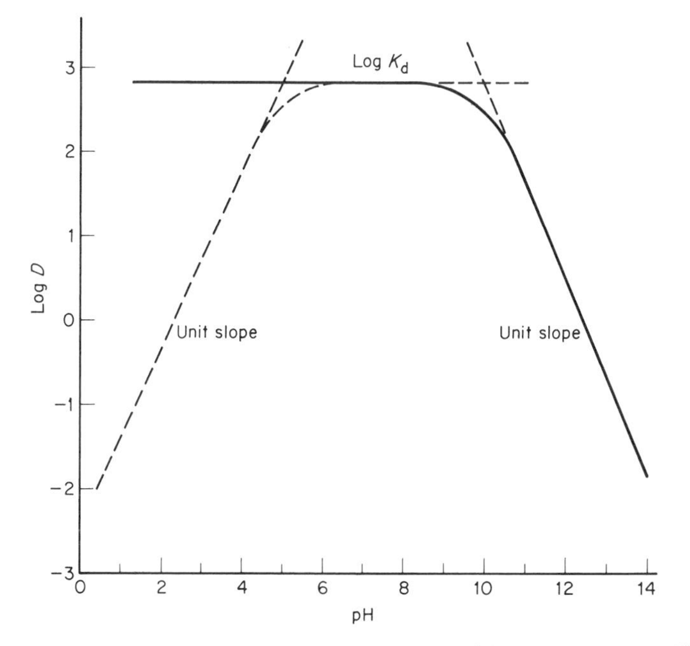

FIG. 3-2. Log D vs pH for a weak acid (RNH_3^+ type) with $K_{HA} = 1.25 \times 10^5$ (---------) and RCOOH type with $K_{HA} = 6.7 \times 10^9$ (———); $K_d = 720$ for both. Typical for 8-hydroxyquinoline.

amines. The uncharged amine base distributes between the two phases. As the hydrogen-ion concentration is increased, increasing amounts of the neutral amine are converted to the positively charged RNH_3^+ ion and the distribution ratio declines. An amphiprotic molecule, such as 8-hydroxyquinoline, shows both acidic and basic behavior.

Since the distribution ratios for organic acids and bases depend markedly on the pH of the aqueous phase, substances with only small differences in acidic behavior may be separated by means of multiple extractions between an organic phase and a buffered aqueous phase. Although not practical for batch liquid–liquid extractions, such separations are quite feasible with extraction trains or by columnar partition chromatographic methods.

Metal Chelate Systems

To convert a metal ion in aqueous solution into an extractable species, the charge has to be neutralized and waters of hydration have to be displaced. Qualitatively, it is to be expected that if metal ions are effectively surrounded by hydrophobic ligands which are able to bring about charge neutralization and occupy all the positions in the coordination sphere of the metal ion, distribution will strongly favor organic solvents. Chelate systems offer one type of metal extraction system, ion association systems another. Factors affecting the formation of metal chelates were discussed in Chapter 2. Although in a strict sense metal chelate systems perhaps belong with the pseudomolecular systems, metal chelate systems constitute such an important class that they will be considered separately.

Extraction Equilibria The equilibria involved in a metal chelate extraction system can be outlined in the following manner:

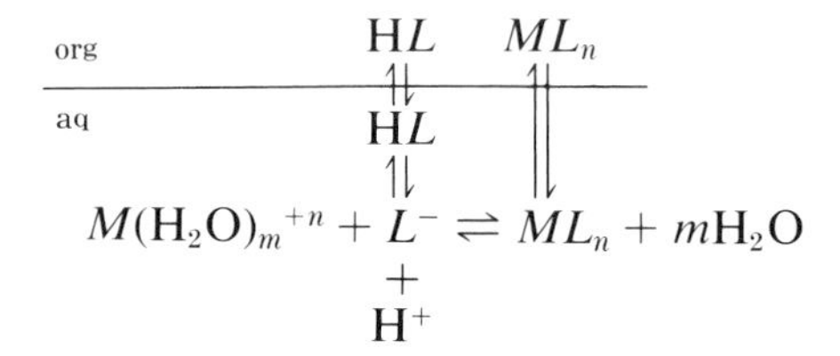

The distribution of the chelating agent, HL, usually a weak, polyfunctional organic acid, between the two phases is expressed by its partition coefficient, $(K_d)_r$

$$(K_d)_r = [\mathrm{H}L]_o/[\mathrm{H}L] \tag{3-28}$$

In the aqueous phase the association of the ligand with hydrogen ions

$$L^- + \mathrm{H}^+ \rightleftharpoons \mathrm{H}L \tag{3-29}$$

is expressed by the appropriate acid association constant, $K_{\mathrm{H}L}$,

$$K_{\mathrm{H}L} = [\mathrm{H}L]/[\mathrm{H}^+][L^-] \tag{3-30}$$

Some of the free ligand ions in the aqueous phase react with the hydrated metal ions to form an extractable chelate, ML_n,

$$(\mathrm{H_2O})_m M^{+n} + nL^- \rightleftharpoons ML_n + m\mathrm{H_2O} \tag{3-31}$$

The formation constant of the metal chelate, K_f, is expressed by

$$K_f = [ML_n]/[M(\mathrm{H_2O})_m{}^{+n}][L^-]^n \tag{3-32}$$

Finally, the distribution of the metal chelate between the two phases is represented by the partition coefficient, $(K_d)_c$

$$(K_d)_c = [ML_n]_o/[ML_n] \tag{3-33}$$

The expression for the distribution ratio of the metal between the two phases is

$$D = [ML_n]_o/([ML_n] + [M(\mathrm{H_2O})_m{}^{+n}]) \tag{3-34}$$

From the equilibrium relations in Eqs. (3-28), (3-30), (3-32), (3-33), and (3-34), it can be shown that

$$\frac{1}{D} = \frac{1}{(K_d)_c} + \frac{K_{HL}{}^n (K_d)_r{}^n [H^+]^n}{K_f (K_d)_c [HL]^n} \tag{3-35}$$

Several simplifying assumptions are involved in the derivation of Eq. (3-35): (1) The absence of species other than ML_n in the organic phase; i.e., no polymers or adducts with solvent or excess reagent. (2) Negligible amounts of intermediate complexes from step-wise equilibria. (3) The absence of hydroxy or other anion coordination complexes with the metal ion in the aqueous phase. The distribution ratio, as shown by Eq. (3-35), is dependent on the nature of the reagent, the reagent concentration, the organic solvent, and the pH of the aqueous phase, but is independent of the initial concentration of the metal.

When all the metal ion is complexed by the chelating agent, $D = (K_d)_c$ and the maximum degree of extraction has been attained (barring polymerization or adduct formation with solvent in the organic phase) regardless of changes in reagent concentration or pH of the aqueous phase. The upper limit of extraction of a metal chelate depends on the relative solubility of the metal chelate rather than on the chelate formation constant or other factors relating to the chelate formation. The effect of changing the solvent on the extractability of a particular metal ion may be inferred from Eq. (3-35) in terms of the changes in the values of the partition coefficients, realizing that such a change may very well be similar for the reagent and its chelate so that

$$\frac{(K_d)_{c,1}}{(K_d)_{c,2}} \simeq \frac{(K_d)_{r,1}}{(K_d)_{r,2}} \tag{3-36}$$

It follows then

$$\frac{D_1}{D_2} = \left[\frac{(K_d)_{r,2}}{(K_d)_{r,1}}\right]^{n-1} \tag{3-37}$$

Thus, a change to a solvent in which a reagent and its multivalent metal chelates ($n > 1$) are more soluble will result in a lower D value.

Solvent extraction offers an opportunity to ascertain species that dominate in each phase over restricted regions of operating parameters. The composition of the complexes formed in the aqueous phase is established from the relationship between the distribution ratio of the metal and one parameter: pH of aqueous phase, the concentration of chelating agent, $[HL]_o$, the concentration of a complexing or masking agent in the aqueous phase, $[X]$, or the concentration of adduct-forming agent, $[B]_o$ over a fairly broad set of conditions. As Rydberg [4] and Schweitzer [5] have shown, the distribution ratio varies differently for various types of complexes when plots

[4] J. Rydberg, *Arkiv. Kemi* **8,** 113 (1955).
[5] G. K. Schweitzer, *Anal. Chim. Acta* **30,** 68 (1963).

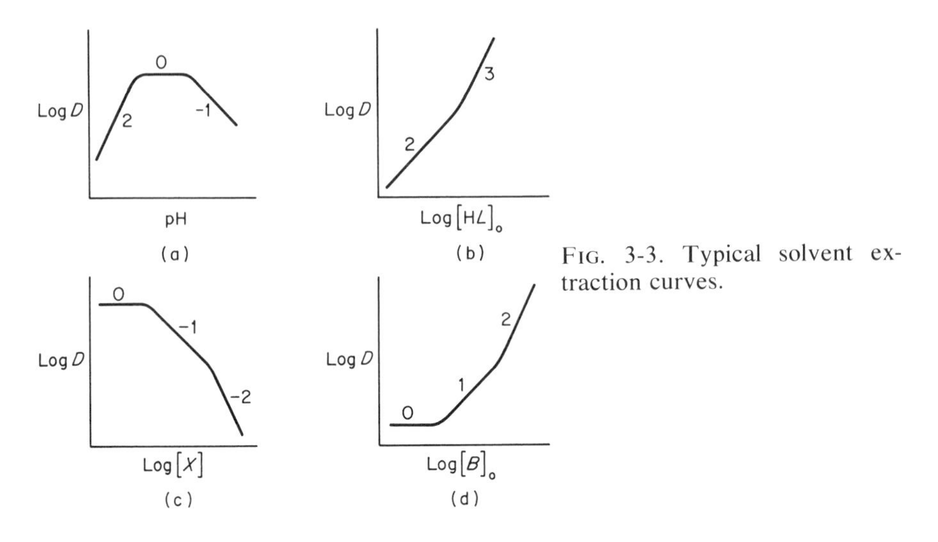

FIG. 3-3. Typical solvent extraction curves.

of the following types are constructed (Fig. 3-3): (a) log D vs pH at constant $[HL]_o$, $[X]$, and $[B]_o$; (b) log D vs log $[HL]_o$ at constant pH, $[X]$, and $[B]_o$; (c) log D vs log $[X]$ at constant pH, $[HL]_o$, and $[B]_o$; and (d) log D vs log $[B]_o$ at constant pH, $[HL]_o$, and $[X]$. Regions of constant slope usually signify conditions under which a single species dominates in each phase. Portions of the curves with changing slopes are usually indicative of the presence of several species in one of the phases.

Hydrogen Ion Dependence In general, the extraction of a simple metal chelate can be described by a plot of log D vs pH at constant reagent concentration $[HL]_o$ in the organic phase (Fig. 3-4). The curve consists essentially of two linear portions. Starting at a low pH, log D increases with a slope of n and then eventually reaches a constant, pH-independent value determined by the partition coefficient of the metal chelate. The slope n is equal to the number of hydrogen ions released per metal ion in the overall extraction reaction.

Over a limited range of conditions, the distribution ratio may usually be given by an expression that involves a singular species in the organic phase and one aqueous-phase species. When $D \simeq (K_d)_c$, the neutral metal chelate, ML_n, dominates in each phase. At values $D < (K_d)_c$, the distribution ratio is given approximately by

$$D \simeq \frac{(K_d)_c K_f [HL]_o^n}{(K_d)_r^n (K_{HL})^n [H^+]^n} \tag{3-38}$$

Equation (3-38) shows that the distribution ratio is dependent only upon the pH of the aqueous phase and the amount of unused chelating agent. All other

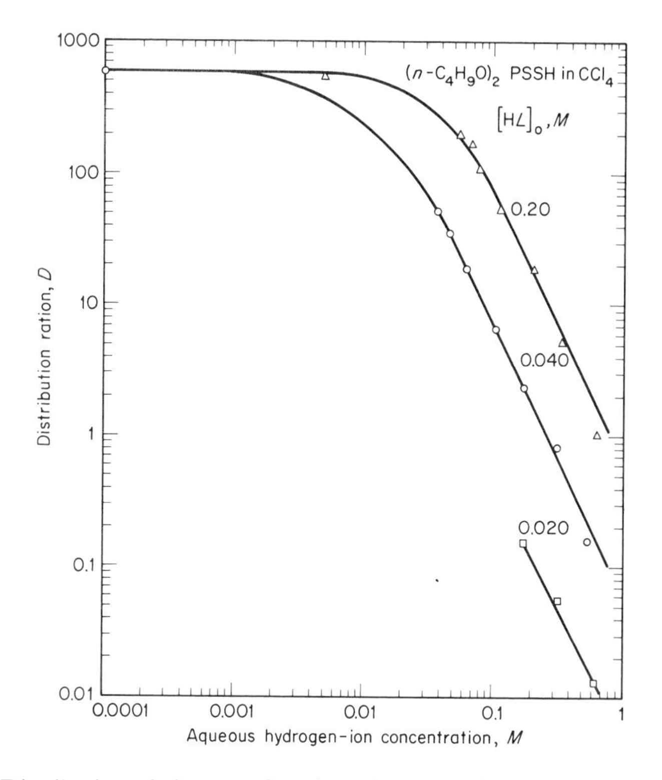

FIG. 3-4. Distribution of zinc as a function of aqueous hydrogen ion concentration for three different concentrations of chelating agent. [From T. H. Handley, R. H. Zucal, and J. A. Dean, *Anal. Chem.* **35,** 1163 (1963). *Courtesy of American Chemical Society.*]

terms are constants for a particular metal chelate-and-solvent pair. Indeed one may write

$$D = K^*[HL]_o^n/[H^+]^n \tag{3-39}$$

where K^* is the extraction constant. The value of the extraction constant can be obtained by substituting the appropriate experimental quantities into Eq. (3-38). From the portions of the log D vs pH plot with zero slope, the value of the partition coefficient, $(K_d)_c$ can be estimated. Values for the acid association constant and the partition coefficient of the reagent are determined separately from a log D vs pH plot for the chelating agent itself. These values then enable the formation constant of the metal chelate to be calculated.

Extraction information is sometimes represented as "percent extraction," denoted by the symbol E, plotted against the pH of the aqueous phase. The extent of extraction of a series of metal ions with a given

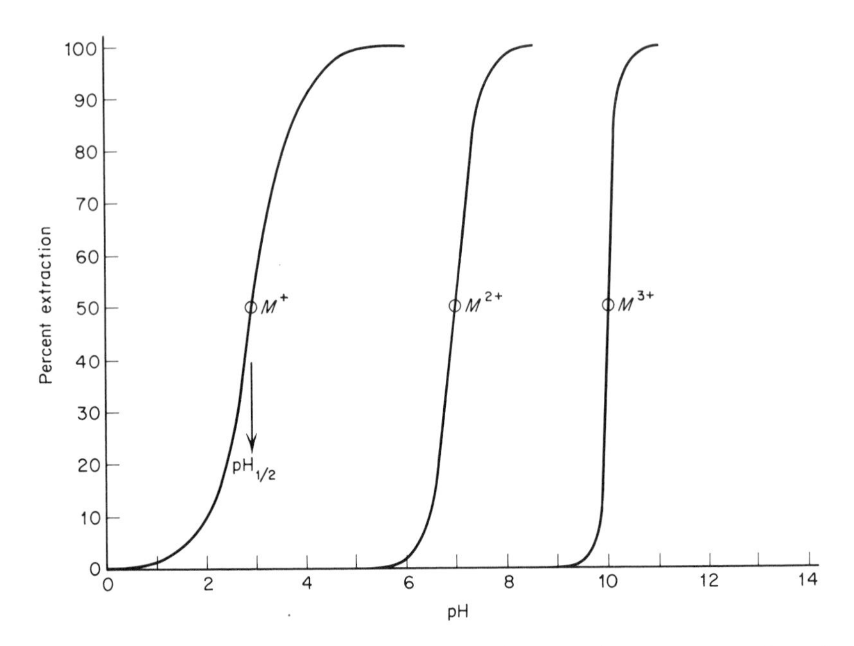

FIG. 3-5. Extent of extraction for metals of different valence states with a given chelating agent. No significance should be attached to the placement of each plot along the pH axis.

chelating agent is shown in this manner in Fig. 3-5. The range between 1 and 99% extraction, when the aqueous phase is shaken with an equal volume of organic solvent, is covered by $4/n$ units of pH. The value of the hydrogen-ion concentration when, for fixed excess chelating agent, the extraction reaches 50% (i.e., $D = 1$), is denoted $pH_{1/2}$. The percent extraction curves do not convey as much information about a system as do the distribution curves. However, they are frequently used in representing extraction data for a series of metal ions when interest centers around the data for removal purposes.

Chelating Agent Dependence The extent of extraction increases with the equilibrium (excess) concentration of chelating agent in the organic phase. At a higher value of the reagent concentration, the entire curve (Fig. 3-4) shifts to a region of lower pH, without alteration of either the slope or the value of the partition coefficient. This fact will prove useful in the extraction of metals from solutions which must be kept sufficiently acid to avoid metal hydrolysis. From a plot of log D vs the logarithm of "free" ligand concentration, i.e., either log $[L^-]_{aq}$ or log $[HL]_o/[H^+]$ (at constant $[HL]_o$), the slope will show the number of reagent ligands incorporated in the extractable complex (Fig. 3-6).

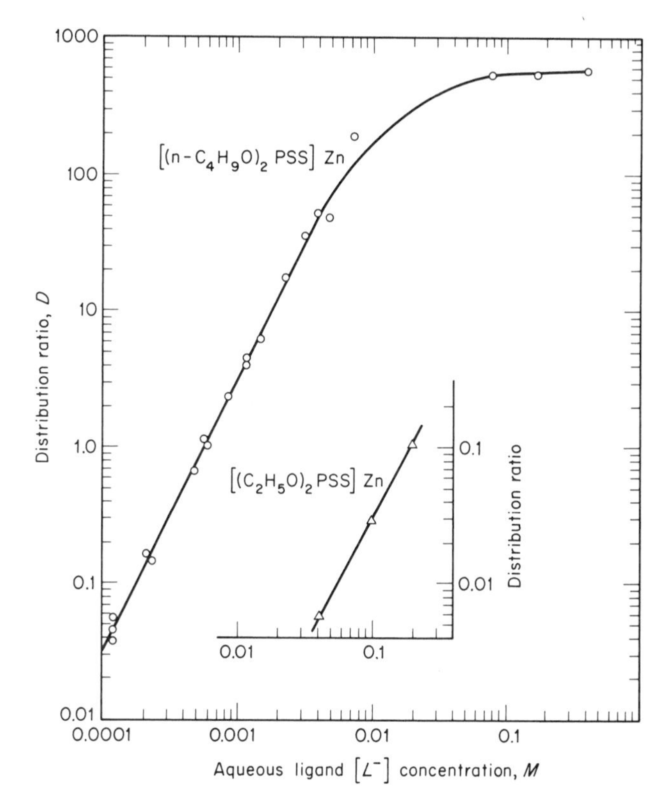

FIG. 3-6. Distribution of zinc as a function of aqueous ligand concentration (or log $[HL]_o/[H^+]$). [From T. H. Handley, R. H. Zucal, and J. A. Dean, *Anal. Chem.* **35,** 1163 (1963). *Courtesy of American Chemical Society.*]

Adduct Formation In the liquid–liquid extraction of the chelates of many transition metal ions whose coordination number is greater than twice the ionic charge, a remarkable increase in their distribution coefficients has been observed when the extraction system contained molecules with atoms that can act as electron pair donors. The enhanced values of the distribution coefficients have been attributed to the formation of chelate adducts between the metal chelate that is an electron pair acceptor and the molecule containing the electron pair donor. This type of adduct formation is a Lewis acid–base interaction in which the metal acceptor atom tends to satisfy its characteristic coordination number by the process of chelation alone. The formation of chelate adducts in general enhances the hydrophobic properties of the metal chelates and increases their extractability into organic solvents. When adduct formation predominates, the value of a in the species $[ML_n{\cdot}aB]_o$ can be determined by a plot of log D vs log $[B]_o$ at constant pH and $[HL]_o$ concentration. The slope of the resulting line is $n + a$.

An example is magnesium 8-hydroxyquinolate dihydrate, $Mg(Ox)_2 \cdot 2H_2O$, which, although insoluble in water, finds the normal coordination energy of its hydrated water to be a barrier to extraction. Substitution of water in the coordination sphere by another more basic ligand that enhances the hydrophobic properties of the metal chelate, such as butyl Cellosolve (denoted S), butyl amine, or excess 8-hydroxyquinoline, increases extractability. The species extracting are represented as $Mg(Ox)_2 \cdot 2S$, $Mg(Ox)_2 \cdot 2C_4H_9NH_2$, and $Mg(Ox)_2 \cdot 2HOx$. In the event that the chelating agent is also involved in adduct formation, an increase in reagent concentration will cause the plateau portion of the extraction curve to shift to a higher value of log D, and the initial linear portion of the curve to shift to a lower pH without changing its slope.

Separation Factor The separation factor for a pair of metal ions, when the value of n is the same, in a given extraction system is given by

$$\frac{D_1}{D_2} = \frac{(K_f K_d)_1}{(K_f K_d)_2} \tag{3-40}$$

Selectivity can be increased by the use of masking agents or competing complexing agents. Addition of a masking agent moves the distribution curve to higher pH values, the shift being greater as the amount of complexing agent is increased and as the magnitude of the formation constant of the masked metal increases. For example, copper(II) is extracted with 8-hydroxyquinoline in $CHCl_3$ at a much lower pH than is uranium(IV); yet, if EDTA is added to the system, the uranium extraction proceeds as before whereas the copper extraction is delayed until the pH is raised by more than 5 units. Often greater selectivity can be achieved through the use of a particular metal chelate, rather than the chelating agent itself, as the ligand source. For example, copper is selectively extracted in the presence of iron(III) and cobalt using lead diethyldithiocarbamate, whereas all three metals would be coextracted with the sodium salt of diethyldithiocarbamate.

When the value of n differs for two metal ions, separation might be possible through adjustment of the pH of the aqueous phase, as was shown in Fig. 3-5.

Ion Association Systems

In ion association systems the inorganic ion, introduced initially as a salt, associates with an oppositely charged ion to form a neutral species, an ion pair, which is extractable into organic solvents. The undissociated ion pair behaves thermodynamically as a neutral molecule. The effect of such factors as ionic radii and the presence of hydrocarbon groups upon the extraction shows that best extraction is attained in the presence of minimum ionic charges at a definite cation-to-anion radius and in the presence of hydrophobic groups in one or both ions. When the coordination sites around the metal ion are incompletely filled by organic ligands, neutral complexes can sometimes be extracted into organic solvents, provided the solvents are of such types that they can coordinate strongly to the metal ion. This leads to a sharp distinction between oxygen-containing solvents such as alcohols, esters, ethers and ketones, and nonpolar solvents such as benzene and carbon tetrachloride.

In this brief treatment, ion association systems will be differentiated into three categories:

(1) Ion association alone.
(2) Chelation and ion association with either cationic or anionic metal chelates.
(3) Simple coordination and ion association—liquid ion exchangers and the "onium" systems.

Dielectric Constant Influence Extraction varies markedly with the dielectric constant of the solvent. A high value of dielectric constant can compensate for the loss of solvation. Thus, if a solvent of higher dielectric constant is used to improve extraction in another respect, the resulting specificity will be reduced. In a solvent with a small dielectric constant, the ions of opposite charge will not be kept apart as effectively as in a higher dielectric medium. The result is the formation of associated aggregates of a relatively bulky cation with a similar bulky anion. On transference to the organic phase, additional aggregation often occurs as all ion pairs tend to polymerize in a highly concentrated solution of small dielectric constant. Of course, polymerization results in a decrease in concentration of the monomeric ion-pair which takes part directly in the distribution equilibrium so that extraction increases.

The effect of temperature is tied in with the magnitude of the dielectric constant. In solvents of relatively high dielectric constant, the dielectric constant decreases with increase in temperature so that association increases. However, in solvents of very low dielectric constant, ion association falls off with rising temperature.

Salting Agents A high concentration of foreign electrolytes decreases the activity of "free" water and so promotes the loss of the hydration shell of the cation. Addition of "salting agents" increases generally the distribution ratio. These are small polyvalent cations which hydrate readily and thus tie up water to an appreciable extent. A secondary effect of the added salt is to decrease the dielectric constant of the aqueous phase and further encourage the formation of ion pairs. Obviously the salting agent should not be extractable.

Simple Ion Association The simplest category of ion association systems involves large and bulky cations and anions whose size and structure are such that they do not have a primary hydration shell. Furthermore, the large size of the ions disrupts the hydrogen-bonded water structure. The larger the ion the greater the amount of disruption and the greater the tendency for the ion association species to be pushed into the organic phase. Dispersion interaction of the solvent molecules in the organic phase is the only force which must be displaced. This force is of the same order of magnitude as that between the ion itself and the nearest organic solvent molecules. However, the dielectric constant of the solvent will play an important role since the solvation energy in the organic phase is primarily electrostatic in nature. Extraction will be greater the higher the dielectric constant of the organic solvent, but the resulting specificity will be reduced.

Relatively simple, stoichiometric equilibria are sufficient to describe these ion association systems. There is dissociation of the ion pairs in the aqueous phase and association in the organic phase. This is true often for the reagent salt as well as for the ion pair involving the principal metal. Examples of the reagent cation include the tetraphenylarsonium (ϕ_4As^+) and tetrabutylammonium (n-Bu_4N^+) ions, present usually as the chloride salts. The formation constant of the ion association complex of tetraphenylarsonium perrhenate

$$(\phi_4As^+,Cl^-)_o + ReO_4^- \rightleftharpoons (\phi_4As^+,ReO_4^-)_o + Cl^- \tag{3-41}$$

is given by

$$K_f = \frac{[(\phi_4As^+,ReO_4^-)]}{[\phi_4As^+][ReO_4^-]} \tag{3-42}$$

Ion pair formation is indicated by placing a comma between the cation and anion which are in association with each other. The distribution ratio is

$$D = \frac{K_f(K_d)_c[(\phi_4As^+,Cl^-)]_o}{K_r(K_d)_r[Cl^-]} \tag{3-43}$$

where K_r and $(K_d)_r$ are the association constant of the tetraphenylarsonium chloride ion pair and the partition coefficient of this ion pair between the two phases, respectively. Higher aggregation of ion pairs in the organic phase often occurs at higher concentrations of extracting species. The formal resemblance of the distribution ratio expression to those developed for metal chelates should be noted.

Chelation and Ion Association Polyvalent cations can be extracted if the cation is made large and hydrocarbon-like. This can be accomplished with certain chelate reagents having two uncharged coordinating atoms such as 1,10-phenanthroline, *IV*, which forms cationic complex chelates with metal ions

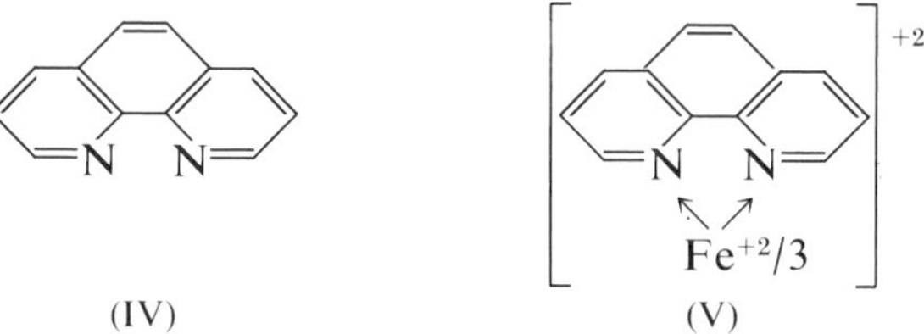

(IV) (V)

such as iron(II), iron(III), cobalt(II) and copper(II). In conjunction with a large anion such as ClO_4^-, *tris*(phenanthroline)iron(II) perchlorate, *V*, extracts moderately well into $CHCl_3$ and quite well into nitrobenzene. By using quite large anions, such as long-chain alkyl sulfonate ions in place of ClO_4^-, the extraction into $CHCl_3$ is improved. Another general method for making small, polyvalent metal cations extractable is to transform them into large negatively charged chelate complexes. An example is the formation of an anionic complex of magnesium with 8-hydroxyquinoline (Ox^-) ion in carbonate solution:

$$Mg^{+2} + 3Ox^- \rightleftharpoons MgOx_3^- \tag{3-44}$$

and extraction in the presence of the large tetrabutylammonium cation:

$$Bu_4N^+ + MgOx_3^- \rightleftharpoons (Bu_4N^+, MgOx_3^-) \tag{3-45}$$

Liquid Ion Exchangers High molecular weight amines in acid solutions form large cations capable of forming extractable ion pairs with a variety of anions. Tri(2-ethylhexyl)amine, methyldioctylamine and tribenzylamine have proved to be quite useful and versatile in the extraction of mineral, organic, and complex metal acids. The role of the amine is not completely clear. It may interact in the organic phase by hydrogen bonding and form micellar aggregates in which the organophilic portions of the molecules face outward to the solvent whereas the ionic portions are shielded from the solvent at the center of the micellar structure. In many of their properties they resemble the resin exchangers (Chapter 5), giving rise to the misnomer "liquid ion exchangers." The anion attached to the ammonium cation may be exchanged for other anionic species. Work is usually confined to extraction from acid solution. Under such conditions the amine is converted to its stable positive species which is practically undissociated in the amine-diluent (aliphatic hydrocarbons) solution. It modifies the solvent, lending polar properties to the solvent mixture and thus making possible the dissolution of excess acid. It provides an ideal method to remove mineral acids from biological preparations under mild conditions. As an illustration of the removal of complex metal acids, let us consider the extraction of zinc as $ZnCl_4^{-2}$ from an HCl solution using a benzene solution of tribenzylamine, R_3N:

benzene

$R_3NH^+,Cl^- \rightleftharpoons$ dimer $\qquad$ dimer $\rightleftharpoons (2R_3NH^+,ZnCl_4^{-2})$

$\upharpoonleft\downharpoonright \qquad\qquad\qquad\qquad \upharpoonleft\downharpoonright$

water

$R_3NH^+,Cl^- \rightleftharpoons Cl^- + R_3NH^+ \qquad 2R_3NH^+ + ZnCl_4^{-2} \rightleftharpoons (2R_3NH^+,ZnCl_4^{-2})$

$\qquad\qquad\qquad \downharpoonleft\upharpoonright \qquad\qquad\qquad \upharpoonleft\downharpoonright$

$\qquad\qquad H^+ + R_3N \qquad\qquad ZnCl_3^- + Cl^-$

$\qquad\qquad\qquad\qquad\qquad\qquad \vdots$

$\qquad\qquad\qquad\qquad\qquad\qquad Zn(H_2O)_4^{+2}$

"Onium" Systems Solvent molecules may participate in ionic reactions as in "onium-salt" formation. Recognition of the hydrated hydronium ion, $H_9O_4^+$, as the cation which pairs with halo-metallic anions in the extractable complex halo-metallic acids, has clarified the role of the oxygen-containing solvent in oxonium extraction systems. This cation must be stabilized by hydrogen bonding to the solvent. Thus the coordinating ability of the solvent (denoted S) is of central significance. In effectiveness, the order appears roughly to be isobutyl methyl ketone > butyl acetate > amyl alcohol > ethyl ether; solvents such as benzene and CCl_4 are ineffective. As an example, let us consider the extraction of iron(III) from HCl solution:

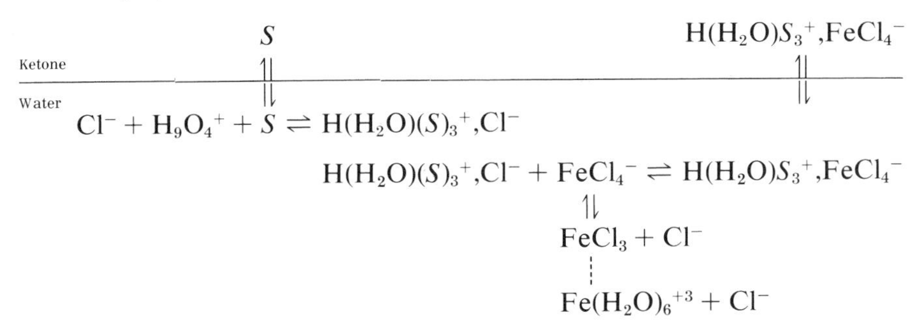

In the aqueous phase chloride ions progressively replace the water molecules in the hydration shell until the ferric ion is tetrahedrally surrounded by chloride ions. Organic solvent molecules are drawn into the aqueous phase where they compete with water molecules for positions in the solvation shell of the proton. The primary species which partitions into the organic phase is assumed to be $[H(H_2O)(Solvent)_3^+,FeCl_4^-]$. In solvents of low dielectric constant, the primary species may aggregate by dipole interaction of the ion pairs; it may also associate with ion-paired hydrochloric acid. The distribution ratio decreases at high HCl concentrations due to organic solvent being pulled over into the aqueous phase and the active competition of ion-paired HCl for the solvent.

COUNTERCURRENT DISTRIBUTION

Liquid–liquid partition methods can be extended to the separation of solutes possessing only small differences in their partition coefficients by a method called countercurrent extraction or distribution. Countercurrent distribution is a multiple partition process with a large number of stages, entirely discontinuous and stepwise in nature. It requires equipment with 100, 200, 300, or even 400 separating chambers in which equilibrations and phase transfers can be carried out automatically. The mode of operation is an automatic mixing of the two phases and the transfer of one phase (moving phase) to the subsequent separation chamber. Compounds with a high K_d value will have the highest migration speeds followed by the compounds with smaller values. Applications lie predominately in the separation of complex mixtures of organic (often natural) substances. Separations are achieved under very mild conditions; thus the method is ideally suited for handling labile materials. No material is wasted, and all material put into a distribution machine can be recovered. The method is often employed when column or gas chromatography fails. It is not successful with compounds of limited solubility, with salts or other strongly polar compounds, or with saturated hydrocarbons.

Equipment

The glass distribution machine, devised by Craig, consists of a train of interlocking glass units, shown in Fig. 3-7. Each unit consists of an extrac-

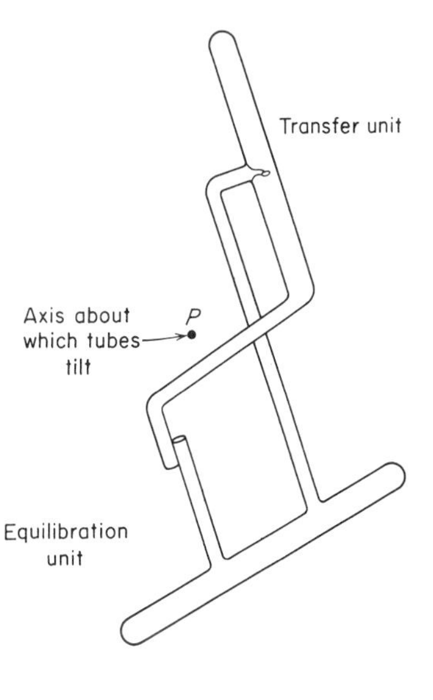

Fig. 3-7. Single interlocking glass unit from a glass countercurrent distribution train (after Craig and Post).

tion tube and a decantation tube. The tubes (often 100 tubes per bank) are mounted on a frame that rocks them back-and-forth a definite number of times during each contacting period, holds them motionless while the phases separate, then tilts them to allow the upper (lighter) phase in each tube to drain into the next tube. Reverse flow is hindered by the constriction in the side arm as it connects to the extraction tube. The entire equilibration and transfer cycle is carried out by an adjustable, automatic mechanism in about 1–2 min.

The operation of a Craig unit resembles in many respects a chromatographic column in the elution method (*q.v.*). First, each tube is filled up to the side arm with heavy-phase (stationary) solvent. The sample is placed in the first tube in the series. For a larger load the sample may be scattered in several of the first tubes. Light-phase (mobile) solvent is then added to the first tube. After the unit has cycled through a contact and separation stage, this lighter phase drains into the next tube and fresh light-phase solvent flows into the first tube. The lighter (or upper) phase gradually travels through the row of tubes. Components of the sample, each with its own partition coefficient, travel at varying rates. The separation ends when the desired components are isolated, each in a separate group of tubes within the series.

In the Ronor column (Fig. 3-8), the separation compartments are cylindrically shaped chambers built from a large number of round glass disks with eccentric openings and a Teflon ring functioning as an outside wall and spacer. Held at an angle of about 45°, the compact column is rotated around

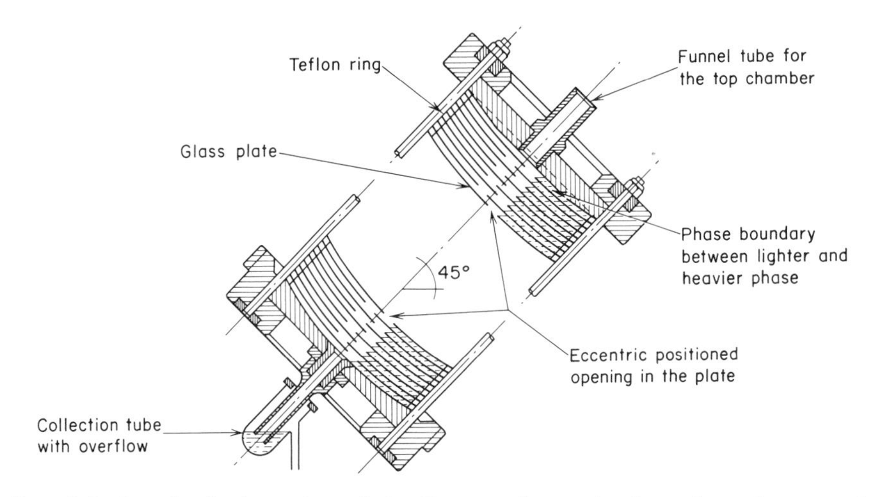

FIG. 3-8. Longitudinal section of the Ronor column: 1, glass plate; 2, excentric opening; 3, Teflon ring; 4, funnel opening to column; 5, phase boundary; 6, collection at column discharge. Courtesy of Brinkmann Instruments.

its longitudinal axis (about 20 rpm). One revolution represents a complete operation cycle—addition of moving phase, mixing and transfer of the moving phase into the next lower compartment. Transfer takes place whenever the eccentric hole in the glass disk reaches the lowest point in its circular motion. In a sense, each compartment represents a theoretical plate (see Chapter 4). Initially the column is filled with stationary phase and subsequently with moving phase. The moving phase will replace a certain amount of stationary phase and form a layer a few millimeters thick on top of the stationary phase. The sample mixture is introduced into the first separation chamber. Fresh moving phase is continuously fed into the upper chamber and, after passing through the column, collected at the discharge end with the help of a fraction collector.

Principle

A systematic extraction scheme for a Craig distribution machine is outlined in Fig. 3-9. Here each contacting phase is represented by a rectangle and numbered serially, starting with zero. The stages or transfer numbers are on the left. The top row gives the state of affairs after one equilibration but no transfers. Solute A is assumed to possess a $K_d = 0.5$; solute B, $K_d = 2.0$. Also assumed are unit quantity of each solute and equal phase volumes. After transfer 1, the fraction of each solute in the four phases is represented by the number inside the appropriate rectangle. Similarly, the distribution of fractional parts after transfers 2, 3, and 4 is shown. Solute A will spread preferentially to the right whether the upper and lower phases are moved alternately, or if only one phase is moved at each equilibration. As each solute migrates it also spreads and proceeds to occupy more tubes. When the transfers are expanded to 9 (10 tubes in the apparatus), the results shown in Fig. 3-10 are obtained.

The fraction of solute in each tube (both phases) can be expressed by the simple binomial

$$(p + q)^n = 1 \qquad (3\text{-}46)$$

where p is the fraction of solute in the mobile phase and q that in the stationary phase. The exponent n is the number of transfers. Since the fraction in each phase at equilibrium is fixed by the distribution ratio and phase volume-ratio, Eq. (3-46) can be written as

$$\left[\frac{D(V_M/V_S)}{1 + D(V_M/V_S)} + \frac{1}{1 + D(V_M/V_S)}\right]^n = 1 \qquad (3\text{-}47)$$

Theoretically there is no limit to the number of transfers which may be effected. In practice, a limit is set by the ability to detect solute in tubes lying well away from the maximum and the diminishing quantity of solute in tubes lying within the broadening band.

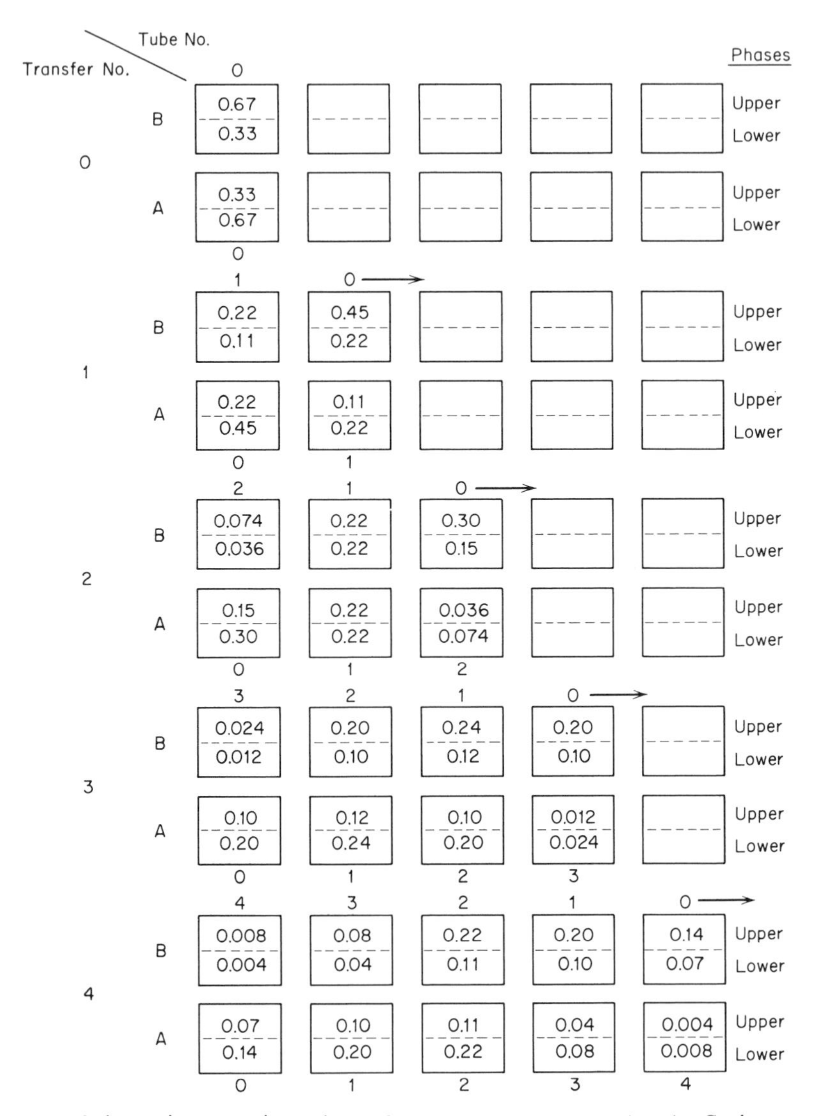

FIG. 3-9. Schematic extraction scheme for four transfers following the Craig countercurrent fundamental procedure. For solute A, $K_d = 0.5$; and for solute B, $K_d = 2.0$.

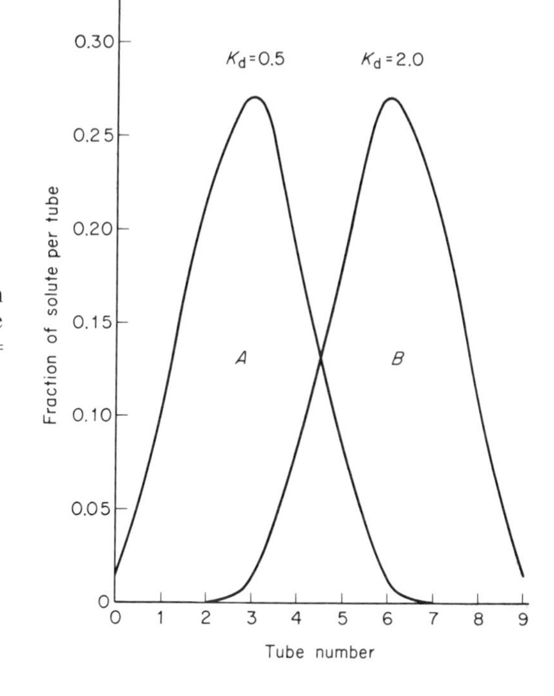

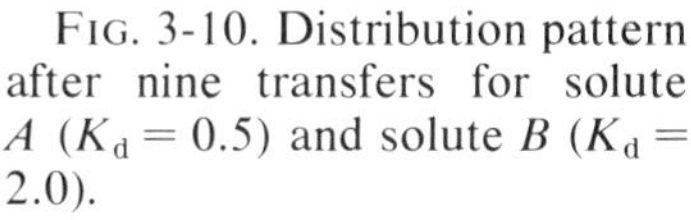
FIG. 3-10. Distribution pattern after nine transfers for solute A ($K_d = 0.5$) and solute B ($K_d = 2.0$).

For distributions involving large numbers of transfers (greater than 20), the mathematics of probability can be applied. The exact position (rth tube) of the peak of the distribution curve and the distribution ratio are related as follows:

$$r_{\max} \simeq np = \frac{nD(V_M/V_S)}{1 + D(V_M/V_S)} \tag{3-48}$$

or

$$D = \frac{r_{\max} + 1}{(n - r_{\max})(V_M/V_S)} \tag{3-49}$$

The distribution ratio also may be calculated from the concentration ratios of neighboring tubes:

$$D = \frac{(r + 1)T_{n,r+1}}{(n - r)(V_M/V_S)T_{n,r}} = \frac{rT_{n,r}}{(n - r + 1)(V_M/V_S)T_{n,r-1}} \tag{3-50}$$

where $T_{n,r}$ is the fraction of solute (both phases) in the rth tube.

The width of the zone containing 99.9% of the solute will extend for 3σ on either side of the zone maximum. For symmetrical zone profiles and unit phase ratio, it is given by

$$W = 6\sqrt{nD}/(1 + D) \tag{3-51}$$

Distribution Patterns One of the first necessary steps in counter-current distribution studies is the determination of the individual distribution ratio for each solute in the mixture. This is possible even though the number and identity of the solutes are totally unknown. With no information available beforehand, the sample must be distributed throughout the extraction train. The number of transfers required can only be estimated from experience. As an example, the distribution pattern for the lower fatty acids using 24 transfers in a 25-tube machine is shown in Fig. 3-11. After the last transfer, the total fatty acid content in each tube can be determined by an acidimetric titration. Results are plotted. If individual peak maxima appear, it is an indication that separation has proceeded to some degree.

The second step involves calculating the individual distribution ratio of each fatty acid, Eq. (3-49) or (3-50). A tube (or tubes) is selected which is most likely to be free of contamination from adjacent solutes. Often a sufficiently accurate estimate may be obtained by successive extractions of one phase from one of the tubes. The total weights of solute in each successive extract are inserted into the equation

$$1 - (w_2/w_1) = 1 - (w_3/w_2) = D(V_M/V_S)/[1 + D(V_M/V_S)] \qquad (3\text{-}52)$$

which is then solved for the distribution ratio.

The third step involves the calculation of the theoretical distribution pattern for each solute tentatively identified (the dashed lines in Fig. 3-11). The fraction of solute (ordinate of the distribution curve) at the maximum,

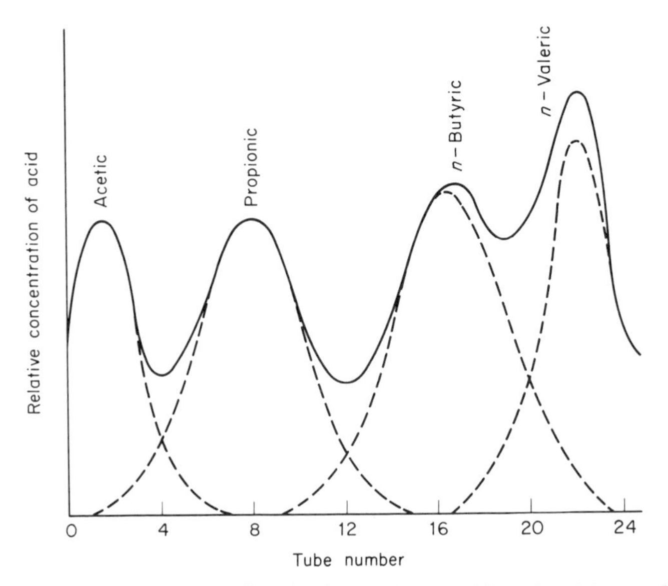

FIG. 3-11. Distribution pattern for the lower fatty acids using 24 transfers. [From *J. Biol. Chem.* **170,** 501 (1947). *Courtesy of Journal of Biological Chemistry.*]

y_o, is given by

$$y_o = 1/\sqrt{2\pi n D(D + 1)^2} \tag{3-53}$$

Since an experimental curve will already be available, the y_o value can be taken directly from it. The other ordinate values are calculated from the y_o value by substituting appropriate value (x) of the number of tubes either to the right or left of y_o in the expression

$$y_x = y_o/\text{antilog } 0.434x^2\sqrt{[2nD(D + 1)^2]} \tag{3-54}$$

Several points on either side of the maximum will provide sufficient points from which to construct the curve.

Finally, the individual distribution curves are summed. If the fit between the experimental and calculated pattern is satisfactory, at least within the limits of experimental error, then it may be presumed that no major component has been overlooked. However, any evidence of skewed experimental bands is strong indication of another solute whose distribution ratio is very similar to that of the major component overlapped.

In the foregoing discussion, the fundamental procedure has been described. It must always be accomplished before proceeding to the other operational variations: single withdrawal, completion of a square, recycling, and partial withdrawal. For operational details on these variations, the articles by Craig should be consulted.

LABORATORY WORK

Batch Extractions

The usual apparatus for batch extractions is a pear-shaped separatory funnel. The solute is extracted from one layer by shaking with a second immiscible phase in a separatory funnel until equilibrium has been attained (either from literature directions or by running a series of equilibrations at several time intervals to ascertain the time for attainment of equilibrium). After equilibration the two layers are allowed to settle out completely, and the layer containing the desired component is removed. Extractions should be performed at constant temperature; both the distribution ratio and the phase volumes are influenced by temperature changes. Simple inversions of the extraction vessel, repeated 50 times during 1–2 minutes, suffice to attain equilibrium in extractions of most organic substances and many metallic salts. Droplets of aqueous phase entrained in the organic phase can be removed by centrifugation or by filtering through a plug of glass wool. Whatman phase separating paper, available in grade 1PS, are silicone-treated papers which serve as a combined filter and separator in place of separatory funnels when solvent extractions are being made from aqueous solutions. The organic phase passes through the silicone-treated paper.

The water phase will be held back at all times in the presence of an immiscible organic solvent. These papers function best when used flat on a Büchner or Hirsch type funnel.

EXPERIMENT 3-1 *Distribution of Iodine Between an Organic Solvent and Aqueous Iodide Solutions*

Prepare these solutions: 1.00*M* iodide: dissolve 166 g KI in 1 liter of 0.001*M* H_2SO_4 (to prevent hydrolysis of I_2 to HIO). Iodine solution: accurately weigh 0.15 g I_2 crystals and dissolve in exactly 500 ml of the assigned organic solvent—cyclohexane, $CHCl_3$, CCl_4, or *n*-heptane. 0.100*N* $Na_2S_2O_3$: dissolve 24.8 g of the pentahydrate salt in 1 liter of distilled water (exact normality need not be known). 0.001*M* H_2SO_4; add 0.1 ml of concentrated acid to 1 liter of distilled water.

Equilibrate individual 25-ml portions of the iodine solution with these aqueous solutions:

Sample No.	1.0*M* iodide, ml	0.001*M* H_2SO_4, ml
1	25.0	0
2	15.0	10.0
3	10.0	15.0
4	5.0	20.0
5	2.5	22.5
6	1.5	23.5
7	1.0	24.0
8	5.0 [a]	20.0

[a] Use 0.1*M* KI solution

After the phases have separated, remove duplicate 10-ml aliquots from each phase. Determine the iodine content of each aliquot by titration with 0.100*M* $Na_2S_2O_3$ delivered from a 1.00-ml microsyringe. Use starch indicator for the aqueous samples. Titrate the organic portions in a glass-stoppered flask in contact with about 10–20 ml of distilled water. Shake well after adding each increment of thiosulfate; the end point is taken to be the disappearance of the violet color from the organic layer.

Estimate the partition coefficient by shaking 25.0 ml of the iodine solution with ten 25-ml portions of 0.001*M* H_2SO_4. Be sure that the phases have completely separated after each equilibration. Combine the aqueous extracts and re-extract with 25.0 ml of the pure organic solvent. Determine the iodine content of the entire organic phase as described previously.

Plot log D vs log $[I^-]$. Extrapolate the linear portion of unit slope to the intersection with the horizontal line of zero slope (the log K_d value). Estimate the value of K_f for the I_3^- species. Also prepare a plot of $1/D$ vs $[I^-]$;

estimate the value of $1/K_d$ from the intercept and the value of K_f/K_d from the slope. At 25°C, these solubilities of I_2 may be used to estimate values of K_d: H_2O, 0.336 g/l; cyclohexane, 2.719 g/100 g; $CHCl_3$, 3.2 g/100 ml; CCl_4, 2.91 g/100 ml. If necessary, correct the iodide concentration for the amount associated with I_2 in the aqueous phase as I_3^-.

EXPERIMENT 3-2 *Iron(III) Determination*

Standard Solutions Approximately 0.1 g of C.P. iron wire is dissolved in a mixture of 10 ml of 1 : 1 HCl and 4 ml of concentrated HNO_3 by gentle heating in a 150-ml beaker. Transfer the resulting solution to a 1-liter volumetric flask and dilute to volume with 0.1*M* HCl.

8-Hydroxyquinoline (oxine), 0.1*M*, is obtained by dissolving 14.5 g of reagent in 1 liter of $CHCl_3$.

PROCEDURE

Take three iron samples containing approximately 30, 60, and 90 μg of iron(III) per 25 ml, adjust the pH to 2.5–2.8, and transfer to 125-ml separatory funnels. Rinse the beakers with 0.001*M* HCl solution and add the rinsings to the separatory funnels. Pipet 25.0 ml of 0.1*M* oxine to each separatory funnel and shake gently for 5 min. After equilibration and phase separation, transfer the $CHCl_3$ layers to glass-stoppered sample bottles. Measure the absorbance at 470 mμ. Prepare a calibration curve from the data and use it to determine the iron(III) content of unknown samples carried through the same procedure.

EXPERIMENT 3-3 *Extraction of Iron(III) Chloride*

The extraction of iron(III) from HCl solution is a simple method of separating iron from the other components of steel alloys. As a series of interrelated experiments, the student might study the influence of $[H^+]$ and $[Cl^-]$ on the extent of extraction and the difference among the several solvents that have been employed—ethyl ether, amyl acetate, and methyl isobutyl ketone. The resultant iron(III) content of the aqueous phase could be determined as described in Experiment 3-2; that in the organic phase by the same procedure after back extraction into an aqueous medium.

At a constant chloride concentration of 6*M*, vary the hydrogen ion concentration from 0.01 to 10*M* with H_2SO_4. Graph the percent extraction vs $[H^+]$. Repeat the series for each type of solvent.

At a constant hydrogen ion concentration of 1*M*, vary the chloride ion concentration from 0.1 to 3*M* with KCl. Graph the percent extraction vs $[Cl^-]$.

To illustrate the influence of masking agents, such as phosphate, repeat the extraction of iron in 6*M* chloride ion over the range of hydrogen ion concentrations from 0.1 to 10*M* in the presence of 1*M* KH_2PO_4.

EXPERIMENT 3-4 *Countercurrent Distribution of Indicators*[6]

Suggested pairs of indicators include methyl red and meta cresol purple, bromcresol green and phenol red, or bromphenol blue and phenol red.

PROCEDURE

Align six 125-ml separatory funnels in a row. To each add 50 ml of *n*-butanol. To funnel 1, add 45 ml of 0.05*M* aqueous Na_2CO_3 and 2.5 ml of 0.04% (*w*/*v*) of each indicator (standard acid–base indicator solution containing about 0.1 g per 250 ml). Shake for 5 min and allow the phases to separate.

Transfer the Na_2CO_3 phase to funnel 2 and add 50 ml of 0.05*M* Na_2CO_3 solution to funnel 1. Repeat the equilibrations with funnels 1 and 2. Transfer the Na_2CO_3 phase from funnel 2 to funnel 3, and from funnel 1 to funnel 2. Add fresh carbonate solution to funnel 1. Repeat the equilibrations and phase transfers until all six funnels are utilized.

Measure the absorbance of each phase at the appropriate wavelength for each indicator. For the pair methyl red–meta cresol purple, these are 430 and 580 mμ, respectively. The appropriate wavelengths are ascertained from absorbance–wavelength graphs over the range 380 to 700 mμ for each indicator in both equilibrated *n*-butanol and aqueous 0.05*M* Na_2CO_3 phases. Graph the results as absorbance vs funnel number.

Problems

1. Starting with 1.000 g of material whose partition coefficient between an organic and aqueous phase is 50, calculate the amount extracted from 100 ml of aqueous phase (a) with three 10-ml portions of the extracting solvent and (b) with one 30-ml portion of this solvent.
2. A 100-ml volume of 6*M* HCl contains 0.2000 g of antimony(V). The partition coefficient for the extraction of $SbCl_5$ from this solution into ethyl ether is 4.25. If 25.0-ml portions of ether are used, how many extractions are required to remove (a) 99.0% and (b) 99.9% of the antimony from the aqueous solution?
3. At 18°C, the partition coefficient of iodine between water and CS_2 is 420. (a) If a solution of 0.018 g of I_2 in 100 ml of water is shaken with 100 ml of CS_2, what weight of I_2 remains in the aqueous phase? (b) If the same aqueous solution of iodine is shaken with two 50-ml portions of CS_2, what amount of iodine remains unextracted?
4. When $D = 10$ for a particular organic/aqueous system, estimate the efficiency of a separation for volume ratio (V_o/V_{aq}) of (a) 10, (b) 1, and (c) 0.1 for a single extraction, and (d–f) for two successive extractions using fresh organic phase in the second extraction.
5. To assess various factors as quickly as possible, a rapid two-phase transfer distribution was carried out using three graduated centrifuge tubes and equal volumes of each phase. After the second transfer and equilibration of the tubes, the percent of original sample in each tube was determined by a suitable procedure with these results. Estimate the distribution ratio for the solutes in each system.

[6] B. Arreguin, J. Padilla, and J. Herran, *J. Chem. Educ.* **39,** 539 (1962).

SYSTEM I

Tube	Percent of sample
0	25
1	50
2	25

SYSTEM II

Tube	Percent of sample
0	27.8
1	44.4
2	27.8

6. When benzoic acid was distributed between benzene and water at a certain temperature, the following results were obtained:

Benzene layer, g/l	86.1	54.2	27.8	16.3	10.0
Aqueous layer, g/l	3.17	2.51	1.82	1.40	1.09

What is the molecular weight in benzene if one assumes a monomer in water?

7. When iodine was distributed between CCl_4 and aqueous potassium iodide solutions (containing $0.001M$ H_2SO_4 to prevent hydrolysis of I_2 to HIO), the following results were obtained at 25°C:

D	1.15	2.28	3.72	10.6	18.5	27.2	50.7
$[I^-]$, M	0.100	0.0500	0.0300	0.0100	0.00500	0.00300	0.00100

The iodide ion concentration represents free, unassociated (as I_3^-) species. (a) Plot the results on double logarithmic paper. (b) Estimate the value of the formation constant of I_3^- from the graphed results by means of successive approximations (i.e., curve fitting) and (c) from the knowledge that the solubility of iodine in water at 25°C is 0.336 g/l and in CCl_4 is 29.2 g/l.

8. When trifluoroacetylacetone was distributed between $CHCl_3$ and water, the following results were obtained:

Distribution ratio	2.00	2.00	2.00	1.40	0.90	0.40	0.20
Equilibrium pH	1.16	2.09	3.25	6.29	6.68	7.40	8.00

Estimate the partition coefficient for the solute and its pK_a value.

9. When diethyldithiophosphoric acid was distributed between CCl_4 and water, the following results were obtained:

Distribution ratio	0.177	0.32	0.70	0.93	1.25
Equilibrium pH	1.14	0.85	0.43	0.26	0.05

Estimate the partition coefficient and the acid ionization constant.

10. The distribution of 2,4-pentanedione between CCl_4 and water is given:

Distribution ratio	3.24	3.20	3.28	0.270	0.090	0.029
Equilibrium pH	1.00	3.00	5.00	10.00	10.50	11.00

Estimate the values for the partition coefficient and the acid ionization constant. Compare these values with those obtained in Problem 8 for the trifluoro derivative.

11. Estimate the partition coefficient and the dimerization constant of di-(2-ethyl hexyl) phosphoric acid between $0.2M$ $HClO_4$ and benzene from these data:

Log $[D2EHPA]_{aq}$	−3.92	−4.11	−4.22	−5.00	−5.60
Log $[D2EHPA]_{org}$	−2.00	−2.30	−2.55	−3.72	−4.32

12. Graph the relationship between log D and pH for the distribution of 8-hydroxyquinoline between $CHCl_3$ and water as a function of pH; $K_d = 720$, $K_a =$

8.0×10^{-6} (hydrogen associated with ring nitrogen atom), $K_a = 1.5 \times 10^{-10}$ (phenolic hydrogen).

13. The distribution of di-*n*-butyl phosphoric acid between methyl isobutyl ketone and $HClO_4$–$NaClO_4$ solutions is tabulated. The initial concentration of solute in the ketone phase was $1.33 \times 10^{-5}M$. Estimate the partition coefficient and the acid ionization constant.

D	pH	*D*	pH
27.3	0.00	10.0	1.30
27.9	0.30	4.55	1.70
24.4	0.48	2.08	2.00
21.8	0.70	1.10	2.30
14.8	1.00	0.42	2.70

14. The distribution of di-*n*-butyl phosphoric acid between $CHCl_3$ and $0.1M$ $HClO_4$ for different concentrations of solute is tabulated. Phase volumes were 15 ml each. Estimate the value of the dimerization constant and the partition coefficient of the monomeric form of the acid. Assume $pK_a = 1.00$, the literature value. Given are initial concentrations of solute, denoted DBP:

D	[DBP], *M*	*D*	[DBP], *M*	*D*	[DBP], *M*
135	0.333	27	1.00×10^{-2}	3.0	1.13×10^{-4}
123	0.200	19.6	5.00×10^{-3}	2.4	5.33×10^{-5}
95.5	0.133	12.5	2.00×10^{-3}	1.7	2.67×10^{-5}
74	0.100	8.8	1.00×10^{-3}	1.6	1.33×10^{-5}
57.6	0.0500	6.5	5.13×10^{-4}	1.4	6.67×10^{-6}
38	0.0200	4.2	2.13×10^{-4}		

15. The addition of iodide ion to aqueous solutions of mercuric iodide causes the formation of HgI_3^- and HgI_4^{-2}. Formulate the expression for the distribution ratio of HgI_2.

16. From measurements made in a nearly saturated solution of HgI_2 in benzene and with $0.01M$ HNO_3 plus $0.02M$ KNO_3 aqueous solution, the partition coefficient of HgI_2 was found to be 47.2. The distribution of HgI_2 as a function of free iodide ion concentration is tabulated:

D	$[I^-]$, *M*	*D*	$[I^-]$, *M*
10.0	7.0×10^{-4}	0.390	0.010
7.00	1.0×10^{-3}	0.290	0.012
4.87	1.5×10^{-3}	0.160	0.016
1.45	4.0×10^{-3}	0.100	0.020

Estimate the formation constants for HgI_3^- and HgI_4^{-2}.

17. The type of predominant species in a metal chelate extraction system can be determined from plots of log D against different experimental variables: pH, log $[HL]_o/[H^+]$, or log $[HL]_o$. For a divalent metal ion, deduce the predominant species in the organic phase and the predominant species in the aqueous phase from the information supplied on the slopes of the respective curves. Assume the absence of solvent adducts and any aqueous anionic complexing agent.

System	A	B	C	D	E	F
Log D vs pH (at constant $[HL]_o$)	2	−1	2	0	−1	2
Log D vs log $[HL]_o/[H^+]$ (at constant $[HL]_o$)	2	−1	2	0	−1	2
Log D vs log $[HL]_o$ (at constant pH)	2	−1	3	1	0	4

18. Deduce the slopes of the plots of log D against the different experimental variables enumerated in Problem 17 for these systems:

System	A	B	C	D	E
Predominant species in organic phase	ML	ML	$ML(HL)$	$ML_3(HL)_3$	ML_3
Predominant species in aqueous phase	M^+	ML_3^{-2}	M^+	M^{+3}	$M(OH)^{+2}$

19. The distribution of Ni^{+2} between $CHCl_3$ and $0.1M$ $NaClO_4$ solutions at 25°C in the presence of dimethylglyoxime is tabulated for several stated *total* concentrations of chelating agent. The pK_a of dimethylglyoxime is 10.46, and its partition coefficient between $CHCl_3$ and water is 0.083. (a) What are the predominant species containing nickel as central ion in the organic phase and in the aqueous phase? (b) Estimate the value of the partition coefficient for the predominant nickel species in the organic phase, and the formation constant of this species.

Log D	pH
0.001M HL	
−2.206	2.45
−1.066	3.09
−0.130	3.55
+0.399	3.88
0.812	4.15
1.200	4.38
1.660	4.75
2.162	5.03
2.400	5.56
2.383	6.17
2.392	7.75
2.432	9.47
2.531	10.32

Log D	pH
0.002M HL	
2.049	4.52
2.049	4.95
2.467	5.48
2.447	6.07
0.0032M HL	
−1.90	2.12
−1.255	2.49
−0.806	2.75
−0.328	2.97

Log D	pH
0.0032M HL	
+0.194	3.24
1.182	3.70
1.864	4.10
2.214	4.44
2.464	4.88
2.521	5.30
2.560	5.96
2.675	7.50
2.709	7.97
2.727	8.74
2.658	9.81

20. The extraction of these four metal ions with 0.01M 8-hydroxyquinoline in $CHCl_3$ adheres to the regular "percent extraction-pH" relationship. After each metal is given the $pH_{1/2}$ value: indium, 2.1; thorium, 4.0, cadmium, 6.3, and silver 8.8. (a) Graph the theoretical percent extraction vs pH curve for each ion. Use the same graph for all plots. (b) For each metal, indicate the point of incipient extraction (assumed to be 0.1%) and the point of complete extraction (99.0%), assuming that the partition coefficients are of sufficient magnitude to meet this requirement in one equilibration. (c) Comment on the feasibility of separating these ions from one another.
21. The distribution ratios for the extraction of metal bromides by ethyl ether are:

M, HBr	1.0	2.0	3.0	4.0	5.0	6.0
D (for Ga^{+3})	0.0020	0.0090	0.015	1.21	29.3	19.0
D (for In^{+3})	0.18	5.76	70.4	1000	166	14.4

(a) For each ion, compute the percent extraction and graph the results on the same plot. (b) At what concentration of HBr is the batch extraction of indium in the presence of gallium most favorable with one equilibration? What amount of gallium would contaminate the indium in the organic phase? Would two equilibrations with fresh organic phase at 2M HBr be better? (c) If one were interested in achieving the minimum contamination from gallium, as in a radiotracer separation, with a moderate yield of "pure" indium, at what concentration of HBr would the batch extraction be conducted? (d) At what HBr concentration is the best separation factor achieved?
22. The extraction of 0.0080M zinc(II) with $(n\text{-}C_4H_9O)_2PSSH$ (for which K_d = 331 and K_a = 0.602) gave these sets of data:

Initial $[HL]_o = 0.200M$

Distribution ratio	590	200	100	20	6.0	1.00
Equilibrium $[H^+]$	0.0050	0.053	0.080	0.20	0.33	0.80

Initial $[HL]_o = 0.040M$

Distribution ratio	590	20	7	0.80	0.20
Equilibrium $[H^+]$	1.0×10^{-4}	0.060	0.10	0.29	0.60

Initial $[HL]_o = 0.020M$

Distribution ratio	0.16	0.055	0.030
Equilibrium $[H^+]$	0.16	0.31	0.40

(a) Ascertain the predominant species in the organic phase and in the aqueous phase from the slopes of the plots enumerated in Problem 17. (b) Estimate the values of the partition coefficient and the formation constant of the zinc chelate. [Hint: Correct $[HL]_o$ for the amount of ligand present as ZnL_2.]
23. From the change in concentration of 8-hydroxyquinoline in $CHCl_3$, predict the power dependency of the reagent concentration in the extraction of indium(III):

$[HL]_o$, M	1.0×10^{-4}	1.0×10^{-3}	1.0×10^{-2}	1.0×10^{-1}
$pH_{1/2}$	3.9	2.9	1.9	0.9

24. The distribution of yttrium between $CHCl_3$-di-n-butyl phosphate and 0.1M HNO_3 is explained by the equilibrium:

$$Y^{+3}(aq) + 3H_2L_2(org) \rightleftharpoons Y(HL_2)_3(org) + 3H^+(aq)$$

From the experimental data, calculate the equilibrium constant for the overall reaction. Phase concentrations of yttrium are given as counts per minute.

$[HL]_o$, M	$[Y^{+3}]_{aq}$	$[Y^{+3}]_o$
0.003	1471	13
0.007	1463	97
0.01	1162	276
0.015	741	634
0.02	457	900
0.03	209	1194
0.05	47	1493
0.1	48	7154

25. In a binary mixture, solute A has a distribution ratio of 0.7 and solute B a distribution ratio of 1.4. Graph the fraction in each tube (a) after 24 transfers and (b) after 100 transfers. (c) In each case, how many tubes contain both solutes?
26. Calculate the distribution pattern for two solutes, $K_d = 0.5$ and $K_d = 2.0$, when separated by the fundamental procedure with a 30-tube train.
27. What degree of separation could be achieved in Problem 25 if the transfers were continued, using the 100-tube machine, until both solutes had emerged completely from the machine?
28. For a solute whose $K_d = 1.0$, calculate the tube possessing the maximum amount of solute, the band width encompassing 99.8% of solute, and the fraction of tubes containing solute to the total number of tubes in the machine after (a) 24, (b) 50, (c) 100, and (d) 200 transfers.
29. Assuming symmetrical zone profiles, graph the elution pattern for these transfers when K_d is 1.0, 5.0, and 10: (a) 50, (b) 100, (c) 500, and (d) 1000.
30. In an ideal situation, how many components could be accommodated in an extraction train at 99.9% separation for these tubes: (a) 100 and (b) 1000?
31. What is the required number of transfers in a countercurrent distribution train to obtain 99% purity, using the fundamental mode of operation, for these separation factors: (a) 1.2, (b) 1.4, (c) 1.6, (d) 1.8, (e) 2.0, (f) 2.2, and (g) 2.4.

BIBLIOGRAPHY

Craig, L. C., and D. Craig, in A. Weissberger (Ed.), *Technique of Organic Chemistry,* Interscience, New York, 1956, Vol. III, pp. 248–300.

Diamond, R. M., and D. G. Tuck, in F. A. Cotton (Ed.), *Progress in Inorganic Chemistry,* Vol. 2, Interscience, New York, 1960, pp. 109–192.

Irving, H. M., and R. J. P. Williams, in I. M. Kolthoff and P. J. Elving (Eds.), *Treatise on Analytical Chemistry,* Part I, Vol. 3, Interscience, New York, 1961, pp. 1309–1365.

Marcus, Y., *Chem. Rev.* **63,** 139 (1963).

Morrison, G. H., and H. Freiser, *Solvent Extraction in Analytical Chemistry,* Wiley, New York, 1957.

Stary, J., *Metal Chelate Solvent Extraction,* Pergamon, Oxford, 1965.

4 *Chromatography–General Principles*

The term "chromatography" embraces a family of closely related separation methods based on the experiments described by Tswett[1] in 1903 and 1906 and by Day in 1897 and 1904. The importance of chromatography lies primarily in its use as an analytical tool, although preparative applications are significant. It serves as a means for the resolution of mixtures and for the isolation and partial description of the separated substances, some of whose presence may be unknown or unsuspected.

In chromatography, the components to be separated are distributed between two phases. One of the phases is a stationary bed of large surface or bulk area, the other a fluid that percolates through or along the stationary bed. The transfer of mass between the mobile phase and the stationary phase occurs either because the molecules of the mixture are adsorbed on particle surfaces or absorbed into particle pores, or they partition into pools of liquid held on surfaces or within pores. The general process is often called sorption.

Separation of the components in a sample is based on the fact that the rate of travel of an individual solute molecule through a column or thin layer of sorbent is directly related to the partition of that molecule between the mobile phase and the stationary phase. The partition coefficient of each component determines how much of it is in the mobile phase at any time and, therefore, the overall time spent in the stationary phase. The latter determines the retention or retardation of a solute. If selective retardation differences prevail, each component will travel through a column (or along a

[1] M. Tswett, *Proc. Warsaw Soc. Nat. Sci., Biol. Sec.* **14,** No. 6 (1903); *Ber. Deut. Botan. Ges.* **24,** 316, 384 (1906), English translation in *J. Chem. Educ.* **44,** 238 (1967).

thin layer of stationary phase) at a rate dependent upon i
tion characteristic. When separation takes place, each co
from the column (or travels the length of a thin layer) a
interval or, if the process is stopped before any of the com
the column bed or thin layer, they will be distributed in sp
tionary phase instead of being distributed in time at the exit. Strain[2] considers the overall process to be a differential migration phenomenon produced by a nonselective driving force, the flow of the mobile phase. The wide choice of materials for both the stationary and mobile phases in chromatography makes it possible to separate substances that differ only slightly in their physical and chemical properties.

Classification of Chromatographic Methods The mobile phase can be a gas or a liquid, whereas the stationary phase can only be a liquid or a solid. This leaves room for various combinations such as liquid–liquid, liquid–solid, gas–liquid, and gas–solid. When the separation involves predominantly a simple partitioning between two liquid phases, one stationary and the other mobile, the process is called *partition chromatography*. If physical surface forces are mainly involved in the retentive ability of the stationary phase, the process is called *adsorption chromatography*. These processes comprise liquid solution chromatography. If the mobile phase is a gas the methods are called *gas–liquid chromatography* and *gas–solid chromatography*. For volatile compounds, gas chromatography (Chapter 11) offers the analytical chemist high resolution, minimal analysis time, and sensitivity in the parts per million range. The fairly extensive range of organic compounds not sufficiently volatile or thermally stable to be handled by gas chromatography can be analyzed by liquid chromatography. Liquid chromatographic methods employ a liquid mobile phase to transport the sample through the partitioning column which is packed with a solid adsorbent (Chapter 7) or a liquid coated on a solid (Chapter 8).

Two other chromatographic methods differ somewhat in their mode of action. In *ion-exchange chromatography* (Chapters 5 and 6), true heteropolar chemical bonds are formed reversibly between ionic components in the mobile and stationary phases. The use of molecular sieves as the stationary phase brings about a classification of molecules based largely on molecular size. The method, actually *exclusion chromatography* (Chapter 12), is referred to as gel-permeation chromatography by polymer chemists and as gel filtration by biochemists.

Sheet methods, such as paper chromatography (Chapter 9) and thin-layer chromatography (Chapter 10), are used to separate the same types of samples as column methods, but are more useful in the qualitative or semi-

[2] H. H. Strain, "Differential Migration Methods of Analysis," in E. Heftmann (Ed.), *Chromatography*, 2nd ed., Reinhold, New York, 1967.

quantitative analysis of unknown mixtures and when handling very small quantities of material.

Gas and liquid chromatography are generally treated individually. Actually there are very few fundamental reasons for their separation. If liquid chromatography is used under operating conditions fully analogous to those for gas chromatography, much the same can be accomplished in analysis. The advent of high-speed, ultra-sensitive, low-noise detection systems for liquid chromatography has reawakened interest in this segment of the column chromatographic methods. In this chapter on general principles, and when examining chapters on specific chromatographic methods, one should be conscious of the similarities in operating conditions and not be misled by merely differences in magnitude of certain parameters.

Chromatographic Behavior of Solutes

The chromatographic behavior of a solute can be described either by its retardation factor (R value) or by a retention volume, V_R (or corresponding elution time, t_R). Either term provides a means for describing the fractional time a solute molecule resides in the mobile phase; i.e., the period of time it moves in the direction of the mobile phase. Each substance will possess a different retardation factor or retention volume for each sorbent and solvent combination. By varying the sorbent–solvent combination and various operating parameters, the degree of retention can be varied over a wide range from nearly total retention to a state of free migration. The first extreme leads to an infinite time of separation, the other extreme fails to produce any resolution.

Partition Coefficient The concentration of solute in each phase is given by the partition coefficient, K_d

$$K_d = C_S/C_M \tag{4-1}$$

where C_S, C_M are the concentrations of solute in the stationary phase and mobile phase, respectively. If the partition isotherm is linear, the partition coefficient will be a constant independent of the solute concentration. Generally this is true at very low concentrations in all chromatographic methods. Adherence to, or deviation from, this ideal situation will be discussed when the specific chromatographic methods are considered.

Retardation Factor The retardation factor, or R value, is the ratio of the displacement velocity of a solute to that of an ideal standard substance (often the mobile phase) which does not dissolve in, or which is not adsorbed by, the stationary phase; i.e., a substance for which $K_d = 0$. The R value then becomes the fraction of time spent by a solute molecule in the mobile phase, or

$$R = t_M/(t_M + t_S) \tag{4-2}$$

where t_M, t_S are the times the molecule remains in the mobile and stationary phases, respectively. In systems where the cross section and the partition coefficient may not be constant along the length of the development path, the ratio of distances traveled by the solute and the mobile phase is called the R_f value;

$$R_f = \frac{\text{Distance traveled by the solute}}{\text{Distance traveled by the mobile phase}}$$

In paper and thin-layer chromatography, particularly, there is a gradient of solvent in the paper or sheet decreasing from the bulk solvent to the advancing front. Since the velocity of the front is greater than that of the following mobile phase, R is greater than R_f by about 15%.

The R value is also an equilibrium property and a function of the partition coefficient because the retention or holdback of zones, which fixes their position in time or space, is determined by the distribution of the solute in the zone between the mobile and stationary phases. When the rate of mass transfer between phases is sufficiently large for a statistical number of transfers to occur, the time average distribution of the molecules approaches the equilibrium distribution. If R is the equilibrium fraction of solute in the mobile phase, then $1 - R$ is the solute equilibrium fraction in the stationary phase. Now the fraction $[R/(1 - R)]$ is the same as the fraction of molecules in the mobile phase divided by the fraction in the stationary phase, i.e.,

$$\frac{R}{1 - R} = \frac{C_M V_M}{C_S V_S} \tag{4-3}$$

where V_M is the volume of the mobile phase and V_S is the volume of the stationary phase. (If adsorption chromatography is involved, a surface area, A_S, should replace the volume, V_S.) Rearranging, and inserting the partition coefficient,

$$R = \frac{V_M}{V_M + K_d V_S} = \frac{1}{1 + K_d(V_S/V_M)} \tag{4-4}$$

It is worth noting the formal resemblance of Eq. (4-4) to Eq. (3-7) which is discussed in liquid–liquid extractions; however, it should be emphasized that the partition coefficient need not be the same in a chromatographic column as in a separatory funnel. The situation is much more complicated in a chromatographic column.

Retention Volume In column elution methods, the significant volume is the amount of mobile phase which has left the column at the instant the maximum of the solute zone emerges from the column. At the appearance of the peak maximum at the column exit, one-half of the solute has eluted in the retention volume, V_R (or V_{max}, as it is sometimes denoted), and one-half remains in the column in the mobile phase, V_M, plus the volume of the stationary phase, V_S. Thus,

$$V_R C_M = V_M C_M + V_S C_S \tag{4-5}$$

Rearranging, and inserting the partition coefficient, we get

$$V_R = V_M + K_d V_S \tag{4-6}$$

The retention volume is related to the retention time through the volume rate of flow of the mobile phase, F_c,

$$V_R = t_R F_c \tag{4-7}$$

and it is also related to the retardation factor

$$V_R = LF_c/R\nu \tag{4-8}$$

where L is the length of the column and ν is the linear flow rate or average velocity of the mobile phase. If the expression for the R value in terms of phase volumes and partition coefficient, Eq. (4-4), is introduced into Eq. (4-8), then

$$V_R = LF_c(V_M + K_d V_S)/\nu V_M \tag{4-9}$$

Now the ratio $[LF_c/\nu]$ is the volume of the mobile phase; therefore, once again Eq. (4-6) is obtained. In Eq. (4-6), the first term of the right-hand member is the total volume of the mobile phase within the column. Often called the bed void or free-column volume, it represents the retention volume of a substance that is unadsorbed by, or insoluble in, the stationary phase (for which $K_d = 0$).

Column Capacity Another term frequently encountered is the capacity factor or partition ratio, k. It is the ratio of the amount of solute in the two phases:

$$k = C_S V_S / C_M V_M \tag{4-10}$$

Now the volumetric phase ratio, β, in a column is designated as V_M/V_S. Thus $k = K_d/\beta$. The interplay of the capacity ratio with column efficiency and resolution is important and, of course, it affects sample size (or column loading).

The partition ratio is related to R by

$$R = 1/(1-k) \tag{4-11}$$

and to the retention volume by the expression

$$k = (V_R - V_M)/V_M \tag{4-12}$$

Temperature Effects Temperature is the most important single parameter in the operation of chromatographic columns. This importance is primarily the result of the marked dependence of the partition coefficient on temperature; secondary effects such as the changes with temperature of gas and liquid volume and gas and liquid diffusivities are also important.

Like all equilibrium constants, K_d changes exponentially with absolute temperature in the following way:

$$K_d = e^{\Delta S^\circ/\mathscr{R}T}\, e^{-\Delta H^\circ/\mathscr{R}T} \simeq a\, e^{-\Delta H^\circ/\mathscr{R}T} \tag{4-13}$$

where ΔS° and ΔH° are the standard entropy and enthalpy of sorption, and $\mathscr{R}$ is the gas constant. The heat of sorption of the solute in the stationary phase is usually negative and ranges from −12 kcal/mole to zero or slightly positive values. Quite often a 20°C temperature rise will halve K_d. This leads to a large increase in the R value [Eq. (4-4)] or a large decrease in the retention volume [Eq. (4-6)]. Such a strong dependence on temperature can be used to advantage in controlling migration, particularly in gas chromatography (*q.v.*).

A correlation is observed between the entropy and enthalpy terms in Eq. (4-13); entropy decreases with an increase in enthalpy. For members of a homologous series, ΔS° tends to vary linearly with ΔH°. This correlation indicates that the decrease in entropy associated with sorption is proportional to the heat of sorption, that is, the greater the heat of interaction with the stationary phase, the greater the degree of order of the solute molecules in that phase.

Temperature also influences resolution, to be discussed later. It suffices to indicate at this point that temperature increases are more often than not detrimental to resolution. Yet the operating temperature should not be reduced indefinitely or separation time may become prohibitive for practical work. The temperature is usually established as a compromise between better resolution (when it is lowered) and higher speed (when it is raised).

Partition Isotherms The relative magnitudes of the partition coefficients, as well as their concentration dependence, are of fundamental importance for chromatographic behavior. Typical partition isotherms, giving the stationary-phase concentration of the species as a function of the mobile phase concentration, are shown in Fig. 4-1. When the mobile phase concentrations are so low that operation extends over only a short part of the isotherm near the origin, the isotherm can be regarded as linear. This situation prevails in conventional gas chromatography and in other types of chromatography when dealing with trace amounts of components. The concentration profiles of components resemble symmetrical bell-shaped peaks of the normal or gaussian error function. Under this condition the concen-

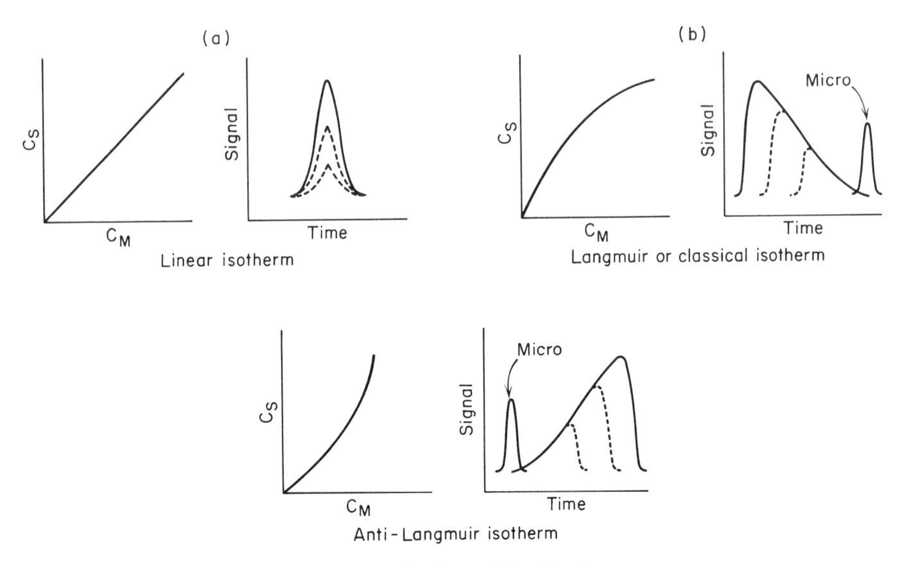

FIG. 4-1. Typical partition isotherms.

tration of the solute in the stationary phase is directly proportional to the concentration of the solute in the mobile phase. The retention time is independent of concentration as shown by the dotted curves.

Peaks with a sharp front and a tailing rear boundary are caused by Langmuir-type isotherms shown in Fig. 4-1(b). This type of peak is often observed in adsorption chromatography. The peak shape is caused by the main portion of the solute band eluting more rapidly than the leading front edge due to a limited number of sites available for adsorption. Retention time may be a function of sample size and normal, narrow peaks are only obtained with micro samples.

Peaks with sloping front boundaries and sharp rear boundaries are caused by the nonlinear isotherm shown in Fig. 4-1(c). This type of peak is often obtained when the solute has low solubility in the stationary liquid phase. The liquid phase becomes modified by the solute molecules causing the main portion of the band to elute slower than the extremities. Retention time is also a function of sample size. This type of isotherm may change slope with temperature.

Molecular Structure The chromatographic behavior of solute molecules is not specific with respect to the molecular skeleton or to particular functional or polar groups. In many cases, however, a relationship between structure and partition coefficient becomes apparent if the quantities $\log (R^{-1} - 1) = R_M$, or $\log V_R$, of members of a homologous series are plotted in a coordinate system versus the number of "homologous structural elements." The points often lie on a smooth curve. The behavior of other compounds belonging to the same homologous series may then be predicted by interpolation or extrapolation. Of course, the contribution of homologous structural elements will prove to be independent from the rest of the molecule only under restricted conditions: If the spatial arrangement offers the same possibility for adsorption or solvation to each structural element and, if, at the same time, the remainder of the molecule behaves identically in each member of the series. For example, a typical series might involve benzene, toluene, and ethylbenzene but not benzene, toluene, and xylene; oxalic, malonic, and succinic acids would not constitute a series.

Chromatography may be utilized in conjunction with specific chemical reagents to determine whether or not particular functional groups are present. Also specific interactions may be intentionally invoked to attain separations. Selective masking will retard the passage of components with specific groups. For example, silica gel or kieselguhr impregnated with silver nitrate retains selectively compounds possessing carbon–carbon double bonds; these same supports impregnated with bisulfite interact with carbonyl groups.

Development of the Chromatogram

The sample is generally introduced at the top of a column or near one edge of a sheet or thin layer of sorbent. Development into separate zones (bands or peaks) takes place when a suitable mobile phase is allowed to percolate through the stationary phase. Chromatographic development may be conducted either by *frontal analysis* (without developer), by *elution analysis* (with a solvent as developer), or by *displacement development* (using a more strongly adsorbed displacing solute). In flow chromatograms the effluent is collected in separate fractions or the emergence of each component from the column is continuously recorded by means of a detector in the effluent stream. This is contrasted with development within the column or thin layer which is discontinued when the fastest moving component, or the solvent front, reaches the column exit or the opposite edge of the sheet or thin layer. Today most workers prefer either elution from columns or discontinuous development in sheets and thin layers.

Frontal Analysis Introduced by Tiselius in 1940, frontal analysis consists of passing the sample solution continuously through an adsorbent. The active centers of the adsorbent are occupied by the more strongly adsorbed components and the least strongly adsorbed component accumulates in the traveling front. What issues from the column first is the pure solvent itself, followed next by the sample ingredient least strongly adsorbed. The concentration of the column effluent, plotted against the effluent volume as

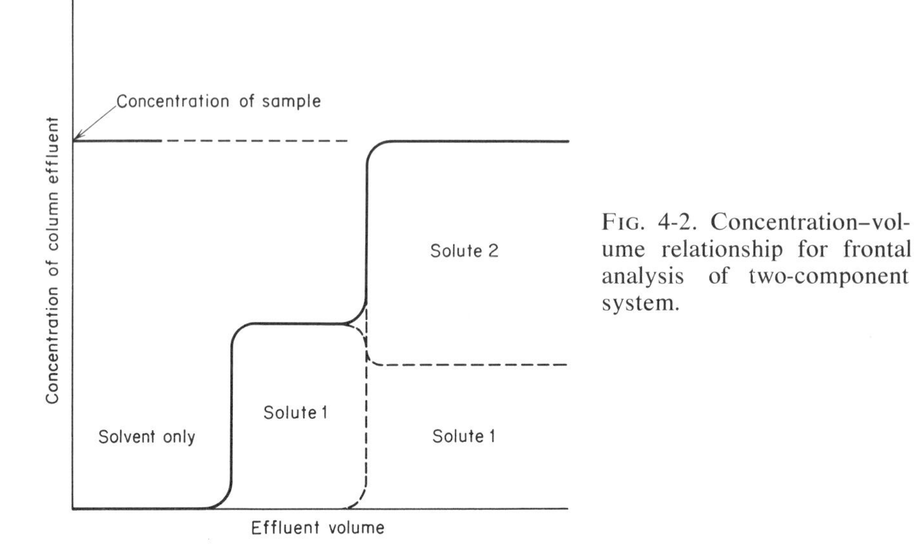

FIG. 4-2. Concentration–volume relationship for frontal analysis of two-component system.

in Fig. 4-2, has a stepped pattern. Successive plateaus indicate the emergence of an additional solute in the effluent.

Frontal analysis leads to resolution only of the least strongly adsorbed solute. Processes of this type are used commercially for the removal of relatively small amounts of undesirable components when these are more strongly adsorbed than the bulk of the material.

The performance of frontal analysis at a number of concentrations provides a rapid and accurate method for obtaining adsorption isotherms. Only a shallow bed of adsorbent and a small volume of sample is required. The application of this technique to the analysis of multicomponent systems and the development of the pertinent theory will be found in papers by Tiselius and Claesson.[3]

Elution Development This is the most widely used technique. A small sample, not occupying over a few percent of the sorbent capacity, is introduced at the upstream end of the stationary phase. Pure solvent, called eluent, is allowed to flow through the system. This leads to a differential migration of the solutes in the mobile phase. Assuming a fixed flow rate for the eluent, the time (and volume) at which each component in the sample will emerge from the column depends on the individual partition coefficients. If these are sufficiently different for the several components, then the individual components of the mixture will split up into separate bands that migrate at different rates, increase their distances from one another, be-

[3] A. Tiselius and S. Claesson, *Arkiv Kemi Mineral. Geol.* **15,** No. 18 (1942); S. Claesson, *Ann. N.Y. Acad. Sci.* **49,** 133 (1948).

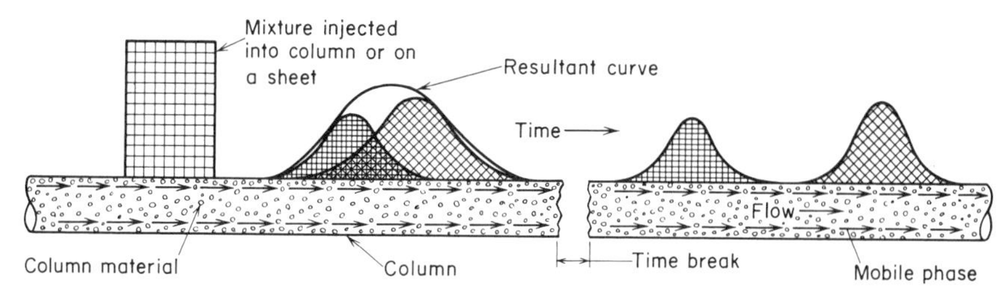

FIG. 4-3. Elution development.

come progressively more diffuse due to zone spreading, and appear in the effluent separated from one another but on a background of the mobile phase (Fig. 4-3).

Displacement Development As in elution analysis, the sample is introduced into the column, or onto a thin layer, as a narrow starting zone. Separation of the sample ingredients is achieved by running a more strongly adsorbed displacing agent into the column. The displacing agent is preferentially adsorbed and concentrates in a zone near the top of the column, gradually spreading downward through the column. All of the sample components are forced off the sorption sites and compelled to move ahead of the front produced by the displacer. In their turn the sample components will displace one another. The net result will be a series of contiguous zones, one for each component, in front of the displacer and in their order of decreasing sorption strength. First to leave the column will be the component least strongly sorbed, followed immediately by the zone of the next least strongly adsorbed solute, and so on, until the front of the displacer emerges [Fig. 4-4(a)].

Diagrams illustrating the theory of displacement development are shown in Fig. 4-4b. Convex adsorption isotherms are a necessary prerequisite. The upper diagram gives the isotherms for the displacing agent, D, and for components 1, 2, 3, and 4. A straight line is drawn through the origin to a point on the isotherm of the displacing agent corresponding to C_D, the concentration of displacing agent selected. Its partition function is given by

$$q_D = f_D(C_D) = V_D C_D \tag{4-14}$$

Assuming a constant flow of the displacing agent, the front of the displacer moves down the column at a rate which depends only on the value of the retention volume ($V_D = q_D/C_D$) of the displacing agent. Components of the sample are compelled by displacement to move at the same rate, and the concentration of each component in its zone becomes adjusted so that

$$\frac{q_D}{C_D} = \frac{q_3}{C_3} = \frac{q_2}{C_2} = \frac{q_1}{C_1}$$

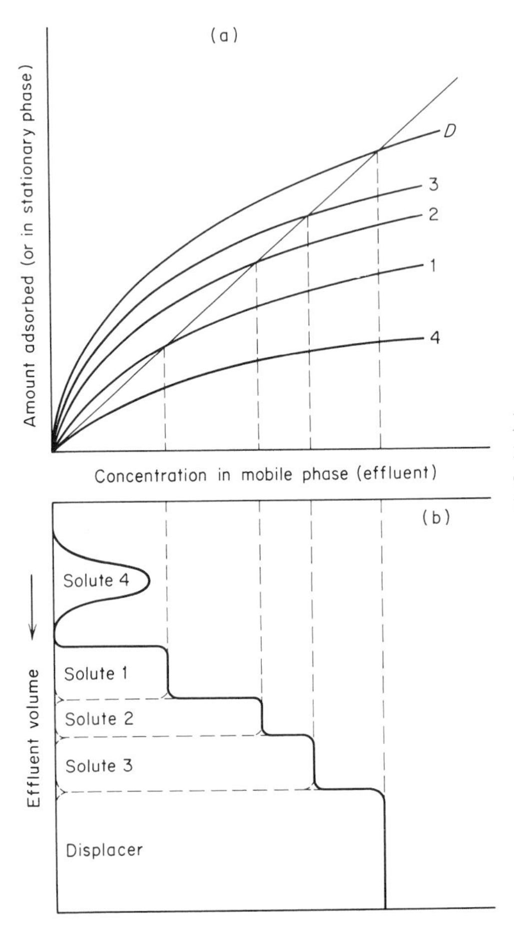

FIG. 4-4. Displacement development: (a) sorption isotherms; (b) concentration–effluent volume profile.

The concentrations of each component in their respective zones are given by the intersection of the line (drawn from the origin to the displacer concentration) with their adsorption isotherms. It is common practice to choose as displacer a substance that is closely related to the substances undergoing separation. Often a higher member of a homologous series is selected.

Figure 4-4(b) shows the concentration at which the components emerge from the column. The amount of each component can be obtained from its concentration and the volume of solution in the corresponding step. The isotherm of component 4 is not intersected by the line drawn from the origin to the displacer concentration. Consequently, component 4 will move faster than the speed of the displacer; actually at a speed characteristic of free elution. Now if the concentration of the displacer is raised to a value C_D', such that the isotherm for component 4 is intercepted, component 4 will appear as the first displacement band.

An advantage of displacement development is that rather heavy loading of the chromatographic column is permissible—this is especially suitable for preparative work. On the other hand, since complete separation of the bands does not take place, the recoveries of each component in pure form are lower than when elution development is employed. However, the bands are self-sharpening, which minimizes tailing. Of course, the technique leaves the stationary phase saturated with the displacing agent after a run. Except, in liquid–liquid partition columns, displacement has no compensating advantages over elution development and is now seldom used.

Gradient Elution In ordinary elution development, there is only a small range of retention volumes (or retardation factors) for optimum separation. The partition coefficients must be sufficiently large so that components eluted early are not pushed off the column as an unresolved series of bands, yet the partition coefficients must be reasonably small if excessive elution times and zone broadening are to be avoided. Gradient elution permits an automatic, gradual attainment of the eluting power required for each component. It counteracts the broadening of zones, and tailing is not so pronounced. Gradient elution development is characterized by the intentional variation of eluting conditions during the course of separation. These include composition of the mobile phase, column pressure, and column temperature. In gas chromatography it is most convenient to vary the partition coefficient values by gradually raising the column temperature (programmed-temperature gas chromatography), although programmed flow has also been employed. In liquid column chromatography, adjustment of the composition of the eluent is the approach most easily achieved.

In liquid column chromatography, gradient elution is conducted with one of the arrangements shown schematically in Fig. 4-5. Fluid with stronger displacing properties is continually approaching from above; zones are forced more closely together and, to compensate, they become narrower. The use of mixing chambers of different size and shape yields elution gradients as shown underneath each arrangement. The shape of the gradient is important. It should be chosen in such a way that it is flat in the region where many components with similar adsorbabilities are crowded together, but steep in the region where adsorbabilities are widely different. The device shown in Fig. 4-5(d) is very flexible and convenient for supplying custom gradients through the automatic switching valve *S*. Snyder's review[4] provides a comprehensive discussion of the technique and of gradient devices.

[4] L. R. Snyder, *Chromatog. Rev.* **7**, 1 (1965).

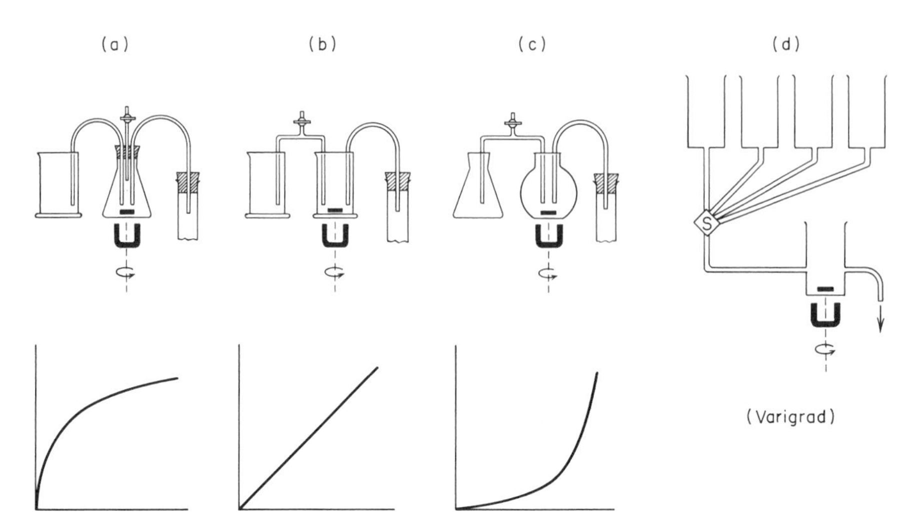

FIG. 4-5. Schematic diagrams of various gradient devices and (below) type of gradient.

Dynamics of Chromatography

The effectiveness of separations in chromatography hinges on two properties—the disengagement of zone centers as they migrate apart and the compactness of the zones. As narrower zones are obtained, more of them can be fitted into a given chromatographic pattern. The partition coefficient expresses the proportionate amount of time spent by each component in each phase. Associated with this figure is the linear dimension, to be denoted plate height, taken in the direction of flow of the mobile phase, during which one partition will have occurred. The partition is affected by the linear velocity of the mobile phase, the diffusion of solute molecules within each phase, and the interchange or mass transfer of solute between the phases. All these enter into expressions for efficiency and resolution. Unfortunately, when considering speed, resolution, and capacity on a single chromatographic arrangement, the choice of one of these desirable operating characteristics precludes the remaining two, although some compromise may be possible to obtain a reasonable performance in all three aspects.

The following sections are not intended to provide a complete survey concerning the selection of optimum parameters for chromatographic systems. This task, because of the diversity of chromatographic methods and experimental requirements, belongs with the more specific coverage of subsequent chapters. Here, only optimal conditions in the general context will be discussed.

Efficiency The measure of column efficiency is expressed by the plate number N,

$$N = 16(V_R/W)^2 \tag{4-15}$$

where W is the width of the elution peak measured in volume units. As shown in Fig. 4-6, it corresponds to the intercept of the front and rear tangents to the inflection points on the elution peak with the base of the peak. An alternate expression is

$$N = 8(V_R/W_e)^2 \tag{4-16}$$

where W_e is the width at C_{max}/e (or $0.368C_{max}$); C_{max} being the concentration of material in the effluent at the peak maximum. Quite naturally, the retention volume may be replaced by the retention time and the width then measured in time units.

An elution peak, or chromatographic zone, approaches a gaussian or standard-error curve in shape under ideal conditions. As for any such curve, the width may be expressed in terms of the standard deviation. As defined, $W = 4\sigma$.

Plate number is a concept carried over from distillation practice. It is somewhat ambiguous but useful in column chromatography. Ideally, it implies that a column can be considered to be divided into a number of segments or plates, with a perfect equilibrium assumed to exist between the solute in the mobile and stationary phases existing in each plate. Except for an infinitesimal rate of flow, changes in the stationary phase composition always lag behind changes in composition of the mobile phase. Thus,

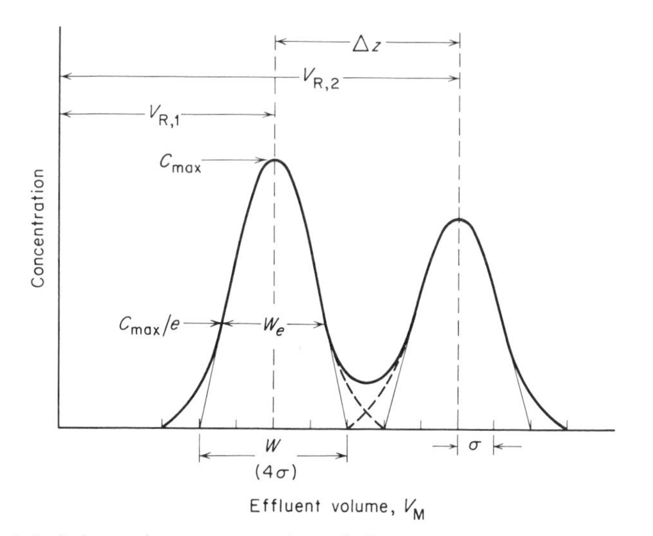

FIG. 4-6. Schematic representation of elution curves and terminology.

the original significance of the theoretical plate is lost. Nevertheless, plate number, and the analogous unit, plate height, is a useful index of column efficiency to describe the extent to which peaks and zones are broadened through the action of physical transport phenomena. Basically, the expression for plate number compares the narrowness of a peak to the length of time ($t_R = V_R/F_c$) the component has been in the column (or on a sheet). Plate height, H, is the length of the column divided by the number of plates: $H = L/N$.

Zone Spreading The theoretical distribution of a solute during a chromatographic development assumes that all of it started in an extremely thin zone at the very top of a column, or as a very small compact spot on a sheet. When a mobile fluid boundary passes through the zone or spot, the rate of sorption and desorption (or exchange) between phases, being finite, cannot quite keep pace with the changes taking place in the mobile phase. Progress down the column, or along the sheet, by individual solutes resembles a random, stop-and-go process.[5] The net forward travel of each component is actually an average value and there is a normal dispersion of values about the mean. Furthermore, when dealing with finite concentrations, the solutes are not initially confined to only the first plate but are spread over a number of plates. The effect will be that a number of chromatograms are started successively. This also leads to increased zone width.

In terms of the standard deviation of the near-gaussian profile of the zone, plate height may be defined as

$$H = \sigma^2/L \tag{4-17}$$

where L is the length of the column traversed by the center of the zone and σ^2 is the variance. Zone spreading is considered to be due to a series of molecular diffusion and local nonequilibrium factors, each factor contributing a certain variance to the gaussian curve.[6] Zone spreading itself originates in the velocity inequalities of the flow pattern, but the extent of spreading is governed largely by diffusion between fast and slow stream paths. Variances arise from ordinary diffusion, eddy diffusion, local nonequilibrium during mass transfer between phases, and the coupling of some of these factors.

Ordinary diffusion within the mobile phase occurs when solute molecules migrate from regions of higher concentration to regions of lower concentration. This takes place in a very complex network of interconnected channels and void spaces whose cross section may be quite constricted or

[5] J. C. Giddings, *Dynamics of Chromatography*, Vol. I, entitled *Principles and Theory*, Dekker, New York, 1965.

[6] J. C. Giddings, *J. Chem. Educ.* **35**, 588 (1958).

tortuous in some locations.[7] The concentration profile after a finite time is gaussian and the variance due to spreading is given by the equation

$$\sigma_d^2 = 2\gamma D_M t = 2\gamma D_M L/\nu \tag{4-18}$$

where D_M is the diffusion coefficient of the solute molecule in the mobile phase, γ is an obstructive factor ($\sim$0.6) indicating the degree to which diffusion is hindered by the granular or fibrous column packing, and t is the time spent by the solute in the mobile phase [$t = L/\nu$ where L is the distance the zone has migrated and ν is the linear flow rate (solution volume per unit cross section of the column and unit time) of the mobile phase]. Under circumstances when the gas is sorbed by the stationary phase, the longitudinal diffusion in the stationary phase also contributes a significant plate height; the variance should be $2\gamma D_M t/(1 + k)$ or $2\gamma D_M t(1 - R)/R$.

A term for eddy diffusion is required because portions of the mobile phase carrying the solute molecules take a longer and, perhaps, a more tortuous and constricted path in the particulate or fibrous medium and consequently lag behind the average, while other portions take shorter and less obstructed paths and thereby move ahead of the average. Superimposed on this is the fact that the mobile phase flow is more rapid in some regions than in others. Variance due to eddy diffusion is expressed by

$$\sigma_e^2 = 2\lambda d_p L \tag{4-19}$$

which implies that zone spreading increases with the migration distance L and the particle or fiber size d_p of the stationary substrate; λ is the packing factor.

Local nonequilibrium effects arise during mass transfer between the mobile and stationary phases. As the mobile phase enters the edge of the zone it brings with it a solute concentration that is smaller than the equilibrium concentration (expressed by the partition coefficient and phase volumes), and desorption occurs until flow and kinetic processes reach a balance. However, forward flow of the mobile phase is continually upsetting the equilibrium. At the leading edge, the reverse prevails. The mobile phase brings with it a solute concentration that is larger than the equilibrium concentration and sorption occurs with kinetic processes attempting to hasten the equilibrium and the forward flow of the mobile phase upsetting the equilibrium. The spreading due to these factors is given by the variance

$$\sigma_N^2 = 2\omega R(1 - R)\nu L t_S \tag{4-20}$$

where t_S is the average residence time of a sorbed solute molecule in the stationary phase, or the desorption time [$t_S \simeq d_p^2/2D_S$, where D_S is the diffusion coefficient in the stationary phase], and ω is a numerical coefficient

[7] J. H. Knox and L. McLaren, *Anal. Chem.* **36**, 1477 (1964).

obtained from rigorous theory[8] [ω is 1/16 when d_p is the diameter of a fiber, 1/30 when d_p is the diameter of a gel or ion exchanger bead, and 2/3 when, in gas chromatography, d_p is the thickness of a liquid film coating a substrate].

Not all the foregoing variances are independent of one another. There is coupling between the term for nonequilibrium in the mobile phase and eddy diffusion; it is expressed by

$$\sigma_c^2 = 1/[(1/2\lambda d_p) + D_M/\nu d_p^2] \tag{4-21}$$

Now from the sum of the several variances and Eq. (4-17), plate height as a function of mobile phase velocity may be written as

$$H = 2\gamma D_M/\nu + \omega R(1 - R)d_p^2\nu/D_S + 1/[(1/2\lambda d_p) + D_M/d_p^2\nu] \tag{4-22}$$ [9]

This relationship can be written in the following simplified form:

$$H = B/\nu + E_S\nu + \frac{1}{1/A + E_M/\nu} \tag{4-23}$$

where A is a constant, B is the coefficient for the diffusion term, and E_M, E_S are the mass transfer coefficients for the mobile and stationary phase, respectively. Equation (4-22) shows that the effective plate height depends on the retardation factor (or the partition ratio) of the species. Thus, in a separation, the plate height is different for the different species. This point is often overlooked. The smaller the partition ratio, the faster the respective zone moves, and the larger is the plate height (at least if $k > 1$, which is usually true).

The terms D, R, and t_s in these variances are influenced by hidden variables, of which temperature is the best example. Harris and Habgood[10] thoroughly discuss the influence of temperature, particularly its role in gas chromatography.

Whether the full coupled expression for plate height, Eq. (4-22), is employed or one of its simplifications, depends on the flow rate of the mobile phase. Actually it is the so-called reduced velocity, $\bar{\nu}$, given by

$$\bar{\nu} = d_p\nu/D_M \tag{4-24}$$

that concerns us, and in this expression the diffusivity term dominates. Diffusivity is some 10^4 to 10^5 times smaller for liquids as compared to gases. Consequently, the reduced velocity is very large in liquid chromatography, and Eq. (4-22) simplifies to

[8] J. C. Giddings, *J. Chromatog.* **5,** 46 (1961).

[9] Rewriting the second term in Eq. (4-22) with the partition ratio in place of the retardation factor, it becomes: $(\omega d_p^2\nu/D_S)[k/(1 + k)^2]$. Except for minor differences in the coefficients, Eq. (4-22) is identical to the van Deemter equation for plate height in gas chromatography and the Glueckauf expression for plate height of ion exchange columns.

[10] W. E. Harris and H. W. Habgood, *Programmed Temperature Gas Chromatography,* Wiley, New York, 1966.

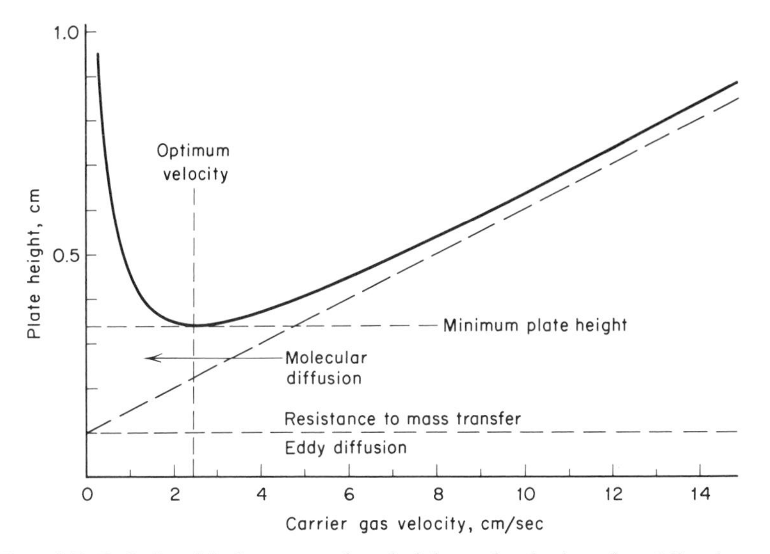

FIG. 4-7. Relationship between plate height and velocity of mobile phase.

$$H = 2\gamma D_M/\nu + 2\lambda d_p + R(1 - R)\nu d_p^2/D_S \qquad (4\text{-}25)$$

when $\bar{\nu} \gg 1$. The range $\bar{\nu} \simeq 1$, described by the low-flow expression,

$$H = 2\gamma D_M/\nu + R(1 - R)\nu d_p^2/D_S + d_p^2\nu/D_M \qquad (4\text{-}26)$$

is encountered only in gas chromatography. When the conditions in gas and liquid chromatography are truly analogous, then the ratio

$$\bar{h} = H/d_p \qquad (4\text{-}27)$$

where $\bar{h}$ is the reduced plate height, will be the same in the two systems, provided comparable conditions exist in both the mobile and stationary phases. The reduced plate height not only transforms various experimental measurements to a common basis, but it also provides a criterion of excellence for the chromatographic system. At moderate values of $\bar{\nu}$ (say 0.2–10) the $\bar{h}$ value for an inherently efficient column should be less than ten and in some cases as low as one or two.[11]

The foregoing equations for plate height enable one to choose operating conditions that decrease the plate height and thereby increase the efficiency of a chromatographic separation. A plot of Eq. (4-23) for any type of linear elution chromatography describes a hyperbola, as shown in Fig. 4-7. There is an optimum velocity of the mobile phase for carrying out a separation at which plate height is a minimum;

$$\nu_{\text{optimum}} = \sqrt{D_M/[R(1 - R)d_p^2/D_S]}. \qquad (4\text{-}28)$$

[11] J. C. Giddings and K. L. Mallik, *Anal. Chem.* **38**, 997 (1966).

It is a compromise velocity which minimizes the nonequilibrium contribution to H and the molecular diffusivity term. It also implies that even with all parameters optimal the plate height can only be reduced to a certain minimum level, usually equal to several particle diameters. This level is fixed by the structural characteristics typical of nearly all chromatographic columnar or sheet material. Since the reduced velocity equals $d_p\nu/D_M$, the optimum velocity is approximated by $\nu_{\text{opt}} \simeq D_M/d_p$. It is very roughly 10 cm/sec in gas chromatography and 10^{-3} cm/sec in liquid chromatography. Often the column will have unneeded resolution capability so that it is possible to trade plates for speed and thus increase the flow rate beyond the optimum value. For example, the extremes of the ratio ν/ν_{opt} extend from about 0.4 to 2.9 with only a reduction to 80% of the maximum resolution.

Other deductions can be made from the equations for plate height. Although no specific particle or fiber size is suggested, in general an improvement comes with smaller particles. Of course, extremely small particles will require high inlet pressures to move forward the mobile phase. High-pressure methods in both gas and liquid chromatography have been described recently.

Resolution The degree of separation of two components is a problem which is common to all chromatographic methods. Exact disengagement between adjacent zones or bands occurs when the trailing edge of the more rapidly moving solute and the leading edge of the following solute just touch or, on a recorded tracing, reach the base line at the same point. In truth, of course, a complete physical separation of adjacent components can never be attained because a chromatographic peak approaches in shape a gaussian distribution. In practical chromatography, however, one is interested only in minimizing the degree of overlap, or cross contamination, between the adjacent zones to a desired experimental level.

Resolution (designated Rs to distinguish it from the R value) is a measure of the degree of separation of zones. In mathematical terms, a fundamental expression applicable to all linear chromatographic methods is

$$Rs = \Delta z/4\sigma \tag{4-29}$$

where Δz is the separation of the zones centers or the gaussian peaks and σ is the standard deviation of the zones. The magnitude of a σ indicates the degree to which the gap between peak centers is filled and cross contaminated by zone spreading (see Fig. 4-6). When resolution becomes less than one, there is appreciable overlapping of one zone by another. When $Rs = 1.0$, there is room for 2σ from each zone, the front of one and the rear of the other. Cross contamination occurs mainly near the midway point in the overlapping half of the zone beyond the 2σ distance; this is about 2%.

To acquire a cross contamination of only 0.1%, resolution must equal 1.5 or more. Glueckauf gives an elegant discussion of cross contamination.[12]

In paper and thin-layer chromatography, where one is concerned with resolution in space, the distance (L) migrated by a zone along the sheet of material is $L = R\nu t$. The gap between adjacent zone centers, ΔR, may be expressed as $\Delta z = \nu t \Delta R$. Since $\nu t = L/R$, $\Delta z = L\Delta R/R$ where the ratio $\Delta R/R$ is the relative velocity difference of the zones. Now from Eq. (4-17), the resolution expression may be written as

$$Rs = \sqrt{L/16H}\,(\Delta R/R) \tag{4-30}$$

In terms of phase volumes and the equilibrium characteristics of columnar liquid chromatography, useful expressions for resolution are

$$Rs = \sqrt{\frac{L}{16H}}\left[\frac{V_S \Delta K_d}{V_M + K_d V_S}\right] = \sqrt{\frac{L}{16H}}\left(\frac{\Delta K_d}{K_d}\right)(1 - R) \tag{4-31}$$

Any value of R near unity for either of the above resolution expressions is injurious. Thus, the achievement of separation is not consistent with high R values. In practical work, conditions should be adjusted so that R values range between 0.2 and 0.5 in liquid partition or adsorption chromatography. Gas chromatography is so rapid that R values as low as 0.01 can often be used without great inconvenience.

For column elution methods where the effluent is examined with respect to time, resolution is best expressed in terms of retention times or the corresponding retention volumes,

$$Rs = (V_{R,2} - V_{R,1})/0.5(W_2 + W_1) \tag{4-32}$$

where the numerator states the separation of the peak maxima and the denominator states the average peak width.

If equal quantities of the two solutes in adjacent zones are assumed, resolution can also be expressed[13] as

$$Rs = \tfrac{1}{4}[(\alpha - 1)/\alpha][k_2/(1 + k_2)]\sqrt{N_2} \tag{4-33}$$

where α is the relative retention ratio $[\alpha = (K_d)_2/(K_d)_1]$, k_2 is the partition ratio, and N_2 the plate number of the second component. In Eq. (4-33), the degree of separation is dependent on three factors: α, k_2, and N_2.

As α increases, the resolution becomes greater. The relative retention ratio pertains only to the relative rates of migration and thus to the peak-to-peak separation. For example, α would represent the relative volatility in gas chromatography, the relative solubility in a stationary liquid phase in partition chromatography, the relative adsorption in adsorption chromatog-

[12] E. Glueckauf, *Trans. Faraday Soc.* **51,** 34 (1955).

[13] An excellent discussion will be found in B. L. Karger, *J. Chem. Educ.* **43,** 47 (1966).

raphy, and the relative exchangeability in ion exchange chromatography. The influence of operating variables, such as temperature and type of stationary and mobile phases, on α, will be considered in later chapters.

The second factor influencing resolution is k_2, the partition ratio, which is given by $k_2 = (K_d)_2(V_S/V_M)$ for partition chromatography, and by $k_2 = (K_d)_2(A_S/A_M)$ for adsorption chromatography where A_S/A_M is the ratio of the cross-sectional area of the solid or liquid surface to the cross-sectional area of the mobile phase. An increase in the partition ratio should improve resolution by causing an increase in the ratio $[k_2/(1 + k_2)]$. Actually this will be true only when k_2 is small since the ratio asymptotically approaches unity with large values of k_2. The amount of stationary phase relative to the mobile phase, either in terms of volume or area, also affects k_2. The ratio V_S/V_M is much smaller in capillary or narrow-bore columns. Even so, resolution may not appear to be significantly changed because k_2 also appears in the expression for plate number. The direction of change in N depends on the relative value of k_2. Because N appears to the $1/2$ power in the resolution expression, the $[k_2/(1 + k_2)]$ term generally predominates over the k_2 expression in the value of N.

In general, the greater the plate number, the better the resolution. One recognized way to increase N is simply to make the column longer. This has its practical limits however. In liquid column chromatography, separation time, often excessive, increases in proportion to length. Super-long columns are a product of the gas chromatography era. Longer columns are feasible only as long as the flow can be maintained through a longer column, perhaps by use of high pressure to force the mobile phase forward, and no extraneous variables, such as evaporation of solvent from a paper strip or a thin layer, are present.

A major problem in the chromatographic separation of complex, multicomponent mixtures arises because the sample components eluted initially tend to be less well separated than components eluted later. Components with small retention volumes move so rapidly that they apparently are unable to avail themselves of the total number of theoretical plates. On the other hand, large partition coefficients may lead to an impossibly long analysis time, large band widths, and small peak heights. Small amounts of strongly held components may escape notice in the final chromatogram. The separation of a multi-component mixture also depends on finding room for the sheer number of the zones present more than it involves selectivity. In this respect, the large number of plates available in capillary or narrow-bore columns and in some thin-layer systems is a distinct advantage.

Choice of Column Length and Flow Velocity After the number of plates required for a desired degree of resolution has been ascertained by means of Eq. (4-33), coupled with Eq. (4-15), one can now proceed to design

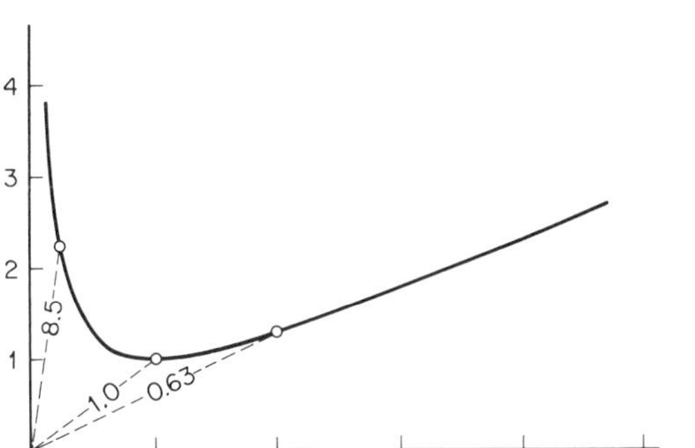

FIG. 4-8. Plate height–velocity curve redrawn as H/H_{min} vs ν/ν_{opt}. Shortest possible separation time for any point on the curve is proportional to the slope of the line joining that point to the origin.

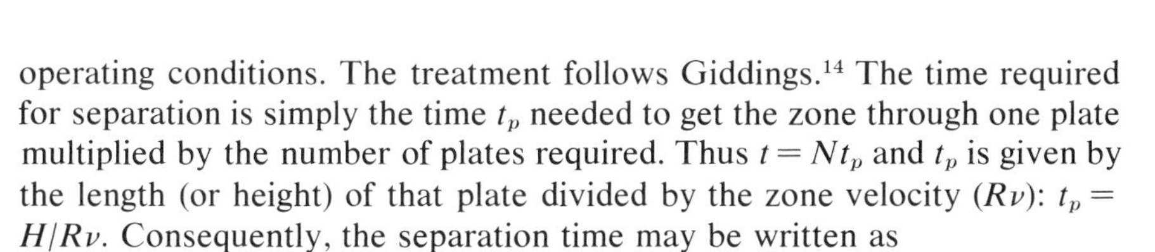

operating conditions. The treatment follows Giddings.[14] The time required for separation is simply the time t_p needed to get the zone through one plate multiplied by the number of plates required. Thus $t = Nt_p$ and t_p is given by the length (or height) of that plate divided by the zone velocity ($R\nu$): $t_p = H/R\nu$. Consequently, the separation time may be written as

$$t = NH/R\nu \tag{4-34}$$

The H/ν ratio can be obtained directly from the experimental plate height–velocity plot—it is simply the slope m of the line drawn from the origin to a point on the graph, as shown in Fig. 4-8. This slope decreases as flow velocity increases, but a point of diminishing returns is reached at a velocity much greater than twice the optimum velocity. Once the velocity has been selected, the column length necessary for the separation is obtained as NH.

A long column is potentially capable of achieving faster separation than a short column because of the decreasing H/ν ratio (see Fig. 4-8). Thus, for practical work one should start with a column which is as long as feasible. Flow velocity is then increased until the resolution is barely adequate. This will provide the utmost speed of analysis.

Quantitative Evaluation and Separation

Analysis directly on the column is infrequently used in column chromatography, except perhaps where substances with low retardation values are separated or occasionally in preparative work. Elution with the mobile phase is continued until the solvent front, or one void volume, has reached the base of the column. Then the entire column is extruded and sectioned by means of a scalpel while using a brush and streaked reagents to locate the individual zones if they are not obvious against the adsorbent. Lixiviation

[14] J. C. Giddings, *Dynamics of Chromatography*, Dekker, New York, 1965; Vol. I, *Theory and Practice*, p. 288.

with a suitable solvent removes the sorbed molecules from the stationary phase.

Elution is generally continued until all the components have left the column. This permits reuse of the column. Analysis of column effluent can be done by any method suitable for continuous analysis—spectrophotometry, polarimetry, refractometry, and radiochemical methods are examples. Frequently, automatic equipment is employed in conjunction with the fraction collection. A recording provides a permanent record. Ideally the property chosen for monitoring on the effluent should be a linear function of solute concentration, and the detector response should also be linearly related to the operative property of the mobile phase (if it varies). Quantitatively, the total amount of a solute eluted is measured by the area under a peak, when a differential detector is used, or by the height of the corresponding step when an integral detector is used. Quantitative evaluation methods are compared in Chapter 11.

Fraction Collection Fraction collectors enable the eluate to be collected in portions as desired and permits one to characterize the various species present in each fraction by independent methods. The apparatus successively places empty vessels under the tip of the column. On the turntable the tubes are arranged in circles or as a spiral. Occasionally the fractions are collected in a back-and-forward pattern which has the advantage of allowing the collecting tubes to be subdivided into racks. These may be removed from the apparatus separately and replaced by others during an elution. Positioning of a new vessel or tube is actuated in one of several ways: collection being based on (1) the lapse of equal intervals of time, (2) constant fraction size (volumetric, gravimetric, or drop counting), or (3) detector output.

Timed flow allows the eluate to drip into each collection tube for a preset period of time. If the flow of the mobile phase is maintained constant, fractions are of uniform volume and equivalent to constant-volume collectors.

Volumetric siphoning cuts the eluate by dripping it into a calibrated siphon that automatically discharges when full. Filling raises an electrolyte solution in the sidearm of the siphon until it touches a pair of detector electrodes, signaling the turntable to present an empty tube. Alternatively the siphon may be mounted on the arm of a balance, which swings over when the siphon is half full. In other cases a beam of light, falling on a photoelectric tube, is interrupted when the liquid in the siphon has reached a certain height. Volumetric siphoning is the basic type of fraction collector and the most economical one. However, a siphon never empties completely and, therefore, each fraction is contaminated with the previous one. With a well-designed siphon in which adequate drainage is assured, the residual contamination is reduced to less than 3%. A siphon-operated collector is essential for volatile solvents which would evaporate under drop-forming conditions.

Drop counting is recommended for extreme accuracy and precision of measurement without holdover of liquid between cuts. A predetermined number of drops is allowed to enter each collection tube with the aid of a mechanical or electronic scaler and with photoelectric detection. Drop size will be constant unless wide fluctuations occur in composition (as in gradient elution) or temperature of the drops. Drops can also be collected on a moving strip of filter paper which then can be processed like a paper chromatogram.

Collection may also be controlled by means of a signal from a detector, located at the exit of the column, so as to shift tubes when the elution of a band is completed. Examples would be flow colorimeters or spectrophotometers, radiometric scanners, and differential refractometers. A disadvantage of using the detector output as a signal for operating the turntable is that baseline drifts may activate the switch and contaminate the desired fractions.

Problems

1. Discuss the merit of the quantity $[N(1-R)^2]$, called the number of effective theoretical plates, as compared with just N.
2. Derive an expression for the minimal column length required for separation in terms of particle diameter. Assume $H \geqslant 2d_p$.
3. In the plate-height equation, assess the influence of particle size on resolution.
4. Derive an expression for the ratio $H/H_{\min}$ in terms of ν/ν_{opt}; graph the expression.
5. Derive an expression for the ratio $Rs/Rs_{\max}$ in terms of ν/ν_{opt}; graph the expression. Shade in the area that represents 90% of maximum resolution.
6. What is the passage time actually spent in motion of active molecules as they migrate through a chromatographic column?
7. A typical liquid chromatrographic system in which D_M is 1.5×10^{-5} cm²/sec and d_p is 0.015 cm (about 100 mesh) will have what mobile phase velocity when the reduced velocity is unity?
8. For a typical laboratory operation on a silica gel partition column, when the particle diameter is 0.015 cm and D_S is 3.4×10^{-5} cm²/sec (e.g., phenol in CS_2), calculate the reduced velocity for column flow rates of 0.001, 0.01, 0.1, and 0.5 cm/sec.
9. To calculate plate height for Problem 8, which expression should be employed?
10. Calculate the plate height for phenol in CS_2 on a silica gel column at the flow velocities given in Problem 8. Assume $R = 0.2$, and that the diffusion coefficient in the mobile phase is not greatly different from that in the stationary phase. Graph the results.
11. With helium as carrier gas, the B-term in the plate-height equation was found to be 0.30 cm²/sec for C_8H_{18} at $T_c = 30$°C. With nitrogen as carrier gas, the term was 0.088 cm²/sec; and with argon, 0.071 cm²/sec. Comment on the difference in values. Estimate the diffusion coefficient of the solute in each carrier gas.
12. Calculate the plate height for the chromatographic separation of glycine on paper whose fiber diameter is 0.0018 cm. Assume $R = 0.3$, and $D_M \simeq D_S = 1.06 \times 10^{-5}$ cm²/sec.

13. A gas–liquid packed column with nitrogen as carrier gas has these constants fitted to the plate height–velocity expression: $A = 0.1$ cm, $B = 0.075$ cm^2/sec, and E terms $= 0.01$ sec. (a) Graph the plate height–velocity expression, and sketch the limiting regions of each term. (b) Calculate the minimum plate height and the optimum carrier-gas flow. (c) To maintain the resolution within 90% of the maximum, the carrier-gas velocity should be restricted to what range?
14. Predict the procedure to be followed to improve each of the separations: (a) Resolution is only 0.2, necessitating a 5-fold increase for baseline separation. How might this best be accomplished? (b) Resolution is 0.8. How might the separation be improved?
15. Derive an expression for the average increase in temperature needed to just double an R value (which is 0.5 or less) in a gas–liquid chromatographic column.
16. Using the expression derived in Problem 15, find the temperature increase that typically doubles the R value (and R is 0.5 or less) when the column operating temperature is 227°C. [Hint: Trouton's rule is still approximately valid.]
17. For close-lying peaks, the relative velocity difference, $\Delta R/R$, is approximately equal to the relative retention time difference, $\Delta t/t$. What is the resolution per unit of relative retention time difference?
18. What is the required number of plates for $Rs = 1.0$ when $(K_d)_1 = 100$, $(K_d)_2 = 110$, and $V_M/V_S = 100$?
19. Express the R values for components 1 and 2 in Problem 18.
20. On a 30% liquid paraffin–Celite packed column, operated at 42°C, these values for plate height and hydrogen gas velocity were obtained using n-hexane:

H (in cm):	0.635	0.510	0.423	0.465	0.552	0.632	0.692	0.749
ν (cm/sec):	0.91	1.51	3.0	4.2	5.55	7.0	8.0	9.0

(a) Graph the data. (b) Estimate the terms in the plate height–velocity expression. (c) What is the optimum linear gas velocity and corresponding plate height? (d) Replot the data using as ordinates H (relative to H_{min}) and as abscissa ν (relative to ν_{opt}). (e) For each graphed point in part (d), calculate the slope of the line representing the H/ν ratio.

21. From data in Problems 4, 5, and 20, plot Rs/Rs_{max} as a function of ν/ν_{opt}. Tabulate the extremes of ν/ν_{opt} for these percentages of maximum resolution: 89, 77, 69, and 62%. Correlate with actual values of linear velocity.
22. On packed columns, 4 meters in length, and containing 13% Carbowax 1000 on Chromosorb R, these data were obtained for nitrobenzene, m-chloronitrobenzene, and p-chloronitrobenzene. Chart speed was 12.5 mm/min.

				Relative retentions (ϕ-NO_2 = 1.0)		Baseline width		
Column	T_c (°C)	F_c (ml/min)	ϕ-NO_2 t'_R (mm)	m-Cl-ϕ-NO_2 t'_R (mm)	p-Cl-ϕ-NO_2 t'_R (mm)	ϕ-NO_2 (mm)	meta- (mm)	para- (mm)
1	210	28	39	1.623	1.792	4.5	7.4	8.4
2	170	98	82	1.742	1.962	8.0	15.3	15.2
3	150	52	183	1.792	2.036	17.8	31.9	33.0
4	150 [a]	33	227	1.960	2.231	24.5	44.0	45.0
5	152 [a]	33	524	1.765	2.025	43.0	70.0	70.0

[a] Different column, same support.

(a) Calculate retention volumes for nitrobenzene, *m*-chloronitrobenzene, and *p*-chloronitrobenzene for each of the columns. (b) Calculate the plate number for nitrobenzene and each chloronitrobenzene on each column. (c) Estimate the resolution for the chloronitrobenzenes on each column. (d) Calculate the zone velocity ($R\nu$) for each component on column 1.

23. In a reversed-phase chromatographic separation on Teflon-6, cyclohexane was used as the stationary phase with 0.5*M* NaCl as the mobile phase. From the values of V_R equal to 18 ml and 62 ml for 3,4-dimethyl phenol and 2,6-dimethyl phenol, respectively, on a column which is 12.5 cm in length and 1.0 cm inside diameter, and which contains 4.0 ml of cyclohexane and 5.0 ml of mobile phase, calculate the distribution ratios and the apparent number of plates. The base widths extended from 11 to 29 ml and from 37 to 97 ml.
24. On the same column used in Problem 23, but whose stationary phase now is cyclohexane containing 6% of 1-hexanol, V_R is equal to 19.5 ml for phenol and 54 ml for *o*-cresol. Corresponding base widths were 13 to 30 ml and 40 to 74 ml. Calculate the distribution ratios and the apparent plate number.

BIBLIOGRAPHY

J. M. Bobbitt, A. E. Schwarting, and R. J. Gritter, *Introduction to Chromatography,* Reinhold, New York, 1968.

H. G. Cassidy, *Fundamentals of Chromatography,* Vol. X in A. Weissberger (Ed.), *Technique of Organic Chemistry,* Interscience, New York, 1957.

J. C. Giddings, *Principles and Theory,* Vol. 1 of *Dynamics of Chromatography,* Dekker, New York, 1965.

J. C. Giddings and R. A. Keller (Eds.), *Advances in Chromatography,* Dekker, New York, 1965– ; 8 volumes released to date.

E. Heftmann (Ed.), *Chromatography,* Reinhold, New York, 1967.

O. Mikeš (Ed.), *Laboratory Handbook of Chromatographic Methods,* Van Nostrand (Princeton, New Jersey), 1961. (Translated English edition, 1966.)

I. Rosenthal, A. R. Weiss, and V. R. Usdin, *Chromatography: General Principles,* in I. M. Kolthoff and P. J. Elving (Eds.), *Treatise on Analytical Chemistry,* Part I, Vol. 3, Interscience, New York, 1961.

5 *Ion Exchange*

Ion-exchange methods are based essentially on a reversible exchange of ions between an external liquid phase and an ionic solid phase. The solid phase consists of a polymeric matrix or crystal lattice, insoluble, but permeable, which contains fixed charge groups and mobile counter ions of opposite charge. These counter ions can be exchanged for other ions in the external liquid phase. Enrichment of one or several of the components is obtained if selective exchange forces are operative. The method is limited to substances at least partially in ionized form. Only resinous ion exchangers are discussed in this chapter; the discussion of other types of exchangers is reserved for Chapter 6.

Chemical Structure of Ion-Exchange Resins

An ion-exchange resin usually consists of polystyrene copolymerized with divinylbenzene to build up an inert three-dimensional, cross-linked matrix of hydrocarbon chains. Protruding from the polymer chains are the ion-exchange sites distributed statistically throughout the entire resin particle. An exchanger might be pictured as a special type of polyelectrolyte gel structure whose ionic units are restricted in motion by attachment to the polymer chains. The charged groups are free to rotate about the position of attachment to the chain, but their translational motion is limited by the immobility of the network as a whole. The ionic sites are balanced by an equivalent number of mobile counter ions. The type and strength of the exchanger is determined by these active groups. Ion exchangers are designated anionic or cationic, according to whether they have an affinity for negative or positive counter ions. Each main group is further subdivided; the cation exchangers into strongly acidic and weakly acidic types, and the anion exchangers into strongly basic or weakly basic types. A selection of commercially available ion-exchange resins is given in Table 5-1. Most

TABLE 5-1. Commercial Ion-Exchange Resins

Name	Type	Effective operating pH range	Functional group	Ionic form	Exchange capacity: dry resin (mEq/g)	Exchange capacity: wet resin (mEq/ml)
Dowex 50	Strongly acidic cation exchanger	0–14	$—SO_3^-H^+$; polystyrene	H or Na	4.3	1.9
Amberlite IR-120		0–14			4.2	1.9
Duolite C-20					5.1	2.1
Wofatit KPS		0–12			4.5	1.3
Zeokarb 225					5.0	2.4
Amberlite IRA-400	Strongly basic anion exchanger	0–14	Quaternary amine	Cl	3.3	1.2
Dowex 1		0–14	$—CH_2—\overset{+}{N}(CH_3)_3$; polystyrene		3.3	1.2
Duolite A-101		0–14			4.0	1.3
Deacidite FF		0–12			3.5	1.5
Amberlite IRA-410		0–12	$—CH_2—\overset{+}{N}(CH_3)_2C_2H_4OH$		3.1	1.35
Dowex 2		0–14			3.3	1.2
Amberlite IR-4B	Weakly or medium basic	0–7	—NHR or $—NR_2$; phenol or polyamine condensation	OH	10	2.5
Wofatit MD					5.7	2.0
Amberlite IR 45		0–7	Polystyrene; —NHR or $—NR_2$		5.0	2.0
Dowex 3		0–7			5.5	2.5
Amberlite IRC 50	Weakly acidic	7–14	Polyacrylic or polymethacrylic acid	H	10	3.5
Duolite CS 101					10.2	3.3
Duolite ES 63	Medium acidic		Polystyrene—$PO(OH)_2$		6.6	3.1
Duolite C 3	Strongly acidic		Phenol condensation; $—SO_3^-H^+$		2.9	1.2

of these are available in a range of particle sizes and with different degrees of cross-linking.

Microreticular Resins For microreticular (gel-like) resins, the synthetic procedure involves a suspension-type polymerization in which styrene is copolymerized with a fixed amount of bifunctional monomer (divinylbenzene). The divinylbenzene causes cross-linking at random positions along the chain of what would otherwise be a linear polymer:

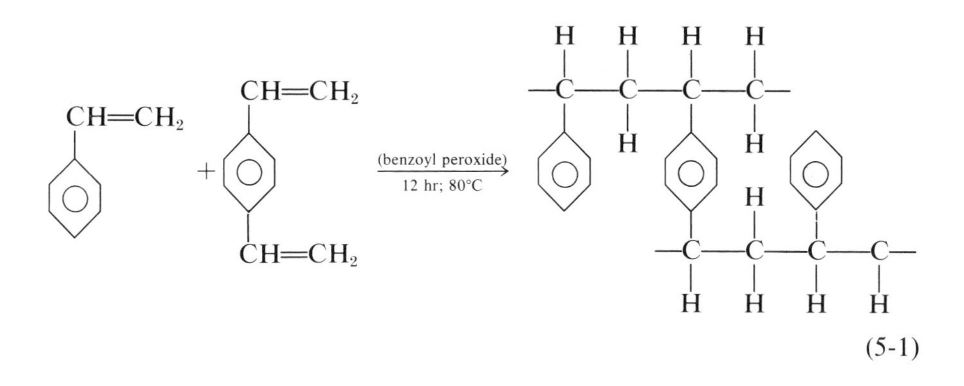

(5-1)

The liquid monomers are vigorously agitated in an aqueous medium containing a suspension stabilizer to prevent agglomeration of the spherical droplets. The size of the individual resin beads depends on the rate of stirring and on the type of stabilizer, and can be varied within wide limits.

The structure of the ion-exchange resin is completed by adding ion-active groups to the copolymer network. To produce an anion exchanger, the resin beads are treated first with a swelling agent, such as trichloroethylene or nitrobenzene, to prevent fracture of the individual spheres and to facilitate the succeeding operations. Next the swollen beads are treated with chloromethyl methyl ether ($ClCH_2OCH_3$) in the presence of a catalyst such as aluminum or zinc chloride to effect a Friedel–Crafts reaction by which chloromethyl groups are attached to the benzene rings in the copolymer. This step must be done carefully so as to attach the methylene chloride groups to the benzene rings without causing them to react further to form methylene bridges. Amination of the chloromethyl group with a tertiary amine produces a strongly basic type of resin with quaternary ammonium groups as exchange sites. Amination with ammonia or aliphatic primary or secondary amines yields a weakly basic type of resin which contains the corresponding primary, secondary, or tertiary amine group at the exchange sites. On the average there is only one anion exchange group for every two benzene nuclei.

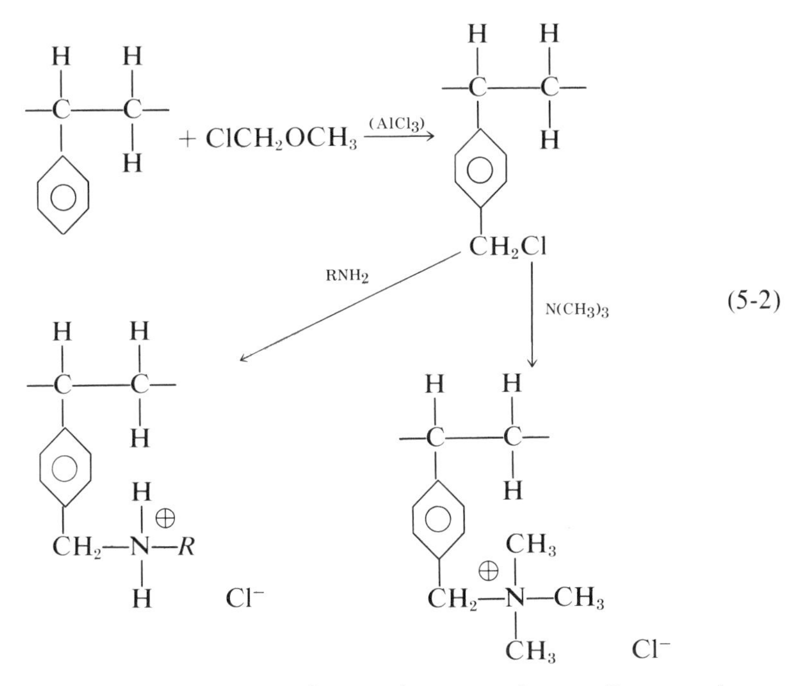

(5-2)

In the production of strong cation exchangers, the swollen copolymer beads are sulfonated with chlorosulfonic acid to introduce the sulfonic acid group into the benzene rings, generally in the para position. Approximately 1.1 sulfonate groups occur per benzene ring.

Weakly acidic cation exchangers are prepared by the copolymerization of methacrylic acid (or its nitrile) and divinylbenzene to form a cross-linked polymethacrylic acid polyelectrolyte structure:

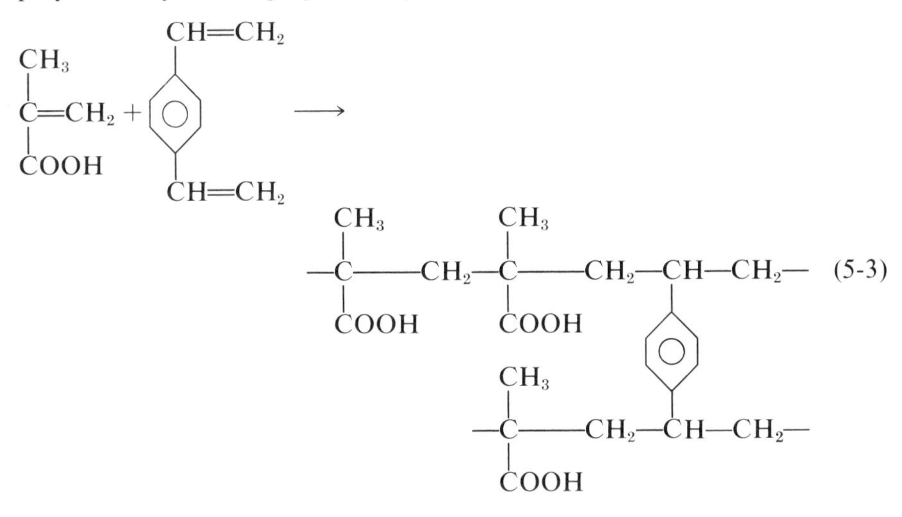

(5-3)

Many other types of functional groupings have been incorporated into resins. Functional groups containing phosphorus are obtained by condensing PCl_3 with the benzene rings of the styrene–divinylbenzene copolymer. Phosphinic acid groups are formed by hydrolysis under mild conditions and phosphonic acid groups by oxidation with chlorine prior to hydrolysis or by oxidation with nitric acid after hydrolysis. The first dissociation in the phosphonic acid resins corresponds to that of a moderately strong acid, while the second dissociation is that of a weak acid. The phosphonous resin shows only the presence of a relatively strong acid group.

A chelating resin, containing as active group iminodiacetate ion, is commercially available. Electron-exchange resins have been prepared by polymerizing vinylhydroquinone with divinylbenzene; these resins act as insoluble reducing and oxidizing agents but one difficulty is their slow rate of reaction.

Macroreticular Resins Macroreticular ion-exchange resins are prepared by a variation of the polymerization technique used for the conventional gel-like resins. An organic solvent which is a good solvent for the monomers, but a poor solvent for the copolymer, is added to the polymerization mixture. As polymerization progresses, the solvent phase is squeezed out by the growing copolymer regions. The pores that remain are several hundred angstroms wide and lie between rigid polymeric structures. Ion active groups are then added to the polymeric structure. The product is an agglomerate of randomly packed microspheres (with normal gel porosity), and extending through the agglomerate is a continuous nongel pore structure. Macroreticular resins possess a high surface area. The channels throughout the rigid pore structure render the bead centers accessible even in nonaqueous solvents, in which microreticular resins do not swell sufficiently. Because of their high porosity and large pore diameters, these resins can handle large organic molecules and will exhibit less fouling when employed in water-softening and deionization applications in commercial installations. These are also resins of choice when aggressive operating conditions are encountered, as when exposed to severe osmotic pressures created during regeneration and rinse operations. In all other respects the macroreticular resins are identical to their microreticular counterparts.

Condensation Resins Resins can also be made by condensation reactions, following the original process of Adams and Holmes.[1] Phenol and formaldehyde are condensed and the product sulfonated to give an exchanger with two kinds of functional groups: sulfonate and phenolic hydroxyl. By condensation of ethylene dichloride or epichlorohydrin with primary and secondary amines, a variety of weakly basic exchangers is

[1] B. A. Adams and E. L. Holmes, *J. Soc. Chem. Ind. (London)* **54T,** 1 (1935).

possible. The sulfonated phenolic-type resins possess noteworthy resistance to radiation damage.

Other Ion-Exchange Media Cellulose ion exchangers are widely used for the chromatography of macromolecules and biologically active materials. These exchangers (Chapter 6) are prepared by imparting anion and cation exchange properties to cellulose by treatment with various reagents.

Permselective ion-exchange membranes, similar in composition to the synthetic ion-exchange resins, find use in membrane separation processes (Chapter 13). Filter paper impregnated with finely divided ion-exchange resins, and paper prepared from cellulose pulp that has been modified by the introduction of ionic exchange sites, finds use in separations requiring only a small amount of sample, such as paper chromatography.

Cross-Linking The cross-linking of a polystyrene-type resin is expressed as the proportion by weight percent of divinylbenzene in the reaction mixture. The symbol "×8" following the resin name and type designates this amount. As the amount of divinylbenzene is increased the cross linkages occur at closer intervals and the effective pore size, permeability, and tendency of the resin to swell in solution are all reduced. At the same time the ionic groups come into effectively closer proximity, resulting in increased selectivity. Intermediate cross-linking, obtained by the addition of 4 to 12% divinylbenzene, is generally used. Low cross-linked resins have a high degree of permeability, come to equilibrium more rapidly, and are able to accommodate large molecules. On the other hand, their wet volume capacity is low and selectivity suffers because the resin phase differs little from the contacting solution in ionic properties. Highly cross-linked resins exhibit properties in the opposite direction. Permeability is low but adequate for inorganic ions; the time necessary for establishment of equilibrium is increased.

Ion-Exchange Equilibria

Ion-exchange methods are based on the distribution of an ionic species between an external solution and a solid resin phase. With modern ion exchangers, exchange reactions are nearly perfectly reversible in most instances, and the position of equilibrium is independent of the direction from which it is approached.

Swelling Microreticular resins are elastic gels that, in the dry state, avidly absorb water and other polar solvents in which they are immersed. While taking up solvent, the gel structure expands, or swells, until the retractile stresses of the distended polymer network balance the osmotic effect. The swollen volume is of interest in chromatography since this gives the volume of resin when immersed in solution. In nonpolar solvents these resins swell hardly at all; their low solvent content and correspondingly small pore width impair diffusion and result in prohibitively low sorption and slow exchange rates. Macroreticular resins, with rigid channels, render the interior of the resin accessible even in nonpolar solvents in which the resin does not swell significantly.

In the swollen state the interior of the resin bead resembles a drop of concentrated electrolyte solution whose concentration may range from 1.2 to 6*M* for different resins and degrees of swelling. The fixed and counter ions tend to form solvation shells with the imbibed solvent, and the highly concentrated solution in the interior tends to dilute itself by taking up additional solvent. The solvent content of the resin particle depends primarily on the activity coefficient of the internal ions and the matrix pore sizes which are 5–50Å, depending on the degree of cross-linking. Volume changes are greater with resins of lower cross-linking and with polar solvents when there exists a strong solvation tendency of the fixed and counter ions. Swelling is also affected by the nature of the counter ions; in water it often follows the same order as the degree of ionic hydration unless ion-pairing occurs.

Whereas the counter ions and ionized groups within the resin determine the extent of hydration of the resin, the skeletal structure determines the stability and life of the resin in actual use. Insolubility is one of the more important properties of an ion-exchange resin. For a resin to be economically useful in cyclic operations, it must resist the solubilizing effects of dilute water solutions, moderate and sometimes concentrated acids, bases, salts, organic solvents, and any drastic volume changes that occur in the transformation from one ionic form to another. The elasticity of the resins is necessary to enable them to withstand these internal strains.

Selectivity If given a choice, all ion exchangers exhibit some degree of preference for one ionic species relative to another. In a batch operation that involves the exchange between the counter ions of the resin (in the hydrogen form) and an electrolyte solution containing M ions of n valence, the following reaction takes place:

$$\underset{\substack{\text{solid}\\\text{phase}}}{H^+R^-} + \underset{\substack{\text{solution}\\\text{phase}}}{1/n\, M^{+n}} \rightleftharpoons \underset{\substack{\text{solid}\\\text{phase}}}{1/n\, M^{+n}R_n^-} + \underset{\substack{\text{solution}\\\text{phase}}}{H^+} \qquad (5\text{-}4)$$

where R^- symbolizes negative exchange sites in the resin matrix. Equilibrium is established when there is no further statistical change in the ratio of M^{+n}/H^+ within the resin phase. At equilibrium the concentration of the ions in solution and in the exchanger is related through the *selectivity coefficient*, $E_H^{M/n}$ (also denoted by the name concentration exchange constant) as follows:

$$E_{\mathrm{H}}^{M/n} = \frac{[M^{+n}]_r[\mathrm{H}^+]}{[\mathrm{H}^+]_r[M^{+n}]} \tag{5-5}$$

where the subscript r refers to the resin phase. Simply stated, the selectivity coefficient expresses the selectivity of the resin for M^{+n} ions in preference to hydrogen ions from a solution containing an equal concentration of each (and for a specified degree of resin loading). A collection of selectivity coefficients is given in Table 5-2 for cations, and in Table 5-3 for anions. Tabulated values refer to equivalents of ion adsorbed (indicated by superscript) from 1 ml of solution per 1 g of dry resin in the H form, for cation exchangers, and in the Cl form for anion exchangers.

TABLE 5-2. Selectivity Coefficients, $E_{\mathrm{H}}^{M/n}$, for Some Metal Ions on Differently Cross-Linked Dowex 50

	4% DVB	*8% DVB*	*16% DVB*
Univalent ions			
Li	0.76	0.79	0.68
H	1.00	1.00	1.00
Na	1.20	1.56	1.61
NH_4	1.44	2.01	2.27
K	1.72	2.28	3.06
Rb	1.86	2.49	3.14
Cs	2.02	2.56	3.17
Ag	3.58	6.70	15.6
Bivalent ions			
UO_2	0.79	0.85	1.05
Mg	0.99	1.15	1.10
Zn	1.05	1.21	1.18
Co	1.08	1.31	1.19
Cu	1.10	1.35	1.40
Cd	1.13	1.36	1.55
Ni	1.16	1.37	1.27
Mn	1.15	1.43	1.54
Ca	1.39	1.80	2.28
Sr	1.57	2.27	3.16
Pb	2.20	3.46	5.65
Ba	2.50	4.02	6.52
Trivalent ions			
Cr	1.6	2.0	2.5
Ce	1.9	2.8	4.1
La	1.9	2.8	4.1

Based on data of Bonner *et al.*, *J. Phys. Chem.* **61,** 326 (1957); *ibid.* **62,** 250 (1958).

TABLE 5-3. Selectivity Coefficients, E_{Cl}^{X}, for Some Anions on Dowex Resins

	Dowex 1	*Dowex 2*
Hydroxide	0.09	0.65
Fluoride	0.09	0.13
Aminoacetate	0.10	0.10
Acetate	0.17	0.18
Formate	0.22	0.22
Dihydrogen phosphate	0.25	0.34
Bicarbonate	0.32	0.53
Chloride	1.00	1.00
Bromate	...	1.01
Bisulfite	1.3	1.3
Cyanide	1.6	1.3
Bromide	2.8	2.3
Nitrate	3.8	3.3
Benzene sulfonate	...	4.0
Bisulfate	4.1	6.1
Phenoxide	5.2	8.7
Iodide	8.7	7.3
p-Toluene sulfonate	...	13.7
Thiocyanate	...	18.5
Perchlorate	...	32
Salicylate	32.2	28

From S. Peterson, *Ann. N. Y. Acad. Sci.* **57,** 144 (1954).

Example 5-1

From the selectivity coefficients listed in the tables, one is able to calculate the selectivity coefficient for the exchange between any pair of ions. For the exchange between a cesium salt solution and a Dowex 50 × 8 cation-exchange resin with 8% cross-linking loaded with sodium ions:

$$E_{Na}^{Cs} = E_{H}^{Cs}/E_{H}^{Na} = 2.56/1.56 = 1.64$$

For the exchange between calcium salt solution and the same resin loaded with sodium ions:

$$E_{Na}^{Ca/2} = E_{H}^{Ca/2}/E_{H}^{Na} = 1.80/1.56 = 1.15$$

or, to account for the unequal charges of the ions:

$$E_{2Na}^{Ca} = [Ca^{+2}]_r[Na^+]^2/[Ca^{+2}][Na^+]_r^2 = (1.15)^2 = 1.32$$

The superscript and subscript on E indicate the kind and number of exchanging ions. As another example, the selectivity coefficient for the exchange between a lanthanum salt solution and the same type of resin loaded with calcium ions is:

$$E_{\mathrm{Ca}/2}^{\mathrm{La}/3} = E_{\mathrm{H}}{}^{\mathrm{La}/3}/E_{\mathrm{H}}{}^{\mathrm{Ca}/2} = 2.8/1.8 = 1.56$$

or

$$E_{3\mathrm{Ca}}^{2\mathrm{La}} = [\mathrm{La}^{+3}]_r^2[\mathrm{Ca}^{+2}]^3/[\mathrm{Ca}^{+2}]_r^3[\mathrm{La}^{+3}]^2 = (1.56)^{3/2} = 1.95$$

The selectivity coefficient is not a constant. It depends on the proportions of the two exchanging ions and on the total concentration of the solution and, for a given type of resin, will vary with the degree of cross-linking. The loading of a resin by an ion is defined by the relation

$$\Lambda = n[M^{+n}]_r/Q_w \tag{5-6}$$

where Q_w is the weight capacity of the resin (in mEq/g). Variation of the selectivity coefficient with degree of cross-linking and resin type is shown by comparing values in Tables 5-2 and 5-3. Variation is greatest with highly cross-linked resins, the environment wherein interionic forces inside the resin matrix are a dominant factor. Even so, differences for cation exchangers are not large; much greater selectivity is found in anion exchangers. Quite often adequate selectivity is achieved only when ion exchange is combined with complex ion formation in batchwise operations, or separations are performed by column chromatography, perhaps in the presence of a complexing agent.

Increasing affinity for the resin shows no exact correlation with any one simple property of the ions. Factors involved are the hydrated ionic radius, which limits the coulombic interaction between ions, and the polarizability of the ions which determines the van der Waals' attraction. Together these factors control the total energy of interaction between cations and anions and hence their tendency to exist as ion pairs in the resin matrix. Few exchange sites are located on the surface of each microreticular resin bead; only such ions as can enter the matrix meshes of the resin particles can participate in exchange reactions. Energy is needed to strip away the solvation shell surrounding ions with large hydrated radii, even though their crystallographic ionic radii may be less than the average matrix mesh size. This explains the position of the Li ion among the alkali metal ions, and fluoride among the halides. Complex relationships apply to exchanges between ions of different net charge; however, the more dilute the solution the more selective the exchange becomes for the polyvalent ion. This holds true for both cationic and anionic exchange.

Ion-Exchange Capacity Exchange capacity is a measure of the counter-ion content, or the ability to take up counter ions, in a unit volume (volume capacity) or weight (weight capacity) of resin. The total available capacity is a theoretical value for the material and can be calculated from the number of functional groups per unit dry weight. However, it is dependent upon the ionic form of the resin. More exchange groups are available per gram of dry resin in the hydrogen form than in the sodium form; the difference is about 0.4 mEq/g. Volume capacity also varies due to the number of water molecules carried by each ion in its hydration shell. Consequently an ion-exchange resin expands or contracts when converted from one ionic form to another. Volume capacity usually refers to a settled bed which is in the hydrogen (or chloride) form, and is fully water-swollen. Weight capacity and volume capacity are related as follows:

$$Q_v = (1 - \beta)\rho Q_w(100 - S/100) \tag{5-7}$$

where $\beta = V_o/V_b$ is the fractional void volume of the packing (V_b is bed volume and V_o is the void or interstitial volume), S is the water content of the resin in weight percent, and ρ is density of the swollen resin.

The actual operating capacity of an ion-exchange resin bed is far more important than the total available capacity. It is the actual capacity under specified operating conditions (flow rate, column temperature, particle size, degree of packing, and concentration of solution feed), and is usually substantially lower than the total capacity. In industrial parlance this is called the break-through or leakage capacity. It represents the exhaustion point in terms of feed volume, after which the adsorbate will leak through into the effluent in gradually increasing amounts which exceed the preset or desired leakage value.

Example 5-2

A strongly basic anion exchanger of 8% cross-linking has a weight capacity of about 3.8 mEq/g (dry). Its density is 1.11 g/ml and the water content is 42–48%. The fractional void volume of a packed bed may be taken as 0.40. The volume capacity is

$$Q_v = (1 - 0.40)(1.11)\left(\frac{100 - 45}{100}\right)(3.8) = 1.4 \text{ mEq/ml}$$

Molar Distribution Coefficient In ion-exchange work it is common practice to describe the distribution of a component using the molar (or weight) distribution coefficient, K_d. The value of K_d is found by batch-equilibration experiments and is defined by the expression:

$$K_d = \frac{[M]_r}{[M]} = \frac{\text{amount of ion on resin per gram of resin}}{\text{amount of ion in solution per milliliter of solution}} \tag{5-8}$$

where M denotes the ion, present in relatively low concentration so that the loading does not exceed 5%. The molar distribution coefficient is analogous to the Nernst partition coefficient in liquid–liquid extraction. For any given conditions, K_d can be calculated if the selectivity coefficient, the resin capacity, and the concentration of the other ion participating in the exchange are known. Assuming the resin is initially loaded with hydrogen ions,

$$K_d = \left[E^{M}_{n\mathrm{H}^+} \left(\frac{[\mathrm{H}^+]_r}{[\mathrm{H}^+]} \right) \right]^n \tag{5-9}$$

For exchanges between ions of unequal charge, the distribution coefficient depends strongly on the concentration of the external solution. For example, if a divalent ion is exchanged for a univalent ion, an increase in concentration of univalent ion will drive the univalent ion into the resin, displacing the divalent ion. This fortunate fact permits the resin to be regenerated by hydrogen or sodium ions (if a cation exchanger) after exhaustion by calcium or magnesium ions as in water softening or deionization operations. Conversely dilute solutions favor sorption of higher valent ions.

For the exchange of an anion of a weakly dissociated acid, the distribution equation must be modified. If the acid is monoprotic, its total quantity both ionized and undissociated is $V_o([A^-] + [\mathrm{H}A])$. The distribution ratio is

$$K_d = \frac{wQ_w[A^-]_r}{V_o([A^-] + [\mathrm{H}A])} \tag{5-10}$$

Here w is weight of resin. Combining this equation with the dissociation constant of the acid and the selectivity coefficient,

$$K_d = \left(\frac{wQ_w[A^-]_r E_B{}^A}{V_o[B^-]} \right) \left(\frac{K_a}{K_a + [H^+]} \right) \tag{5-11}$$

Thus, acids of different strength, or the anions corresponding to them, can be separated from each other at appropriate values of pH. Caution must be exercised in the use of buffers to control the pH of the external solution. The ions introduced may themselves displace the adsorbate from the resin and thus alter the distribution ratio. This disturbance is not significant if the buffer concentration is very small and if the anion of the buffer has a small affinity for the resin.

Batch Separations In batch operations the molar distribution coefficient alone does not give any information about the proportion of an ion adsorbed or not adsorbed by an ion exchanger. For more specific information, one must know the amount of resin (in grams) and the volume of the contacting solution (in milliliters). Then

$$\frac{\text{Total amount of ion in resin phase}}{\text{Total amount of ion in solution phase}} = \frac{[M]_r w}{[M]V} \tag{5-12}$$

Batch separation involves simply placing the resin into the contacting solution and allowing equilibrium to be established by appropriate standing and stirring. The solution phase is then removed by centrifugation.

The extent to which exchange takes place in a batch separation is limited by the selectivity of the resin at equilibrium. Complete uptake of a particular ion by a batch of resin can be achieved when $K_d > (V/w)10^3$. Under practical conditions an ion will be adequately adsorbed if $K_d > 10^4$ and will remain in solution if $K_d < 10^{-2}$. Where applicable, a batch operation offers the advantage of an extremely simple technique for chemical separation.

Example 5-3

If a cation exchanger in the hydrogen form (with a capacity of 3.0 mEq/g) is shaken with 50 ml of a 0.001*M* cesium salt solution, what percentage of the cesium ions will be adsorbed by 1.0 g of the resin? From Table 5-2, $E_H^{Cs} = 2.56$, thus

$$K_d = [Cs^+]_r/[Cs^+] = 2.56([H^+]_r/[H^+])$$

Now the maximum amount of cesium which can enter the resin is 50 ml × 0.001*M* = 0.050 mEq. It then follows that the minimum value of $[H^+]_r =$ 2.95 mEq, and the maximum value of $[H^+] = 0.001M$. The minimum value of the distribution ratio is

$$K_d = (2.56)(2.95)/0.001 = 7550$$

From Eq. (5-12)

$$\frac{\text{Amount of Cs in the resin}}{\text{Amount of Cs in the solution}} = 7550(1/50) = 151$$

Thus, at equilibrium the 1 g of resin has removed 99.34% of the cesium ions from solution. If desired, the calculations could be repeated with more refined values for $[H^+]_r$ and $[H^+]$.

If the amount of resin were increased to 2.0 g, the amount of cesium remaining in the solution would decrease to 0.33%, half the former value. However, if the stripped solution were decanted and placed in contact with 1 g of fresh resin, K_d becomes 1.2×10^6 and the amount of unadsorbed cesium drops to 0.004%. Two batch equilibrations would effectively remove the cesium from the solution.

Ion-Exchange Chromatography

In column elution chromatography a small amount of sample is introduced as a thin band at the top of the ion-exchange column; then the adsorbed ions are gradually moved down the column by a suitable eluent in a series of desorption–sorption steps. If the selectivity coefficients of the sample ions differ sufficiently, each will travel down the column at different rates and emerge as a distinct band. Since the theory of chromatographic columns is discussed in Chapter 4, it will suffice here to limit the discussion to those aspects that apply to ion exchange and to review their most important practical consequences.

Retention Parameters In column operations, a new parameter, the volume distribution ratio, D_v, is useful. It is defined as

$$D_v = \frac{\text{Amount of ion in 1 ml of resin bed}}{\text{Amount of ion in 1 ml of interstitial volume}} = \frac{[M]_r/V_b}{[M]/V_o} \quad (5\text{-}13)$$

The volume distribution ratio is related to the weight distribution coefficient by

$$D_v = K_d\beta \quad (5\text{-}14)$$

where β is the fractional void volume (V_o/V_b) of the settled column. Lacking other information, V_o, the liquid volume in the interstices between the individual resin beads, is approximately 0.4 of the total bed volume, V_b, for uniform, close-packed spheres. The volume distribution ratio differs from the molar partition or distribution coefficient in that $[M]_r$ is the amount of ion in the ion exchanger per unit volume of the bed (including the interstitial volume) rather than per unit weight or volume V_r of the swollen ion exchanger alone. It is strongly dependent on the degree of cross-linking in the resin particle.

The capacity or partition ratio, k, as used in other types of chromatography, provides information concerning the amount of the ion in each phase, viz.

$$k = \frac{[M]_r V_r}{[M] V_o} \quad (5\text{-}15)$$

Important relationships exist between the volume distribution ratio and the volume of eluent needed for obtaining the maximum concentration of an eluted ion in the effluent, $V_{\max}$ (or V_R):

$$V_{\max} = V_b(D_v + \beta) \quad (5\text{-}16)$$

and also between the molar distribution coefficient and the volume needed to elute the peak of an adsorbate band from a resin column:

$$V_{\max} = K_d V_o + V_o \quad (5\text{-}17)$$

These relationships are strictly applicable only when the column loading is less than about 3–5% of the total column capacity.

Columns containing the same type of resin, but in different amounts or settled bed volumes, can be interrelated through the multiples of the interstitial volume that are required to elute the peak of the band. Rearranging Eq. (5-17) as follows:

$$(V_{\max} - V_o)/V_o = K_d \quad (5\text{-}18)$$

shows that the division of $V_{\max}$ by the interstitial volume expresses the peak maximum in terms of numbers of free column volumes. One interstitial volume is subtracted; it represents the interstitial volume present in the column prior to the passage of eluent.

Efficiency The number of effective sorption–desorption steps, or plate number, N, in a resin bed may be expressed by the relation

$$N = 16(V_{max}/W)^2 \tag{5-19}$$

where W is defined as the number of solution-volume units in the effluent as measured between intercepts of tangents to the front and rear flanks of the peak and the base line (or 4σ), as indicated in Fig. 5-1. The bandwidth depends critically on the overall exchange rate, but only slightly on the nature of the rate-controlling mechanism. The plate number is given also by

$$N = L/H \tag{5-20}$$

where L is the column length and H is the plate height.

The plate height can be expressed by four terms:

$$H = \underset{\text{eddy diffusion}}{1.64r} + \underset{\text{particle diffusion}}{\frac{D_v}{(D_v+\beta)^2}\left(\frac{0.142r^2\nu}{D_r}\right)} + \underset{\text{film diffusion}}{\left(\frac{D_v}{D_v+\beta}\right)^2 \frac{0.266r^2\nu}{D_f(1+70r\nu)}} + \underset{\text{longitudinal diffusion}}{\frac{D_f\beta\sqrt{2}}{\nu}} \tag{5-21}$$

where r is the radius of the resin beads (in cm), ν is the velocity of the mobile (eluent) phase (in cm/sec), and D_r, D_f are diffusion coefficients of the solute within the resin and liquid phases, respectively. Equation (5-21) shows that the plate height depends on the volume distribution ratio of the ionic species. Thus, in a separation, the plate height is different for each species.

For ion-exchange systems not involving a chemical step, the rate-controlling process at high concentrations (0.1M or larger) of ions in the external solution is diffusion into and within the resin bead. At low concentrations (0.001M or less) the diffusion of ions across the film or layer of solution

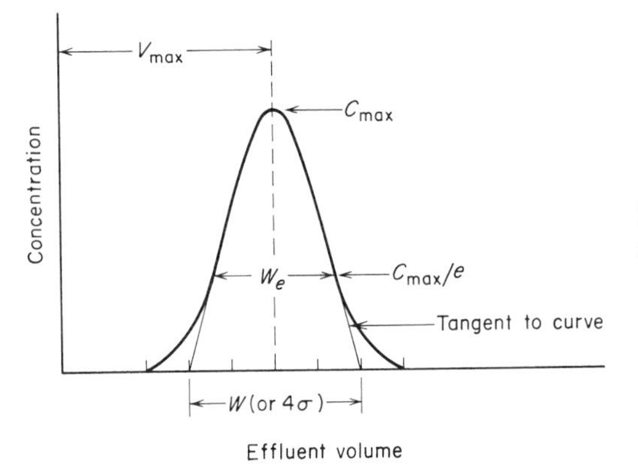

FIG. 5-1. Shape of ideal elution curve.

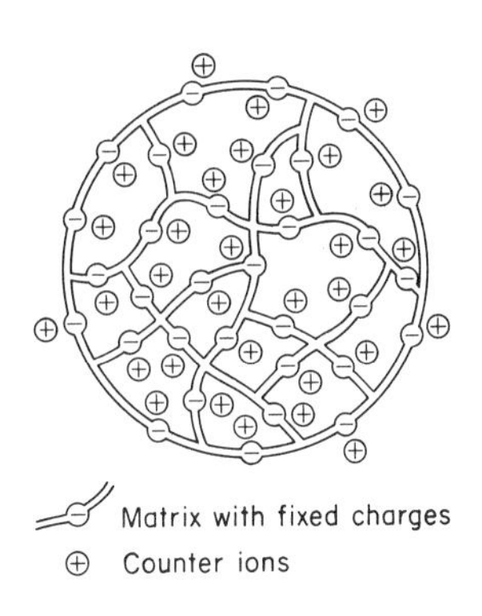

FIG. 5-2. Schematic structure of an ion exchange resin bead.

surrounding the resin bead is rate-controlling. The size of the resin beads determines the area of the surface available for film diffusion, and also the size of the liquid interstices between the individual beads and thus the distance an ion in the external solution must be transported to reach the particle. The reasons for the two limiting cases become apparent upon examining the diagrammatic sketch of a resin particle (Fig. 5-2). Only a small fraction of the total exchange sites exist within the surface film. Consequently, only if the external solution has a low ionic concentration will there be sufficient exchangeable sites available on the surface to handle all of the ions that are able to diffuse into the film, coupled with concurrent diffusion of ions from the interior to the surface to replenish the exhausted sites. As the ionic concentration of the external solution increases, a concentration is reached eventually when the rate of surface exchange becomes equal to the rate of replenishment of the surface ions by diffusion within the particle to the surface film. For many ion-exchange systems, a mixed rate mechanism will be involved.

For flow rates that are slow enough to permit equilibrium conditions to be approximated, the effective plate height is the diameter of the resin beads. However, in practical situations with a finite flow rate, the effective plate height is 5–10 times the mean diameter of the resin beads. This increase in plate height is caused by the lag in attainment of equilibrium between the mobile and stationary phases. While an exchanging ion is participating in a lateral diffusion into the surface film and interior of the resin bead and any replaced counter ion is diffusing outward, the mobile phase is inexorably moving forward. In general, the best compromise between minimum plate height and flow rate of the mobile phase for analytical purposes is obtained by working in the region where the first three terms of Eq. (5-21) are about equal. Longitudinal diffusion contributes little to total plate height unless the flow rate is extremely slow; as a result, this term is often neglected.

Example 5-4

The calculation of optimum operating conditions is illustrated for the separation of 0.05 mEq each of chloride and bromide ions using KNO_3 as eluent on a column of Dowex 2 × 6, cross section 1.0 cm², with resin particles of radius 0.01 cm, and weight capacity 3.3 mEq/g. Assume $\beta = 0.6$, $D_r = 3.7 \times 10^{-7}$ cm²/sec, and $D_f = 2 \times 10^{-5}$ cm²/sec for both ions.

Optimum operating conditions are achieved if the first three terms in Eq. (5-21) are equated to one another, and the fourth term neglected. It is wise to double the first term to take into account inevitable irregularities in column packing. In this way the plate height is

$$H = (2)(3)(1.64)(0.01 \text{ cm}) \simeq 10r = 0.10 \text{ cm}$$

The mean volume distribution ratio for the halide ions is given by equating the particle diffusion term of Eq. (5-21) with the film diffusion term:

$$D_v = \frac{0.142D_f}{0.266D_r} = \frac{0.142(2 \times 10^{-5} \text{ cm}^2/\text{sec})}{0.266(3.7 \times 10^{-7} \text{ cm}^2/\text{sec})} = 29$$

The required eluent velocity, to be calculated in a later section, is so low, $70r\nu \ll 1$, that the correction in the denominator of the film diffusion term is negligible. Now from Eq. (5-14), the mean molar distribution coefficient is calculated:

$$K_d = D_v/\beta = 29/0.6 = 48$$

From Table 5-3, the individual selectivity coefficients are calculated:

$$E^{\text{Cl}}_{\text{NO}_3} = \frac{1}{E_{\text{Cl}}{}^{\text{NO}_3}} = \frac{1}{3.3} = 0.30$$

$$E^{\text{Br}}_{\text{NO}_3} = \frac{E_{\text{Cl}}{}^{\text{Br}}}{E_{\text{Cl}}{}^{\text{NO}_3}} = \frac{2.3}{3.3} = 0.70$$

from which the average selectivity coefficient of the halide ions, $E^{X}_{\text{NO}_3}$, is 0.50.

Now from the mean value of K_d and the weight capacity of the resin, the concentration of KNO_3 required for elution can be calculated from Eq. (5-9):

$$C_{\text{KNO}_3} = \frac{Q_w E^{X}_{\text{NO}_3}}{K_d} = \frac{(3.3 \text{ mEq/g})(0.50)}{48} = 0.035M$$

The optimum velocity of eluent is obtained by equating the first (doubled) term of Eq. 5-21 with the third term:

$$\nu = \frac{3.28(D_v + \beta)^2 D_f}{0.266 D_v^2 r}$$

$$\nu = \frac{(3.28)(29 + 0.4)^2(2 \times 10^{-5} \text{ cm}^2/\text{sec})}{(0.266)(29)^2(0.01 \text{ cm})} = 0.025 \text{ cm/sec}$$

With a column cross-section of 1.0 cm², this corresponds to an eluent flow of 1.5 ml/min.

Finally, the individual distribution coefficients can be calculated. They are for the bromide ion:

$$K_d = \frac{Q_w E_{NO_3}^{Br}}{[NO_3^-]} = \frac{(3.3 \text{ mEq/g})(0.7)}{0.035M} = 66$$

and for the chloride ion:

$$K_d = \frac{Q_w E_{NO_3}^{Cl}}{[NO_3^-]} = \frac{(3.3 \text{ mEq/g})(0.3)}{0.035M} = 29$$

The corresponding volume distribution ratios are 40 and 17, respectively.

A different approach to the calculation of optimum operating conditions may be necessary in separations involving counter ions of different valences and in elution with complexing agents where the volume distribution ratios depend strongly on the solution concentration and, in some cases, on the pH. Under these conditions, it may be advantageous to operate under conditions where the relative retention ratios are most favorable.

Resolution Resolution is the ratio of peak separation to average peak width, that is,

$$Rs = \frac{(V_{max,2} - V_{max,1})}{0.5(W_1 + W_2)} \tag{5-22}$$

where the numerator states the separation of the peak maxima and the denominator is the average band width of the two peaks, a term which is linked to column efficiency through the plate height. It was brought out in Chapter 4 that when $Rs = 1.0$, the extent of cross contamination is about 2%. The extent of cross contamination can also be estimated from the experimental elution curves (Fig. 5-1) in the manner described by Glueckhauf.[2] Let us assume that the effluent from the column is divided into two portions to give products of equal percentage purity, i.e., $\eta_1 = \eta_2$ where $\eta_1 = \Delta m_2/m_1$ and $\eta_2 = \Delta m_1/m_2$. The ratio $\Delta m/m$ represents a chosen degree of purity. When the products are unequal in amount, the fractional impurity must be multiplied by the factor: $(A_1^2 + A_2^2)/2A_1A_2$, where A_1, A_2 are the areas under the elution curves.

The relative peak retention ratio, $\alpha = V_{max,2}/V_{max,1}$, is related to the volume distribution ratio or the molar distribution coefficient by:

$$\alpha = \frac{(D_v)_2 + \beta}{(D_v)_1 + \beta} = \frac{(K_d)_2 + 1}{(K_d)_1 + 1} \tag{5-23}$$

The interrelation of product purity, relative peak retention ratio, and number of effective plates is shown in Fig. 5-3. This graph can be used for deter-

[2] E. Glueckhauf, *Trans. Faraday Soc.* **51**, 34 (1955).

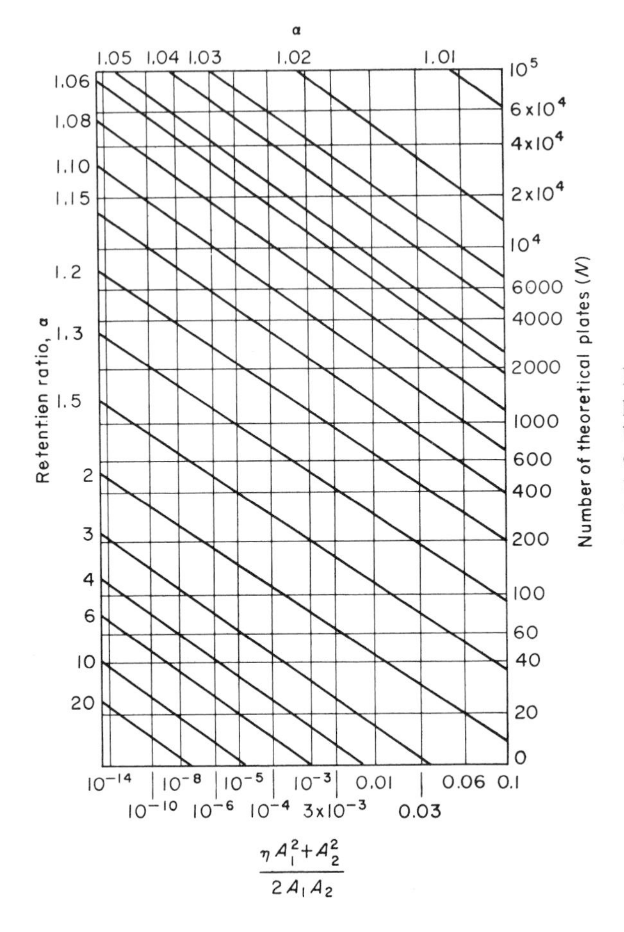

FIG. 5-3. Interrelation of product purity, relative retention ratio, and number of effective plates in elution development. After E. Gleuckhauf, *Trans. Faraday Soc.* **51,** 34 (1955).

mining the number of effective plates required for attaining the desired purity of products.

Example 5-5

If the product purity of the halides discussed in Example 5-4 is specified to be 99.8%, then $\eta = \Delta m/m = 0.001$, since the amount of chloride and bromide is equal. From the optimum values of the distribution coefficients (or the volume distribution ratios), the relative peak retention ratio is

$$\alpha = \frac{(K_d)_2 + 1}{(K_d)_1 + 1} = \frac{66 + 1}{29 + 1} = 2.2$$

Referring to Fig. 5-3, the plate number required is 60. From the plate height of 0.1 cm, estimated in Example 5-4, the required height of the resin bed is

$$L = NH = (60)(0.1) = 6.0 \text{ cm}$$

These values should be regarded as "minimum values." In practice, one should overdesign his experiment, that is, use a somewhat longer column to handle deviations from equilibrium conditions.

From the column dimensions, the bed volume is

$$V_b = (1.0 \text{ cm}^2)(6.0 \text{ cm}) = 6.0 \text{ cm}^3$$

and the void or free-column volume, V_o, is 3.6 ml.

The chloride peak will emerge at the effluent volume

$$V_{\max} = V_b(D_v + \beta) = 6.0(17 + 0.6) = 106 \text{ ml}$$

and the bromide peak at

$$V_{\max} = 6.0(40 + 0.6) = 244 \text{ ml}$$

The corresponding band widths are obtained from the plate number expression,

$$W = 4V_{\max}/\sqrt{N}$$

and are 55 ml for the chloride band and 126 ml for the bromide band. The total duration of the separation is, from the optimum eluent flow rate,

$$t = (106 \text{ ml} + 244 \text{ ml})/(1.5 \text{ ml/min}) = 233 \text{ min}$$

Two difficulties should be immediately apparent from Example 5-5. The separation is lengthy and the concentration profile is quite flat, particularly for the bromide band. These difficulties stem from the extreme purity specified and the small difference in selectivity coefficients. If purity were specified to be only 99%, the plate number required would decrease to about 32, and the column length could be shortened to 3.2 cm, or roughly half the former length. The values of $V_{\max}$ would be correspondingly less, as would be the band widths. Both bands would sharpen significantly. Further improvement could be attained by using a more concentrated eluent to displace the bromide band.

Example 5-6

If after the chloride band had been eluted completely, the eluent concentration had been increased from 0.035*M* to 0.60*M*, the bromide band would elute as a very sharp band. Under this new eluent condition, the distribution coefficient for bromide ion becomes

$$K_d = \frac{Q_w E_{\text{NO}_3}^{\text{Br}}}{[\text{NO}_3^-]} = \frac{(3.3 \text{ mEq/g})(0.7)}{0.60M} = 3.9$$

from which, using the 6-cm column,

$$V_{\max} = 3.6(3.9) + 3.6 = 18.7 \text{ ml}$$

Thus, at 167 ml, ($V_{\max}$ + 42) for the chloride ion plus 19 ml, and ignoring the movement of the bromide band down the column while the chloride band was being eluted, the bromide peak would emerge from the column. At about 175 ml the bromide band would be completely eluted. Each band would have been separated in a purity of 99.8% over an elapsed time of 116 min.

The efficient separation of many mixtures by ion-exchange chromatography requires that the concentration of eluent be changed in the course of the elution. The change in concentration may be continuous (as in gradient elution) or may be stepwise.

Applications of Ion Exchangers

Ion-exchange resins are particularly useful and convenient to handle. Columns can be made in any desired size, from a few cubic millimeters up to preparative columns of many cubic meters. The diameter of the column depends on the amount of material to be treated; the length depends on the difficulty of the separations to be accomplished. Since the range of applications of ion exchange is very great, the choice of examples must of necessity be subjective.

Column Filtration The column filtration technique may be divided into continuous and discontinuous methods. In the continuous method the solution is fed continuously into the top of the ion-exchange column. The ions of the same electrical charge as the counter ion of the exchanger are retained in the resin phase if their selectivity coefficients exceed that of the counter ion. Oppositely charged ions, as well as uncharged chemical species, ordinarily do not adsorb on the ion exchanger and remain in the external solution phase. Gradually the original counter ion is forced out of the column by the species in the feed solution which have more affinity for the resin. This displacement process continues until the column capacity approaches exhaustion. The net result is the removal from the feed solution of all ions more strongly adsorbed by the resin than the counter ion, and their replacement in the effluent by an equivalent amount of the original counter ion. In this manner, undesirable cations may be removed by a cation exchanger and, similarly, unwanted anions by an anion exchanger. Examples, to be discussed later, include deionization, water softening, and removal of interferent ions in various analytical methods.

If interest centers around an ionic species present in relatively low concentration, the sample is introduced onto the top of the column, as before, but now a washing solution of an appropriate composition is passed through the column to remove the unadsorbed material from the interstitial volume (and resin interior) of the column. After the unwanted material has been washed from the column, the adsorbed component, which has been held near the top of the column, is desorbed by an appropriate eluent and washed out of the column. The species of interest is obtained in a relatively small volume of effluent that is free of all other ions present in the original sample. In this manner trace elements present in waters from springs and rivers can be concentrated and later determined.

Removal of Interfering Ions Foreign ions are frequently the cause of interference in many types of analytical work. Undesirable cations may be removed by passing the sample through a cation-exchange column in the hydrogen form or some other innocuous ionic form. In this manner cations such as calcium, iron and aluminum, which interfere in the determination of phosphate, are easily separated from phosphate by passage of the solution through a bed of sulfonic acid resin in the hydrogen form. Iron and copper, which interfere in the iodimetric determination of arsenic, are separated similarly, as can be the metals which interfere with determination of sulfate gravimetrically. In a different mode of operation, passage of the solution containing alkali metals to be analyzed through a cation exchanger adsorbs the alkali metal ions, permitting interfering ions such as phosphate to pass through; the alkali metals are subsequently eluted.

By the use of an anion exchanger, usually in the chloride form, interfering anions can be removed prior to the determination of a cation. Thus phosphate is removed before the determination of calcium and magnesium by titration with EDTA.

Exchange resins may be employed to effect complete and rapid removal from solution those ions which, because of their color, taste, odor, or toxicity, are undesirable in many processes or products. Metallic ion residues from crop protection chemicals may be removed from crushed fruit sirups. Iron may be removed from phosphoric acid pickling baths by use of a strongly acidic cation exchanger. Calcium, responsible for hard curd formation in the digestion of milk, may be replaced by sodium.

Special chelating resins, containing iminodiacetate active groups, are useful in removing traces of heavy metals from a wide range of product streams and in separating various heavy metals. For example, this type of resin will remove traces of iron, copper, or zinc from concentrated solutions of alkali and alkaline earth metal salts. Exceptionally high selectivity makes possible the separation of closely related species; for example, nickel can be separated from cobalt and copper from nickel. Heavy metals are easily eluted by mineral acids.

Although the least glamorous, the treatment of water with ion-exchange materials reigns as the major application. To soften natural waters, it is necessary to replace small quantities of Ca^{+2}, Mg^{+2}, and other polyvalent cations by Na^{+}. At the same time, or in successive columns, the SO_4^{-2} (or HSO_4^{-}) and CO_3^{-2} (or HCO_3^{-}) and other polyvalent anions are replaced by Cl^{-}. Unless the cations are eliminated, sticky and insoluble soap curds may appear in washing operations. From dilute solutions the polyvalent ions are preferentially taken up because of their selectivity coefficients, but in the regeneration step with highly concentrated brine solution, the sodium ion is preferred.

Ion exchange can be used to remove completely all electrolytes, including weak acids such as silicic acid, carbonic acid, boric acid, and phenols. True nonelectrolytes, however, cannot be removed; therefore, one cannot assume sterility and freedom from pyrogens. Complete removal of all electrolytes from solution is accomplished by combining the use of a cation exchange resin in the hydrogen form with an anion exchange resin in the hydroxide form. The strong acid (sulfonic acid) and strong base (quaternary amine) resins are required. These may be used in series or as a regenerable mixture. Macroreticular resins are used to deionize substances in nonpolar solvents. Deionization is capable of producing water with a specific resistance of 250,000 to 1,000,000 ohm cm, equivalent to as little as 2 parts per million of total electrolyte content. The quality of ordinary distilled water can be improved by "polishing" with passage through a mixed resin bed. Small cartridge units are available for laboratory use. In the monobed method a virtually neutral solution pH is maintained during the entire exchange operation; this is important when handling pH-sensitive materials.

Concentration and Recovery of Traces By passing extremely dilute solutions through a short column of an ion exchanger, it is possible to concentrate traces of ionic constituents from large volumes of solutions or from matrix elements which are not adsorbed on the exchanger. To do this successfully, the exchanger must have considerable selectivity for the ion sought, or else the solution must contain few competing ions. Subsequent elution of these adsorbed ions with a small amount of an appropriate eluent results in the production of a concentrated solution of the adsorbed ions in a small volume of solvent. In this way quantitative recovery and determination of traces of copper from sea water or from rayon and pickling wastes may be performed with a carboxylic type of resin. Valuable ionic materials present at low concentration can be recovered from processing solutions; examples include citric and ascorbic acids from citrus wastes, fatty acids from soap wastes, tartaric acid from wine residues, and nicotinic acid from vitamin production wastes. Stream pollution may be mitigated with the recovered metals partially offsetting the cost of the waste disposal method.

In the pharmaceutical field, the antibiotic streptomycin is adsorbed from the filtered fermentation broth onto the salt form of the carboxylic cation resin and is eluted, highly purified and at a high concentration, with dilute mineral acid.

Through prior concentration on an exchanger, the applicability of many spectrophotometric methods is extended. For example, copper in milk occurs in the range of 10^{-4} $\mu g/ml$; after concentration on a cation exchange resin, the effluent will be 0.1 $\mu g/ml$, a thousand-fold concentration.

Stoichiometric Substitutions An analyst is frequently concerned with the determination of the total concentration of cations (or anions) without any breakdown into individual cations. Passage of the sample through a column of resin in the hydrogen form results in the retention of the cations and the release of an equivalent amount of hydrogen ions to the effluent. Rinsing the column and titration of the effluent and rinsings with a standard base gives the total cation content of the sample. Any "free" acid or original alkalinity in the sample is determined by a separate titration. An analogous procedure with exchangers of the strongly basic type permits the determination of the total anion concentration. A stoichiometric substitution is often much simpler than the direct determination of many ions when only a single ion is present in solution. For example, phosphorus in phosphate rock is determined by passage through a sulfonic acid resin in the hydrogen form which converts phosphates and chlorides to the respective acids, which then may be titrated to the proper end points potentiometrically or by the use of two indicators.

The ability of ion exchangers to trade their counter ions for other ions in solution makes possible the preparation either of acids from salts or of a desired metal salt from the corresponding salt of another metal. In the pharmaceutical field conversions of one salt into another include potassium–sodium penicillin, vitamin bromide–chloride, and streptomycin chloride—sulfate transformations. Silicic acid, in non-flocculent form, is prepared by passage of sodium silicate through the hydrogen form of a cation exchanger to yield the free acid. Preparation of carbonate-free hydroxide solution is accomplished by passing a known weight of dissolved sodium chloride through a strongly basic anion exchanger in the hydroxide form. This conversion of salts to their corresponding bases (or acids) is particularly advantageous for standardizing solutions that may be prepared from pure salts, but salts which cannot be dried to a definite weight or whose composition is not known exactly.

Chromatographic Separation of Cations and Anions Chromatographic separation of similarly charged ions can be carried out by means of ion exchange columns on the basis of the different distribution coefficients. The separation factor is determined chiefly by the selectivity of the resin, and the efficiency of the separation by the number of theoretical plates in the ion-exchange column. The elution chromatographic separation of lithium, sodium, and potassium ions on a hydrogen-form cation-exchange column can be carried out with $0.1M$ HCl solution (Fig. 5-4). These ions appear in

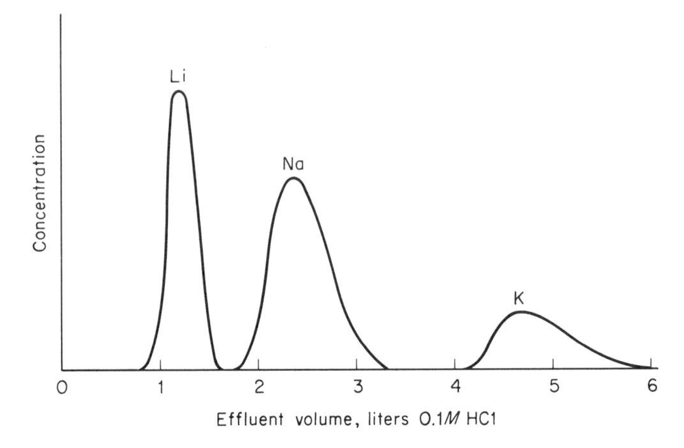

FIG. 5-4. Elution curves of alkali ions from Wofatit KPS-200 (120 cm × 10 mm) column; flow rate, 3–5 ml/min.

the order anticipated: Li, Na, and finally potassium. They can be determined in the collected fractions by flame photometry. A separation is very lengthy and takes about 6–8 hr. However, the time could be shortened if the eluent solution were altered after the emergence of the lithium peak and again after the emergence of the trailing edge of the sodium peak.

The efficiency of the separation of cations can be markedly increased by the use of complexing agents in the eluent. Buffered solutions containing lactate, citrate, EDTA, or α-hydroxy-isobutyrate as the complexing agent will bring about separations if some of the cations are converted into neutral or negatively charged ionic species. The extent of conversion into such species will be a function of pH and complex formation constants as well as dependent upon selective masking action. Remarkable separation of rare-earth ions has been accomplished on a commercial scale. Complexing agents can also be used to separate elements into groups. In this manner 36 radioactive metal ions in fission products were separated into groups by 6 different eluting solutions containing different complexing agents.[3] However, separation of metals by cation exchange through differential complex formation is in general not as effective as anion-exchange separation of metal complexes.

The analytical separation of aldehydes and ketones and the separation of these substances from alcohols is based upon the fact that the carbonyl group of the former forms relatively stable addition compounds with the bisulfite ion which are, in turn, strongly adsorbed by anion-exchange resins. Alcohols do not form the sulfite-addition products and are therefore not adsorbed by the bisulfite salt of the anion-exchange resin. The mixture of aldehydes and ketones that are adsorbed may be separated simply; the

[3] W. J. Blaedel, E. D. Olsen, and R. F. Buchanan, *Anal. Chem.* **32,** 1867 (1960).

ketones are eluted by hot (75°C) water, whereas the aldehydes are not, but the latter may be eluted subsequently with a NaCl solution. The association of sugars with borate counter ions in anion exchangers has long been used for their chromatographic separation from one another. Sugars, as with polyhydroxy compounds in general, form with borate ions complexes of various stabilities and dissociate to various degrees. Disaccharides, having a lesser tendency to form complexes, can readily be separated from monosaccharides which form more stable complexes. Moreover, monosaccharides can be separated from each other within the classes of hexoses, pentoses, and tetroses.

Acids can be separated according to their strengths on strong anion-exchange columns, with the weakest acids emerging first, either by displacement development with a strong acid or, since acid dissociation depends on the pH, by gradient elution with buffers of decreasing pH. Amino acids, which add protons to form cations in the pH ranges below their isoelectric points, can be separated on cation-exchange columns by gradient elution with buffers of increasing pH; here, the most acidic components emerge first and the most basic last. An automatic amino acid analyzer is available commercially.

Advantage can be taken of weakly basic anion-exchange resins to separate acids of different strengths. Acids having a small dissociation constant are retained only to a slight extent and, if the dissociation constant of the acid is less than the base constant of the exchanger, virtually no retention occurs. In this manner, HCN, carbonic, silicic, and boric acids can be separated from phosphoric, sulfuric, and hydrochloric acids.

Anion Exchange of Metal Complexes Practically every metal may, to a substantial extent, be converted to a negatively charged complex through suitable masking systems. This fact, coupled with the greater selectivity of anion exchangers, makes anion exchange a logical tool for handling metals. The complexes, negatively charged, are adsorbed by the exchanger and eluted by changing the concentration of complexer in the eluent sufficiently to cause dissociation of the metal complexes or a decrease in the fraction of the metal present as an anionic complex in solution. If interconversion of complexes is fast, control of ligand concentration affords a powerful tool for control of adsorbability, since ligand concentration controls the fraction of the metal as adsorbable complex.

The adsorption of stable negatively charged complexes on anion exchangers is similar in principle to the adsorption of simple nonmetallic anions. The selectivity coefficient of the anion exchanger for a complex MX_p^{-q} in a medium containing the ion X^- from the eluent is given by

$$E_X^{MX_p^{-q}/q} = \frac{[MX_p^{-q}]_r[X^-]^q}{[MX_p^{-q}][X^-]_r^q} \tag{5-24}$$

where p is the maximum coordination number of the metal ion. Assuming that only one complex species is adsorbed, the volume distribution ratio is

$$D_v = \Phi \frac{[X^-]_r^q}{[X^-]} E_X^{MX_p^{-q}}/q \tag{5-25}$$

where Φ is the ion fraction of metal in the complex anion; that is

$$\Phi = [MX_p^{-q}]/C_m \tag{5-26}$$

and C_m designates the total concentration of metal ion whether complexed or not. For each metal and complexing ligand there is a characteristic curve of log D_v vs molarity of complexer. Examples are given in Fig. 5-5 for metals which form chloride complexes. Similar studies have been reported from fluoride solutions,[4] nitrate solutions,[5] and many other systems.

In some fortunate cases one or a group of the metal ions in a mixture will not be transformed into a negatively charged species and will not be bound by the resin, whereas the remainder of the metal ions will be converted to anionic complexes and selectively adsorbed. For example, on a column containing EDTA and at pH 7, the alkaline earth cations form stable complexes and are adsorbed, while the alkali metal ions do not form complexes and pass through the column without retention. Another favorable case arises when one of the similarly charged ions bound on the column can be eluted with an eluent concentration which has practically no effect on the other anionic complexes. An example of this selective desorption is silver and lead which can be adsorbed on the column from $2N$ HCl and subsequently desorbed with $10M$ HCl; most other chloroanions are retained.

It is possible to devise a large number of separation schemes in which a group of metals is adsorbed on a resin from a concentrated solution of the complexer, then each metal in turn is eluted by progressively lowering the complexer concentration. Thus a cation which forms no (or only a very weak) anionic complex is readily displaced by several column volumes of the complexing agent, while its companions are complexed and retained in the column. For separating two ions, it is advisable to choose a complexer concentration for which the separation factor is maximal, yet, at the same time, it is important that the volume distribution ratio of the ion to be eluted first not be higher than unity. When $D_v = 1$, the peak maximum emerges within two column volumes approximately. The separation of nickel, manganese, cobalt, copper, iron(III), and zinc ions can be carried out as shown in Fig. 5-6. On an anion-exchange column, previously washed with $12M$ HCl, the mixture of cations in $12M$ HCl is poured onto the column bed. The nickel is washed out with several column volumes of $12M$ HCl. The re-

[4] J. P. Faris, *Anal. Chem.* **32,** 520 (1960).
[5] J. P. Faris and R. F. Buchanan, *Anal. Chem.* **36,** 1157 (1964).

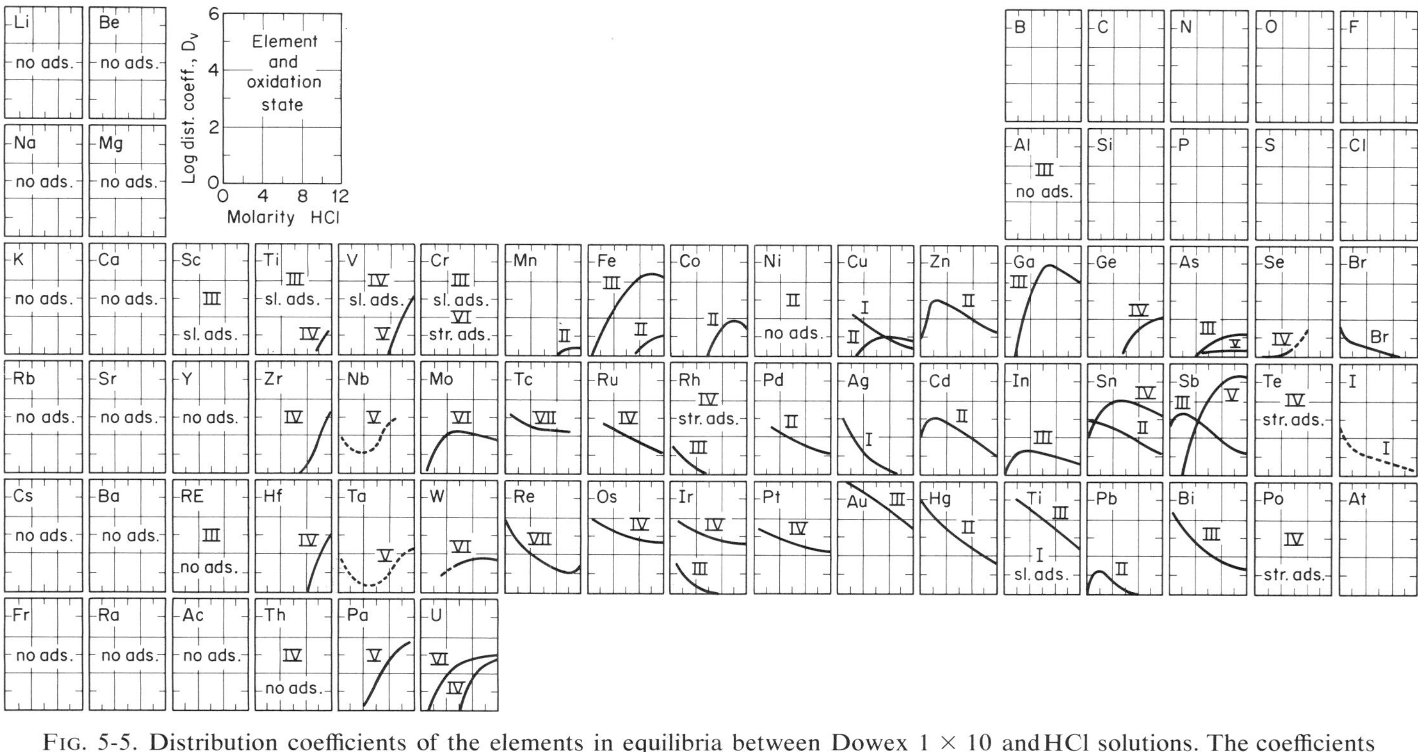

FIG. 5-5. Distribution coefficients of the elements in equilibria between Dowex 1 × 10 and HCl solutions. The coefficients are shown as a function of the HCl molarity. Abbreviations: no ads., no adsorption, 0.1 < 1*M* HCl < 12; sl. ads., slight adsorption in 12*M* HCl ($0.3 \leqslant K_d \leqslant 1$); str. ads., strong adsorption, $K_d \gg 1$. From K. A. Kraus and F. Nelson, *Proc. Intern. Conf. Peaceful Uses Atomic Energy, Geneva* **7,** 113 (1956).

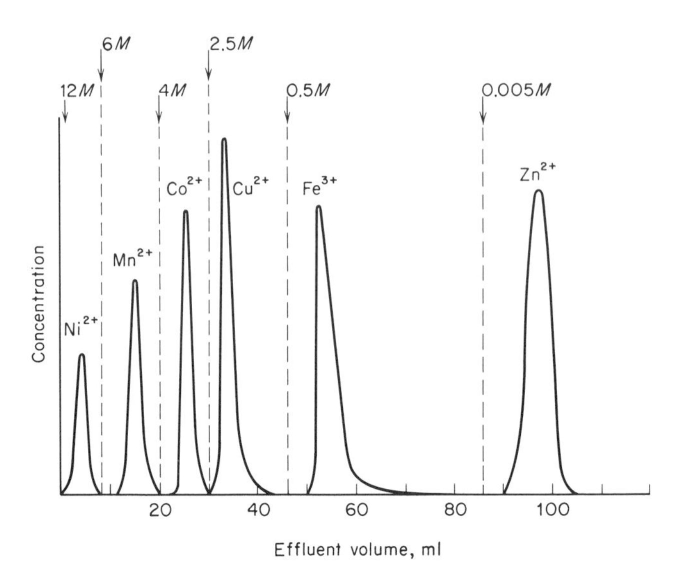

FIG. 5-6. Separation of transitional metals on an anion-exchange column. After K. A. Kraus and G. E. Moore, *J. Am. Chem. Soc.* **75,** 1460 (1953).

ceiver is changed and the manganese ions washed out with 6*M* HCl, the cobalt ions with 4*M* HCl, the copper ions with 2.5*M* HCl, the iron(III) ions with 0.5*M* acid, and finally the zinc ions with 0.005*M* acid, changing the receiver between the various fractions.

Ligand Exchange Chromatography In this method a cation-exchange resin saturated with a complex-forming metal, such as Cu(II), Fe(III), Co(II), Ni(II), or Zn(II), acts as a solid adsorbent. Ligands which may be anions or neutral molecules such as ammonia, amines, amino acids, or olefins are removed from the liquid phase by formation of complexes with the metal attached to the resin and subsequent displacement of water or other solvents coordinated to the metal ion. The advantage of ligand exchange chromatography lies in the high selectivity of coordinative complex formation. Ligands instead of counter ions are exchanged; those with stronger complexing tendency are more strongly retained. It is an efficient way to separate ligands from non-ligands. Elution development uses a ligand in the eluent that complexes with the metal less strongly than the ligands of the mixture.

As an example of the method, amines may be separated. On a cation exchanger carrying copper(II) ions, these ions are retained on the resin in the form of the amine complex. When the amine mixture is introduced at the top of the column, ammonia molecules are displaced and the amine molecules coordinate with the copper ions. If an aqueous ammonia solution is now passed down the column, the amines are gradually displaced downward and eventually emerge from the bed in the sequence of increasing complexing tendency toward copper(II) ions. The selectivity of the metal–resin column can be adjusted and even reversed at will by varying the concentration of eluent ligand, and even by displacing the ligand with a buffer of a pH such that the metal–ligand bonds are broken but the metal–resin bonds are not.

Nonaqueous Solvent Systems In principle, the general theory of ion exchange applies to any solvent. Naturally the nature of the solvent affects the solubility, dissociation, and solvation of solute molecules in solution and the swelling and dissociation of the ion exchanger. Consequently, drastic changes in ion-exchange selectivity and rates of exchange should be anticipated in nonaqueous solvents.

In a nonaqueous medium of high dielectric constant, exchange behavior is similar to that in aqueous solutions. Ionization of the electrolyte portion of the resin and of the solute in solution is enhanced. In the absence of an ionizing medium, the resin will not swell, thereby decreasing the diffusion rate throughout the resin particle and the exchange process will be limited essentially to the exposed surface area. Moreover, in a nonpolar solvent, a considerable degree of adsorption due to van der Waals' type of interaction can occur between the hydrophobic backbone or skeletal matrix of the resin and the hydrophobic parts of a solute molecule.

Most microreticular resins are ineffective in nonpolar solvents if the moisture content of the resin phase is less than 1% of its equilibrium moisture capacity. At a 10% level, the ion-exchange materials are almost as effective as they are when fully saturated with water. Thus, it is possible to conduct some ion-exchange reactions in nonpolar media by introducing moisture into the resin phase (with mutually miscible liquids and water), by employing long contact times, and by using finely divided ion-exchange resins. When these possibilities are inadmissible, macroreticular resins, by virtue of their high surface area and true porosity, enable one to conduct exchange reactions in nonpolar and nonswelling solvents at excellent rates using large particles that are in a completely anhydrous state.

The high surface area and unique pore structure of macroreticular ion-exchange resins have led to their use as sorbents in nonpolar solvents, particularly in systems where small traces of water are objectionable. Macroreticular exchangers are an ideal sorbent for removing ammonia or amines from hydrocarbons, iron from glacial acetic acid, traces of copper from hydrocarbons, and traces of alkali from various organic solvents or reaction mixtures. The anion-exchange resins are useful for removing mercaptans and phenols from nonpolar solvents, and HCl, HCN, and other acids from various solvents or reaction mixtures. Basic and acidic gases, such as NH_3, CO_2, H_2S, and SO_2, can be effectively removed from gas streams using the appropriate dry exchange resins.

LABORATORY WORK

Resin Pretreatment

Unless one uses specially purified, analytical-grade resins, which are available commercially at a premium price, fresh resins or reused resins must be pretreated before preparing a column.

For cation-exchange resins, the cycle of treatments involves placing the resin particles on a Büchner funnel and washing with 3*M* HCl, 1*M* NaOH, and ethanol, with a water rinse (10 bed volumes) of deionized water following each reagent. Apply two volumes of reagent solution for each volume of resin bed at a flow rate of about 0.05–0.1 bed volume per minute. In some instances, impurities may have to be removed from reused resins with complexing agents.

Anion-exchange resins are conditioned by treatment with hot water, 1*M* HCl, 0.5*M* NaOH, and ethanol, with a deionized water rinse following each step. The cycle of treatment should be repeated several times. For weakly basic resins, 0.5*M* aqueous ammonia or sodium carbonate should be substituted for the sodium hydroxide.

The exchangeable ion of the resin should be the same as one ion of the eluent. Conversion to the desired ionic form is accomplished with 5–10 bed volumes of 0.01*M* solutions at a rate of 0.1–0.2 bed volumes per minute when univalent ions are to be replaced by polyvalent ions. For replacing an ion of higher valence by one of lower valence, 1*M* solutions are most effective. Before use the column is rinsed with several bed volumes of the eluent. The effluent should give no test for the original ion.

Column Preparation

A tube 20–25 cm in length and 10–12 mm internal diameter is convenient. Laboratory supply houses stock columns made specifically for ion-exchange chromatography and constructed from glass tubes of various diameters by fusing into the tube a sintered glass filter and drawing down the tube at the bottom. Ordinary laboratory burets can be used by inserting a small plug of glass wool in the restriction above the stopcock and covering with a layer of clean sand. Two arrangements are shown in Fig. 5-7.

The selected resin, usually of particle size 100–200 mesh, is swirled in a beaker with deionized water and transferred to the column through a funnel and with washing. Exacting separations will require wet-screening the resin before use to secure a narrow range of sizes. Finer particles are

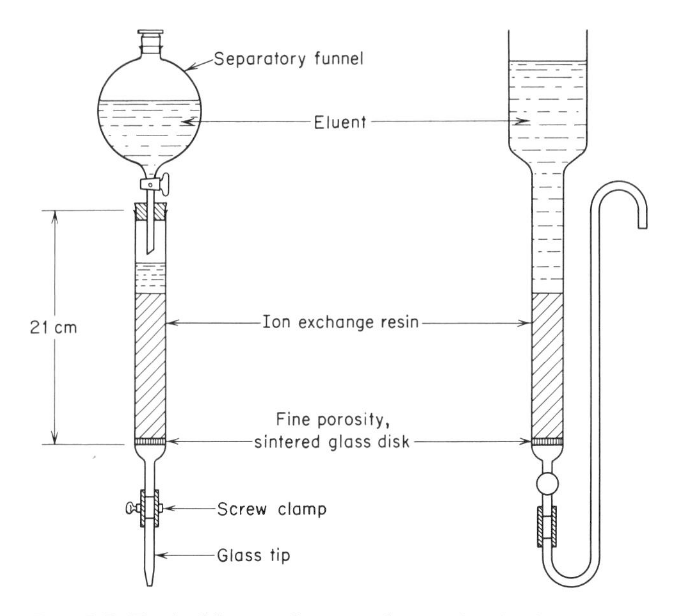

FIG. 5-7. Typical ion-exchange columns for the laboratory.

more desirable in critical chromatographic separations but they retard flow rates through columns when time in a laboratory period is a premium.

The height of the resin bed required depends upon the scale of operations and operating parameters. For many applications, a bed volume of 10–15 ml is adequate. This amount of resin of 8% cross-linking holds about 20–30 mEq of cations on cation exchangers, and about half this quantity of anions on anion exchangers. Longer or narrower columns are used for difficult separations or when small amounts of material are handled. The column is never filled more than about half way with resin. NEVER FILL A COLUMN WITH DRY RESIN. Swelling pressures may burst the glass column.

In columns with sintered-glass disks, attach a water line to the bottom of the column and pass a stream of deionized water upward through the column. This releases air pockets and floats the resin particles. Maintain the water flow until all air pockets are removed and all particles have achieved mobility. Drain off the excess water. NEVER ALLOW THE LIQUID LEVEL TO DRAIN BELOW THE TOP OF THE RESIN BED.

The sample is added to the top of the column dropwise without disturbing the resin bed. The amount of adsorbables should not exceed 3% of the total column capacity. After the sample is drained into the resin bed, the sides of the tube are rinsed with small portions of eluent, each portion being drained to the level of the resin bed before the next one is added. It is convenient to feed the eluent to the column by means of an adjustable reservoir and siphon arrangement, or a 125-ml separatory funnel, inserted into the top of the column through a 1-hole stopper.

Flow rate of eluent can be adjusted by means of a screw clamp, stopcock, or needle valve. When gravity does not suffice, liquid is forced through the column under pressure of nitrogen gas from a small cylinder. Never use a partial vacuum from a water aspirator as this creates air pockets in the column. A recommended flow rate, if not computed as outlined in the chapter, is 0.5–2.0 ml per min per 1 cm^2 column cross section.

The effluent volume is computed from the point of addition of the sample to the column. Automatic fraction collectors are convenient.

EXPERIMENT 5-1 *Equilibrium Distribution Coefficients*

Batch Equilibrations Batch equilibrations have the advantage of lending themselves to simultaneous measurement of a large number of samples and require little time per experiment since they consist simply of equilibrating small known amounts of resin and solution followed by analysis of the two phases after attainment of equilibrium.

Equilibrations are carried out with batches of resin dried in air to constant weight or in a vacuum desiccator at 60°C over Anhydrone for 24 hr. The water content of the "dried" resin is determined by drying a sample of known weight at 105°C to constant weight. With this information it is possible to calculate the distribution coefficients on a dry-weight basis. Weigh accurately approximately 1 g of "dried" cation-exchange resin into a 125-ml glass stoppered flask. Pipet into the flask 5 ml of 0.05*M* metal salt solution and 45 ml of the appropriate solvent mixture. Stopper the flask and shake for 24 hr. Pipet an aliquot from the supernatant liquid, evaporate the solvent to near dryness (if necessary) and determine the metal ion content by an appropriate analytical method. As a check the amount of cation in the resin can be determined after ashing the resin and weighing the metal sulfate, whenever this is possible.

For anion-exchange resin, use 2.0-g samples. Specific directions follow for the chloride–nitrate system. Convert 2 g of strongly basic anion-exchange resin to the chloride form with 50 ml of 1*M* NaCl by pouring the chloride solution through the resin in a column. Wash free from chloride with deionized water. Transfer the resin to a 250-ml flask, add exactly 200 ml of 0.1*M* $NaNO_3$ solution. After standing for 24 hr with intermittent shaking, remove 25 ml of the solution by pipet and titrate with 0.1*M* standard $AgNO_3$.

Column Equilibrations This first method is suitable for medium values of D_v. Continuously pour onto a short column of resin the eluting electrolyte and solution containing adsorbate ions in low concentration. Ascertain the volume of effluent required to attain the 50% breakthrough concentration of adsorbate. In an ideal case the latter is equal to V_{max}, from which D_v can be calculated.

If the value of D_v is very low, pour a solution of known composition that contains the eluent ion and the adsorbate ion through the column until the composition at the inlet and outlet is identical. Separate the solution from the resin by filtration and gentle suction on a Büchner funnel. Determine the amount of ions remaining in the resin by eluting them with a suitable regenerant. Calculate the value of D_v from a knowledge of the column volume, the concentration of ion found in the resin phase, and the concentration of the solution phase.

When the value of D_v is very high, greater than 10^3, saturate the resin phase with the adsorbate ions. Then wash out the resin with a known volume of eluent and determine the amount of adsorbate ions in the eluent. From a knowledge of the amount of adsorbate ions originally present in the resin (approximately the resin capacity), the value of D_v can be calculated.

EXPERIMENT 5-2 *Determination of Total Salt Concentration*

To a column of approximately 10–20 ml of a sulfonic acid exchanger in the hydrogen form, add a sample of the solution to be analyzed. The sample should contain 1–4 mEq of electrolyte (less than one half capacity of the resin bed) at a concentration of 0.1*M* or less. Collect the effluent in a 250-ml flask. When the sample has completely entered the resin bed, rinse the column with 80–100 ml of deionized water (5–6 bed volumes) and collect the effluent in the same flask. Titrate the contents of the flask with 0.1*M* standard NaOH to pH 8.5; if bicarbonates are present, to pH 4. Also titrate a second aliquot of the unknown sample, without treatment by ion exchange, to the same end point. This number of milliequivalents is added to the number obtained on the column effluent.

Suggested samples include tap water or river water, and known weights of salts such as KNO_3 or NH_4Cl. This is one way to standardize solutions which are prepared from salts that are pure but of unknown composition or unstable to drying.

EXPERIMENT 5-3 *Fractionation of Metals as Chloride Complexes*

Slurry 25 g of quaternary type anion-exchange resin with 100 ml of 9*M* HCl and wash by decantation with 200 ml of deionized water. The conditioned resin is resuspended in 9*M* HCl and then transferred to the column using 9*M* HCl as the transfer medium. Rinse the column with 25 ml of the same acid.

Add a sample containing approximately 2–25 mg of each of the metals dissolved in 9*M* HCl. Transfer carefully to the column using small increments of $9M$ HCl to rinse the sample container and the inner surfaces of the column above the resin bed. Suggested metals are Ni, Cu, Co, Fe(III), and Zn.

Pass $9M$ HCl through the column at a rate of 4 ml per min until a volume of 125 ml is collected. This first eluate contains the nickel.

The eluate is now changed to $4M$ HCl and 125 ml of effluent is collected in a fresh flask. This fraction contains the cobalt. Copper is eluted with 125 ml of $2.5M$ HCl, iron(III) with $0.5M$ HCl, and zinc with $0.005M$ HCl.

Each fraction may be titrated by means of complexometric procedures. Spot-test methods may be employed to ascertain the breakthrough and cessation points for each cation.

It is instructive to assign separate metals to individual students to elute under specified conditions.

EXPERIMENT 5-4 *Elution Curves of Chloride and Bromide*

Prepare a 20-ml resin bed with a quaternary type of anion-exchange resin (e.g., Dowex 1 × 8), 100–200 mesh, in the nitrate form. Add exactly a known amount of $0.1M$ standard NaCl and $0.1M$ NaBr solution (10 ml of each is recommended) to the column and allow to flow into the beads. Elute the column with $0.05M$ KNO_3 solution. Collect 10-ml aliquots of effluent (starting from the time the sample solution was added) until about 130 ml of eluent has passed through the column. Continue the elution with $0.6M$ KNO_3 until 150 ml of eluent have passed through the column.

Titrate the effluent fractions with $0.1M$ standard $AgNO_3$, delivered if necessary from a microsyringe buret. Plot the results as milliequivalents of $AgNO_3$ solution vs eluent volume. Compute the total number of milliequivalents of chloride and bromide found experimentally and compare with the known amount present.

From the experimental value of V_{max}, and the void fraction, estimate the values of K_d and D_v, and the number of plates in the column.

Other Suggestions: Another pair of anions is oxalate and bromide (see Problem 23). Individual students could investigate the elution curves of single anions as a function of eluent concentration and nature of the eluent. The influence of eluent flow rate upon the peak sharpness is another possibility.

Problems

1. If an exchanger has 1.65 mEq/cm^3 of column bed, how many exchange groups are there per cm^3 of bed? If these groups are uniformly distributed, what number will any cylindrical segment of the column ($1\ cm^2 \times 0.01\ cm$) contain?
2. A column, $1\ cm^2 \times 10\ cm$, contains 5 g resin (dried basis, density = 1.515 g/cm^3). Calculate the bed volume, the resin volume, and the void column volume.

3. Dowex 50 × 10, in the hydrogen form, has a density of 1.25 g/cm^3 and a water content of 47 weight percent. Calculate the volume capacity.
4. Dowex 50 × 2, in the hydrogen form, has a density of 1.09 g/cm^3 and a water content of 77 weight percent. Calculate the volume capacity.
5. Suppose that 1 g of resin, in the hydrogen form, is equilibrated with 10 ml of 0.25*M* HCl, and solutes *A* and *B*, which are added to separate bottles in concentrations of $10^{-4}M$. After equilibration 58.84% of *A* and 32.26% of *B* remain in the solution phase. (a) Calculate the distribution coefficient for solutes *A* and *B*. (b) How many plates would be needed in a chromatographic column to separate the two cations with 99.9% purity in each solute fraction?
6. Using the column described in Problem 2, calculate the volume at which each peak emerges from the column for solutes *A* and *B* of Problem 5. Also state the peak location in terms of number of column void volumes.
7. The actual elution curves of solutes *A* and *B* (Problem 5) are illustrated in Fig. 5-8. Estimate the number of plates in the column. Since insufficient plates are available for the required purity, suggest a remedy. The molar ratio of *A* to *B* is 1.7.
8. Calculate the molal pK_a value of the ionogenic groups in polymethacrylic acid resin. End point on the titration curve occurs at 9.2 mEq/g. At the point of half conversion, the external solution has pH = 6.4 in the presence of 0.1*M* NaCl. Assume the water content is 60 weight percent.
9. For these experimental operating conditions: column length, 108 cm; particle radius, 2.4×10^{-3} cm; D_v for aspartic acid, 1.57 at pH 2.9; $D_r = 3.3 \times 10^{-7}$ cm^2/sec; $D_f = 1.0 \times 10^{-5}$ cm^2/sec; void fraction = 0.4; calculate the plate height and plate number, and plot the plate height vs eluent velocity at these velocities (in cm/sec): 0.00783, 0.0335, 0.0685, 0.0983, 0.132, 0.173.
10. Estimate the plate height for a Dowex 50 × 8 column, 102 cm in length by 9.0 mm in diameter, packed with particles 0.0125 cm in radius, during the elution of sodium with 0.7*M* HCl flowing at 0.22 $ml/cm^2/min$. $D_v = 6$; other data in Problem 13.
11. To separate approximately equal amounts of Rb^+ and Cs^+ on an 8% cross-linked resin (particle diameter 0.05 mm), how many plates would be required for 0.1% cross contamination, and what column length would be required?

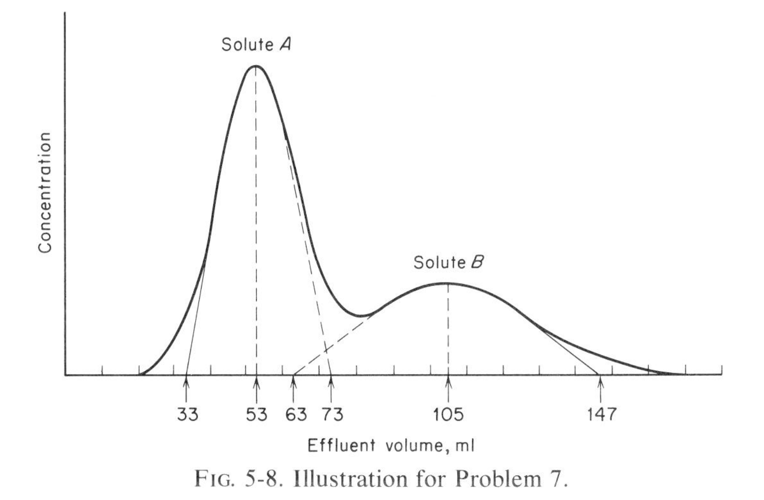

FIG. 5-8. Illustration for Problem 7.

12. To separate Rb^+ and Cs^+, present in approximately equal amounts on an 8% cross-linked resin, how many plates would be required for 1.0% cross contamination? What column length is required if $H = 0.025$ mm?
13. Find optimum operating conditions for the separation of trace quantities of Na^+ and K^+ by elution development with HCl on Dowex 50 × 8. The desired purity is 99.8%. The following data are given: $r = 0.025$ cm (35 mesh), $Q_v =$ 1.2 mEq/ml, void fraction = 0.4, $D_r = 2 \times 10^{-6}$ cm²/sec, $D_f = 2 \times 10^{-5}$ cm²/sec, and column cross section is 1.1 cm².
14. Since the operating time is excessive for separating Na^+ and K^+ in Problem 13, redesign the operating conditions using 270–325 mesh (0.0025 cm) resin particles.
15. Redesign Problem 13 using Dowex 50 × 16, 35 mesh, all other conditions unchanged.
16. For either Problem 13, 14, or 15, sketch the elution peaks of Na^+ and K^+, locate V_{max}, W, and the 2σ and 3σ values on each side of V_{max}. Is the overlap of curves 0.1% or less?
17. On a column of Amberlite IRA-400 (10–35 μ particle diameter), 2.50 cm × 0.360 cm², the eluent peaks obtained for Tb, Gd, and Eu are shown in Fig. 5-9. (a) Estimate the plate number for each solute band. (b) From the average plate number, calculate the plate height. (c) For adjacent bands, calculate the resolution obtained and the plate number required for 99.9% purity. (d) Assuming the void volume is 0.30 cm³, estimate the distribution coefficient for each solute.
18. An incomplete separation of propylene glycol and *meso*-butylene glycol occurred on a column of Dowex 1 × 8, 70.7 cm × 2.28 cm², with 0.29*M* sodium borate as eluent. The interstitial volume was 74 ml. From the elution graph, V_{max} were 370 and 441 ml, and the half width at C_{max}/e were 25 and 29 ml, respectively. (a) Determine the plate number and the partition coefficient. (b) What length of column would achieve a separation within 1.0%, assuming equal quantities of each glycol? (c) On the original (70.7 cm) column, where would cuts be made to secure a portion of each glycol at 99.8% purity?
19. On a column of Amberlite IRA-400 (10–35 μ particle diameter), 4.20 cm × 0.36 cm², the elution graph obtained is shown at two operating temperatures in Fig. 5-10. (a) Estimate the plate number for each band. (b) For adjacent bands, calculate the resolution obtained and (c) the plate number required for 99.8% purity. Assume the Cs peak represents an unadsorbed solute.

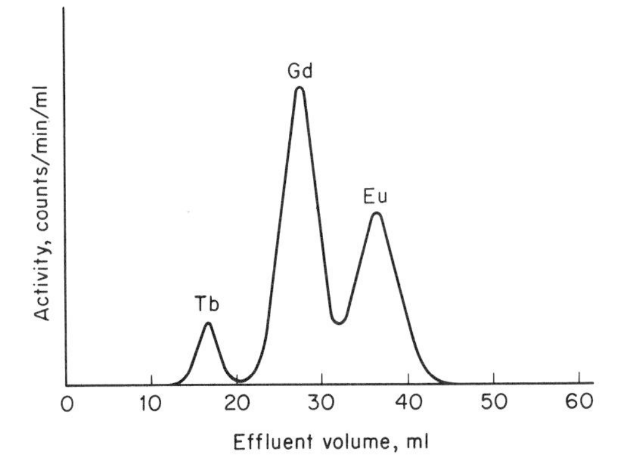

FIG. 5-9. Illustration for Problem 17.

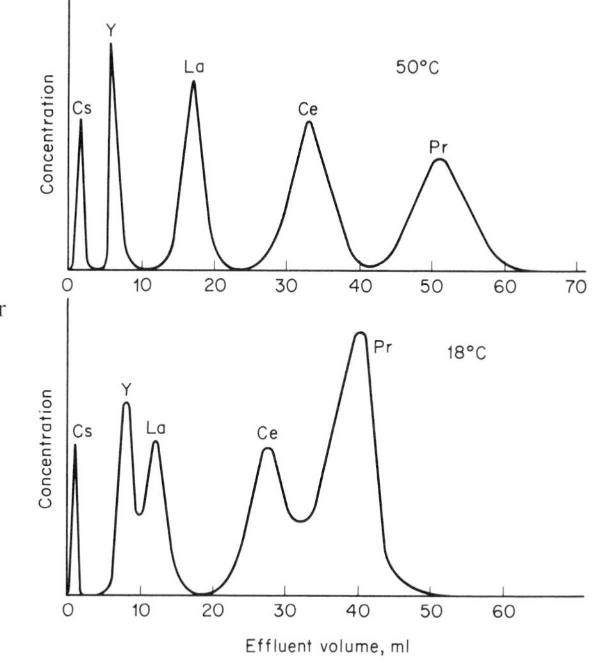

FIG. 5-10. Illustration for Problem 19.

20. A 0.5-g sample of silicate was dissolved, concentrated, and transferred to a column of Dowex 50 × 8, 25 cm × 2.4 cm² (particle radius = 0.005 cm, Q_v = 1.7 mEq/ml) and eluted with 0.7*M* HCl at a flow rate of 0.55 cm/min. Ascertain whether the separation of Li, Na, K, and Mg from each other is complete within 0.1% cross-contamination by calculating for each V_{max}, the break-through volume and the cessation point for each band.
21. If one were only interested in the separation of K^+ and Mg^{+2} on the column employed in Problem 20, (a) show that the Mg–K elution curves overlap when the eluent is 1.0*M* HCl, and (b) show that if the eluent were 10*M* HCl, the Mg^{+2} band would theoretically precede the K^+ band and be adequately resolved.
22. Verify the correctness of these laboratory directions: Samples of the alkali halides, not more than 0.3*M* in any one halide, were pipetted (2 ml) onto a bed of Dowex 1 × 8 resin (6.7 cm × 3.14 cm², r = 0.0055 cm, Q_v = 1.5 mEq/ml) in the nitrate form, previously equilibrated with 0.5*M* $NaNO_3$. Chloride was eluted by the passage of 35 ml of 0.50*M* $NaNO_3$ at a flow rate of 1.0 cm/min. At this point the eluent was changed to 2.0*M* $NaNO_3$. The bromide was collected after 28 ml and the iodide after 51 ml. Time required was 28 min.
23. Elution of 0.2 mEq each of oxalate and bromide through a column, 6.7 × 3.8 cm², of Dowex 1 × 8 (r = 0.0055 cm, Q_v = 1.5 mEq/ml) with 0.05*M* $NaNO_3$ (at an unspecified flow rate) is reported to be unsatisfactory. Suggest a method to improve the separation. Calculate new values for V_{max}, W, and 3σ break-through and cessation values. Approximate values, read from the published elution graphs for the unsatisfactory separation were: oxalate broke through at 325 ml, peaked at 520 ml, and ceased eluting at 735 ml; corresponding values

for bromide were 500 ml, 780 ml, and cessation value not shown. Assume the breakthrough and cessation points represent the base width for 6σ.

24. Suggested operating conditions for the separation of Ta and Nb on 2.8 g of oven-dried Dowex 1 × 10 (200 mesh, $\rho = 1.67$ g/ml, $Q_w = 3.0$ mEq/g) in a column 20 cm × 0.63 cm (i.d.) included a flow rate of eluent at 8–10 ml/hr. The following values of V_{max} (in ml) were observed:

Eluent, *M* HF	24	20	16	12	8
V_{max} (Ta)	65	97	130	190	>350
W (Ta)	15	23	36	55	~100
V_{max} (Nb)	98	150	152	122	160
W (Nb)	25	45	60	45	48

(a) Calculate V_b, V_r, and V_o. (b) Estimate the plate number for each element. (c) Graph the elution curves for each pair at each eluent concentration. (d) Estimate the resolution of Ta–Nb for each eluent concentration. (e) What eluent concentration would you recommend? State reasons for your choice.

25. For the adsorption of gallium(III) from HCl solutions, these are the observed volume distribution coefficients on a strong base anion exchanger:

HCl, molality	D_v	HCl, molality	D_v
3.0	100	8.0	2.4×10^5
3.95	1.0×10^3	10.0	1.6×10^5
5.0	1.0×10^4	12.0	8.4×10^4
6.4	9.6×10^4	13.0	6.4×10^4
7.0	1.6×10^5	14.0	4.6×10^4
7.5	2.25×10^5	16.0	2.5×10^4

Plot the data as log D_v vs molality HCl. Assuming that a linear relationship exists between log E/G (where G are the activity coefficients), calculate the ion fractions: $\Phi = [GaCl_4^-]/C_{Ga}$, and plot the results on the same curve.

26. These experimental values of V_{max} were reported [*Anal. Chem.* **33,** 460 (1961)], all obtained on Dowex 50 × 12 ($Q_v = 2.5$ mEq/ml), bed volume was 5 ml:

	V_{max} (in ml)					
HCl, *M*	Hg^{+2}	Cd^{+2}	Zn^{+2}	Mn^{+2}	Fe^{+3}	Ni^{+2}
0.5	4.0	23	197	265	950	235
1.0	4.0	4.5	38	62.5	97	58
2.0	4.0	...	5.1	14	12	...
4.0	4.0	2.35	2.4	4.9	3.1	5.8

For each concentration of HCl, calculate the theoretical D_v values for exchange involving hydrogen ion and metal ion. Comment on the discrepancies between calculated and observed values.

27. From the information in Problem 26, deduce a separation scheme using a resin bed, 5.0 cm × 1.0 cm^2, of beads 0.005 cm in radius.
28. Devise a scheme for the separation of Fe(III), Al, Co, Ni, and Mo(VI) using both cation and anion exchanger columns—Dowex 50 × 8 and Dowex 1 × 2.
29. In the preceding Problem 28, aluminum and nickel are not retained on the anion exchange column. The effluent, after evaporation and adjustment of acidity, is added to a column of Dowex 50 × 8, 57.5 cm × 1.76 cm^2, 100–200 mesh. What concentration of HCl should be used as eluent, and when should the cuts be made for nickel and aluminum? Assume $Q_v = 1.7$ mEq/ml. $E_H^{Al/3} = 1.73$.
30. Suggest ion-exchange methods for the recovery of these constituents in industrial operations: (a) tin from electro-tinning solutions; (b) chromium from plating rinse baths; (c) silver from photographic rinse waters; (d) zinc from mine waters; (e) amino acids from meat and fish by-products; (f) copper from ammoniacal "blue water" and spinning acid from the rayon cuprammonium process.
31. Suggest suitable ion-exchange methods to handle these problems: (a) removal of silica from waters used in high-pressure boilers and steam turbines; (b) adsorption of ammonia from acetone; (c) removal of formic acid from formaldehyde; (d) removal of malic (and its bitter taste) from apple syrup; (e) removal of HCl and sulfuric acid from lactic acid; (f) removal of sulfate ions from ethylene glycol; (g) removal of electrolytes from colloids; (h) separation of small amounts of aluminum in steel; (i) separation of Zn, Cd, and Pb ions from Fe(III), Mn, Cu, Co, Ni, and Al.

BIBLIOGRAPHY

Helfferich, F., *Ion Exchange,* McGraw-Hill, New York, 1962.

Inczedy, J., *Analytical Applications of Ion Exchangers,* Pergamon, London, 1966.

Marinsky, J. A., *Ion Exchange: A Series of Advances,* Dekker, New York, 1966; Vol. 2 has appeared.

Reiman, W., and A. C. Breyer, "Chromatography: Columnar Liquid–Solid Ion-Exchange Processes," in I. M. Kolthoff and P. J. Elving (Eds.), *Treatise on Analytical Chemistry,* Part I, Volume 3, Chap. 35, Interscience, New York, 1961.

Samuelson, O., *Ion Exchange Separations in Analytical Chemistry,* Wiley, New York, 1963.

6 *Specialized Ion-Exchange Systems*

The separation of particular types of substances sometimes requires the development of specialized ion-exchange media for the stationary phase, with structures and properties particularly suited to the systems. The use of ion exchangers with hydrophilic matrices, inorganic ion exchangers, and ion-retardation resins will now be discussed.

Ion Exchangers with Hydrophilic Matrices

Interaction between an adsorbate and ion-exchange resin based on styrene-divinylbenzene copolymers was restricted essentially to ionic forces because of the hydrophobic nature of the resin matrix. Labile macromolecules, which normally exist in a predominantly hydrophilic environment, ran the risk of being denatured by the disorientation produced by short-range (van der Waals') forces acting between hydrophobic regions in any part of the adsorbate molecule and the resin matrix. Size is another problem. The separation of macromolecules by ion exchange usually requires the use of ion-exchange materials that either have the majority of the functional groups on the outside of the polymer matrix or have a structure sufficiently porous to allow the macromolecules to penetrate.

These limitations have been overcome by the introduction of ion-exchange media with a hydrophilic type of matrix. Suitable exchangers can be prepared by reactions which introduce ionizable groups into polyacrylamide gels or carbohydrate polymers such as cellulose or dextran. The mutual distance of the charged exchange sites in these hydrophilic matrices is approximately 50 Å, in contrast to conventional ion-exchange resins with about 10 Å distance between charged groups. This low density of exchange sites permits exchange of delicate molecules of a biological origin under mild conditions.

Cellulose-Base Ion Exchangers Ion-exchange celluloses are made by reacting the free hydroxyl groups of cellulose with functional groups containing either basic or acidic properties. The cellulose matrix consists almost entirely of anhydroglucose units joined in ether linkage through the 1,4-carbon atoms. The exchange groups are bound to the fibrous surface of the cellulose matrix by ether or ester linkages through the alcohol groups of the anhydroglucose rings. Several types of ion-exchange groups are described in Table 6-1. The more common anion exchangers are usually prepared by treatment of cellulose with epichlorohydrin and an amine. To obtain cation exchangers, sulfonic, carboxylic, or phosphoric acid groups are introduced.

Exchanger materials are rod- or thread-like particles with an average

TABLE 6-1. Classification of Ion-Exchange Groups in Hydrophilic Matrices

Type of exchanger	*Literature designation*	*Abbreviation*	*Exchange group*
Anion exchanger:			
Strongly basic	Guanidoethyl	GE	$—O—C_2H_4—N(H)—C(NH_2)=NH_2^+$
	Triethylamine	TEAE	$—O—C_2H_4—\overset{+}{N}(C_2H_5)_3$
Weakly basic	Diethylaminoethyl	DEAE	$—O—C_2H_4—NH^+(C_2H_5)_2$
	Aminoethyl	AE	$—O—C_2H_4—NH_2$
	Triethanolamine	ECTEOLA	$—\overset{+}{N}(CH_2—CH_2OH)_3$
Cation exchanger:			
Strongly acidic	Sulfoethyl	SE	$—O—C_2H_4—SO_3^-$
	Sulfomethyl	SM	$—O—CH_2—SO_3^-$
Intermediate acid	Phosphonic	P	$—O—P(O^-)(O)(O)$
Weakly acid	Carboxymethyl	CM	$—O—CH_2—COO^-$

diameter of 18 μ. The lengths vary from 20 to 300 μ, with an 80 μ average. Aggregates of carbohydrate chains, with areas of higher and lower orientation, and described as crystalline and amorphous regions, respectively, are randomly orientated along the fiber axis. Within and between these areas, which are stabilized and cross-linked by hydrogen bonding, are located "holes" with a wide size range and available ion-exchange sites (Fig. 6-1). Exchange capacities are not particularly high, ranging from 0.2 mEq/g for types SE and ECTEOLA to 0.9 mEq/g for strongly basic types. The ether derivatives of cellulose (types CM, SM, SE, TEAE, DEAE, and AE) are reasonably stable and may be used under a wide range of conditions. The ester derivatives, however, are subject to hydrolysis at pH values less than 2 and greater than 10, especially above room temperature.

Polymeric Gel Exchangers Ion exchangers can be made by incorporating functional groups into the series of porous gels produced originally for gel permeation (filtration) applications (Chapter 12). A list of these gels and their properties is given in Table 12-1.

The extent of uptake of a macromolecule will depend on such factors as the charge density of the macromolecule, the porosity of the gel, and the pH and the ionic strength of the solution. The gel porosity is governed by the degree of cross-linking. In general, maximum uptake of a macromolecule will be facilitated by using a gel that has a pore size larger than that of the macromolecule. Type P-2 is suitable for the separation of low molecular weight materials. Generally, for solutes of molecular weight less than 10,000, the P-30 or G-25 type will be the best choice, whereas the G-50 or P-100 will have the highest available capacity for solutes of a molecular weight above 10,000. Very high molecular weight solutes that are unable to penetrate the gel structure are adsorbed at the surface of the beads only.

Exchangers from polyacrylamide and polysaccharide gels are strongly hydrophilic and swell in water and salt solutions. They are supplied in the

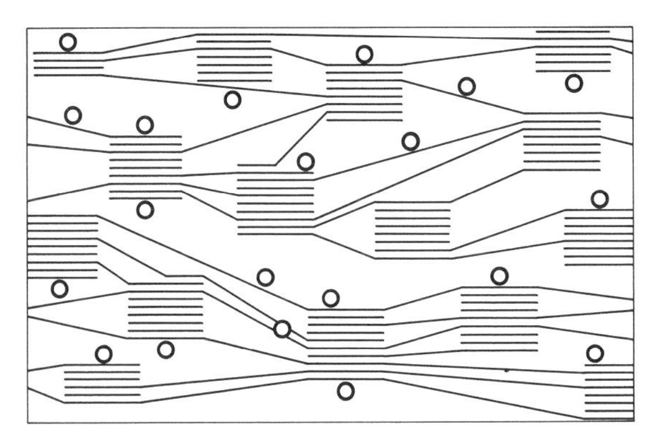

FIG. 6-1. Diagrammatic microstructure of the fibrous ion-exchange celluloses. The solid lines represent the aggregates of carbohydrate chains. The open circles are the ion-exchange sites.

form of small spherical beads. The porosity of the swollen bead depends on the ionic strength of the eluent used; the hydrated bed volume changes as well. The lower the capacity and the lower the initial porosity, the less change there will be in bed volume with changes in ionic strength. This same general statement holds true for changes in pH. Exchange capacity ranges from 2–6 mEq/g.

Exchanger Characteristics Although inorganic ions seem to follow normal ion-exchange processes, small changes in the ionic strength of an eluent can cause large shifts in the adsorption equilibria of macromolecules. These often have a larger effect than pH changes. With large molecules, such as proteins, both the adsorbate and the ion exchanger are polyelectrolytes and are therefore capable of interaction at a number of sites. What is involved during a columnar separation is the making and breaking of "an ion-exchange site – macromolecule complex" rather than normal multiplate ion exchange on the polystyrene–divinylbenzene exchangers. It is possible to elute a protein molecule by changing the pH so as to reduce the number of charges on the protein, or, alternatively, by increasing the salt or buffer-ion concentration of the eluent so as to compete for the charges on the ion exchanger. Stronger eluting conditions are usually necessary for large molecules, as compared with small molecules, owing to the greater number of bonds which can be formed with the ion exchanger.

The pH values of buffers chosen for elution should be selected to lie within the pH range where all the functional groups are completely or almost completely ionized. The most useful pH range is just within to somewhat beyond the titration range of the exchange group (Figs. 6-2 and 6-3). Outside this range the ion exchanger can be used only at the cost of reduced capacity. Ion exchangers possessing the carboxyl group (type CM) will have their maximum capacity at pH values above 6, whereas type SE and

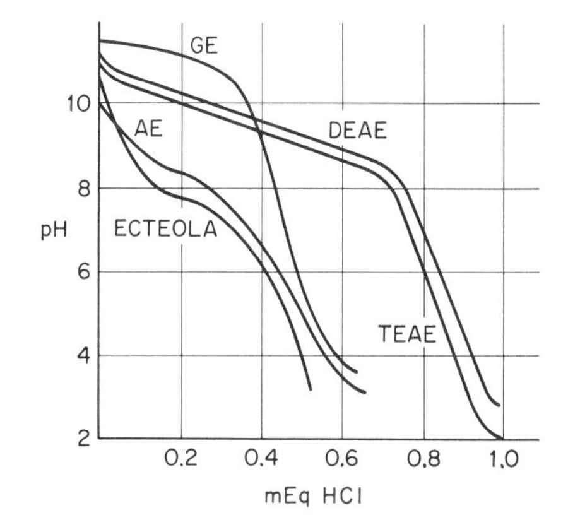

FIG. 6-2. Typical titration curves with basic cellulose ion exchangers.

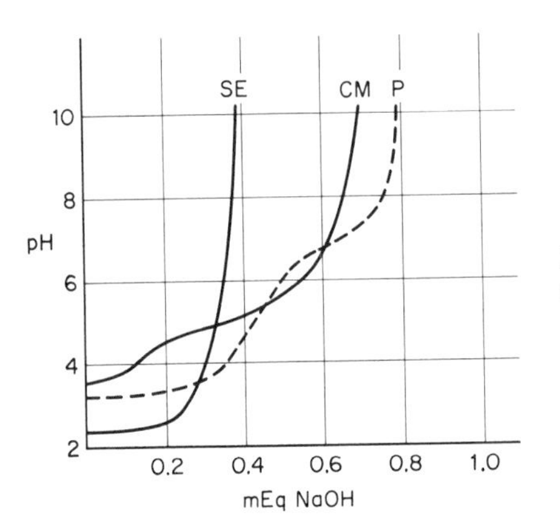

FIG. 6-3. Typical titration curves with acidic cellulose ion exchangers.

SM exchangers are fully dissociated below pH 10. For type DEAE the situation is complicated because the exchanger contains not only the diethylaminoethyl groups ($pK_a = 9.5$), but a small amount of diethylamine groups ($pK_a = 5.7$). The buffering ion should not react with the column material and should be chosen to have a charge of the same sign as the ionized groups on the exchanger; i.e., cationic buffers for anion exchangers and anionic buffers for cation exchangers.

Inorganic Ion-Exchange Crystals

Ion-exchange crystals are synthetic inorganic microcrystalline aggregates with ion-exchange capabilities similar to those of conventional resins. Inorganic exchangers, with their rigid lattices and identical repeating groups, possess considerably higher selectivities than conventional organic resins with their randomly oriented and flexible matrix structures.

Zeolites Zeolites are naturally occurring inorganic minerals; sodium aluminum silicates are synthetic analogs. The crystal structure possesses cation-exchange properties. In the crystal lattice some SiO_4^{-4} groups placed in the tetrahedral arrangement are substituted by AlO_4^{-5} groups. Every AlO_4^{-5} group means a free negative charge which can bind uni- and divalent cations. These zeolites also have a porous, cross-linked structure (see Fig. 12-9). Cations in the pores between the lattice are readily exchanged for other positively charged ions of similar size. Neither the natural or synthetic material has high capacity nor good chemical and mechanical resistance. They frequently find use in water softening units due to their relatively low cost.

Zirconium Acid Salts Inorganic cation exchangers can also be prepared by combining group IV oxides with the more acidic oxides of groups

V and VI. Non-stoichiometric microcrystalline gels of zirconyl phosphate with different $ZrO_2:P_2O_5$ ratios are available. Containing configurations such as

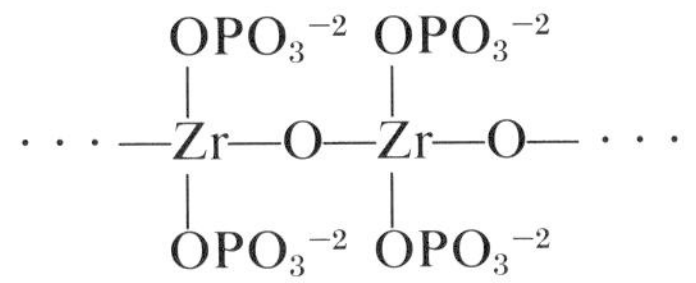

they may be pictured as consisting of an insoluble matrix of zirconium oxide which holds the cation-exchange group—phosphate. Molybdate and tungstate may take the place of phosphate, and titanium, thorium, and tin that of zirconium. Exchange capacity rises from 1 mEq/g at pH 1.5 to 4.5 mEq/g at pH 9.5.

Hydrous Zirconium Oxide At pH levels below the isoelectric point of zirconium oxide, the positively charged polymer functions as an anion exchanger with high selectivities for polyvalent anions over monovalent anions. It has an exceptionally high capacity for fluoride ion since this ion forms a complex with zirconium. In basic solutions the negatively charged polymer becomes a cation exchanger.

Applications Ion-exchange crystals can be used in areas not covered by conventional ion-exchange resins. Their ability to perform separations at temperatures up to 300°C and to withstand high levels of radiation without breakdown makes these materials useful for the separation of selected radionuclides. They also exchange effectively in nonaqueous solutions.

The distribution ratios for 60 metal ions with hydrous zirconium oxide, zirconyl phosphate, zirconyl tungstate, and zirconyl molybdate have been reported from nitrate media over the pH range of 1 to 5.[1] Zirconyl phosphate in the hydrogen form adsorbs the alkali metals from acidic solutions leaving the alkaline earth metals in the effluent. With the exchanger in the ammonium form, the alkali metals are eluted first with ammonium chloride, after which the alkaline earth metals can be selectively eluted with $0.14M$ ammonium sulfate in 60 volume percent methanol (for magnesium) followed by the elution of calcium with $0.2M$ ammonium nitrate plus $0.005M$ nitric acid, and strontium with $1M$ ammonium nitrate.[2]

Microcrystalline ammonium 12-molybdophosphate exhibits high selectivity for individual alkali ions; the larger alkali ions can be selectively adsorbed from solutions containing large amounts of the smaller alkali metal ions. Adsorbed ions can be recovered by elution or by simply dissolving the ion exchanger with a dilute alkaline solution. Potassium hexa-

[1] W. J. Maeck, M. E. Kussy, and J. E. Rein, *Anal. Chem.* **35,** 2086 (1963).
[2] M. H. Campbell, *Anal. Chem.* **37,** 252 (1965).

cyanocobalt(II)-ferrate(II) finds use for the isolation and concentration of highly radioactive Cs^{137} from fission product waste solutions that arise from the processing of nuclear fuels.

Ion Retardation

Ion retardation is based upon the reversible adsorption of electrolytes by amphoteric exchangers which contain both acidic and basic functional groups. These special resins are prepared by saturating a quaternary base resin with acrylic acid, or by saturating a sulfonic acid resin with vinylpyridine, and then polymerizing the latter. The result is a linear polymer with charges of one sign trapped inside a network polymer having charges of the opposite sign. These are the "snake-cage" polyelectrolyte resins. The linear chains of the poly-counter ions are so intricately intertwined with the cross-linked matrix that they cannot be displaced by other counter ions. Each group is the counter ion of the other (Fig. 6-4); the bond between them is indicated by dotted lines.

Ion-Retardation Process The anionic and cationic adsorption sites in ion-retardation resins are so closely associated that they partially neutralize each other's electrical charges. However, the sites still have an attraction for mobile anions and cations and can associate with them to some extent. When an aqueous solution containing a salt and a non-electrolyte is fed onto a column of the resin (until the ion-adsorbing capacity of the bed is utilized as completely as possible), the resin phase will hold back the dissolved salts and allow the non-electrolytes to emerge from the column first since the latter are not retarded. By this method, non-ionogenic substances can be recovered in a salt-free state. The adsorbed ions are then eluted by rinsing the bed with water. Separations of this type are especially effective with rather large molecules of a hydrophilic nature. For example, ion-retardation resin will remove both acids and salts from substances such as amino acids, polypeptides, proteins, enzymes, nucleic acids, polyglycerides, and sugars. There is less degradation of labile molecules during desalting since salt adsorption occurs in a neutral environment, as compared with conventional mixed-bed resins.

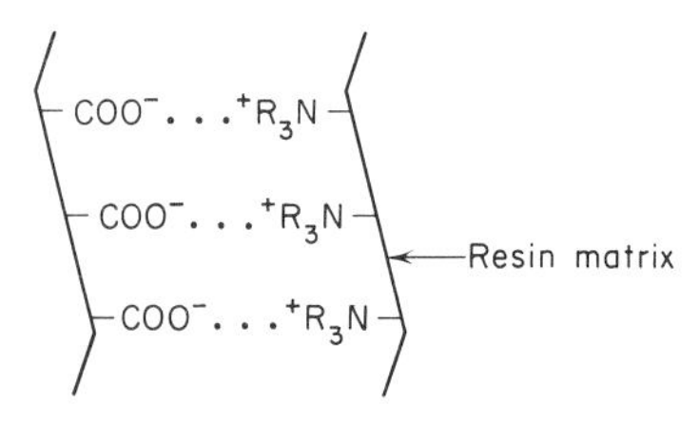

FIG. 6-4. Schematic structure of an ion retardation resin. The solid curved line represents the organic polymer matrix; the dotted lines, ionic bonds.

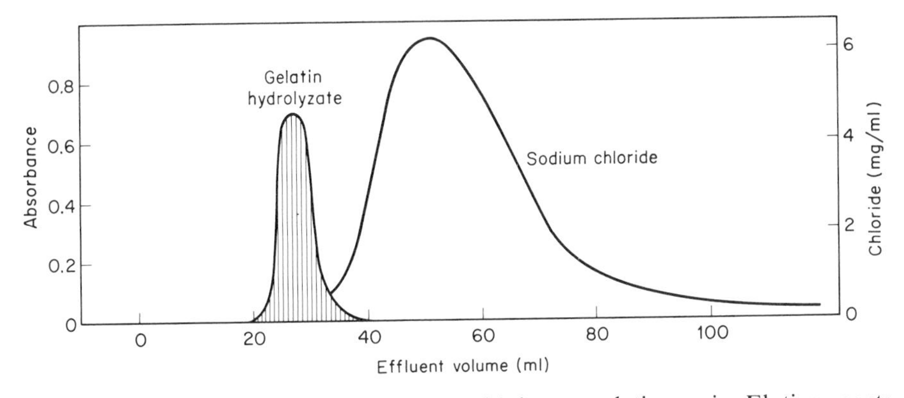

FIG. 6-5. Desalting a protein hydrolyzate with ion retardation resin. Eluting agent: water.

The removal of sodium chloride from gelatin hydrolysate (Fig. 6-5) is depicted below:

$$\underset{\text{(resin)}}{(COO^- \cdots {}^+NR_3)} + \underset{\text{(solution)}}{Na^+Cl} \rightleftharpoons \underset{\text{(resin)}}{(RCOO^-Na^+Cl^-\ {}^+NR_3)} \qquad (6\text{-}1)$$

The strength with which the retardation resin adsorbs an ion varies with the type of ion although, in general, the order of selectivity parallels that of the corresponding Dowex 1 and Dowex 30 resins. Hydrogen ion is so strongly held by the carboxyl group that it is not eluted with water. Some other ions which are chelated strongly by polyacrylic acid are also difficult to remove. Adsorbed salts move down the column during water elution because the fixed exchange groups compete with the mobile salt ions to become self-adsorbed; Eq. (6-1) progressing from right to left. A flow of water washes away the mobile ions driving the reaction to the left. As the salts move down the column they are repeatedly adsorbed and desorbed, and thus retarded relative to non-electrolytes.

LABORATORY WORK

General Directions for Work with Cellulose-Base, Polyacrylamide, and Dextran Exchangers

Packing the Column The amount of ion exchanger required for a particular experiment depends on the available capacity towards the compounds to be separated. So many factors are involved, the available capacity should be quickly estimated by micro-scale batch experiments.

A bed height-to-diameter ratio of 10:1 is recommended. Column heights are generally 10 to 20 cm. Longer columns show no advantage. Any kind of laboratory column may be used. However, coarse sintered-glass disks are not recommended as bed supports because they easily become clogged by small particles of ion exchanger. A layer of sand over the porous disk is satisfactory.

The ion exchanger is stirred into the buffer to be used at the beginning of the elution. After 1–2 minutes with cellulose-base exchangers, the slurry is decanted into a tall graduate cylinder and allowed to settle for 45 min. The turbid supernatant liquid is discarded; the sediment is used for pouring the column as a 5% slurry while the liquid is allowed to drain freely from the exit of the column.

The gel-type exchangers must be allowed to swell in a large excess of initial eluent buffer for the specified period of time. The supernatant liquid should be replaced several times by fresh buffer solution. The fully swollen gel slurry is poured carefully into the column and allowed to settle under gravity for 5 min, then the outlet is opened to allow a gentle flow.

Microbial growth in columns can be prevented by using 0.02% sodium azide for cation exchangers and 1% butanol for anion exchangers. These reagents are added to the prepared columns prior to periods of storage.

If the exchanger cannot be used in the form supplied, the counter ion must be replaced with the desired ion. DEAE-exchanger is allowed to swell in distilled water and then is washed repeatedly with 0.5*N* NaOH on a Büchner funnel until free of chloride ion. The excess NaOH is removed by rinsing with water. Then the exchanger is treated with a solution of an acid containing the desired counter ion (but not above 0.5*M* concentrations for strong mineral acids). The excess conversion acid is removed by rinsing with water. Finally the exchanger is washed with the initial eluent buffer until equilibrium is attained (both pH and counter-ion concentrations in the washings).

CM- and SE-exchangers are converted to other forms by first washing with 0.5*N* HCl and then with a base containing the desired counter ion.

Column Chromatography Samples can be applied to ion-exchanger columns in any manner the worker finds satisfactory, but it is best if the sample is in the same solution that will be used for the initial elution. If it is necessary to transfer the sample from one solution to another this can be done with gel permeation (filtration) methods (*q.v.*) or with dialysis.

Eluent flow rates are 5 to 25 $ml/cm^2/hr$. Such low rates are necessitated by the rate at which equilibrium is attained with macromolecules. To maintain a low and constant flow rate, the use of a pump is recommended. Coupling the effluent to the flow-cell of a recording spectrophotometer provides a convenient method for following an elution and preparing a permanent record.

Reuse of cellulose exchangers requires washing with 4 column volumes of 0.1*M* NaOH and 4 column volumes of water. Anion exchangers should receive a subsequent washing with 2 column volumes of 0.1*M* HCl to remove carbon dioxide which may cause artifact peaks.

After a completed experiment with gel exchangers, retained impurities should be removed from the bed by continuing the elution with solutions of increasing concentration, until an ionic strength of about 2 is reached. Then the gel is removed from the column and washed with distilled water to remove the excess of salt. It is then ready to be equilibrated for the next experiment.

Choice of Eluent The most common method of working with cellulose and porous gel exchangers is to shift the ionic strength of the eluent. This can be done stepwise or as a gradient. As a rule adsorption equilibria on these ion exchangers are much more sensitive against changes in salt concentrations than against change in pH although in many cases shifting the pH will also separate the components of a mixture. Theoretical considerations lead to the following rules for the proper choice of eluent: [3]

(1) Use cationic buffers (tris-HCl, piperazine-$HClO_4$) with anion exchangers; anionic buffers (phosphate, acetate) with cation exchangers.
(2) With anion exchangers use falling pH gradients, with cation exchangers use rising pH gradients.
(3) Avoid using buffers whose pH lies in the neighborhood of the pK_a value of adsorbent.

If it is planned to isolate the chromatographed fractions by evaporation of the solvent, the use of volatile buffers should be attempted, such as carbonic acid, carbonates, acetates, and formates of ammonium trimethylammonium ion.

Usually the composition of the eluent is changed during elution. One-step procedures are used if only one substance with established elution behavior is to be purified routinely. If, however, a complete analysis of a complex mixture of strongly differing components is intended, gradients covering a wide range of buffer concentrations must be applied. Gradient mixing devices are described in Chapter 4.

EXPERIMENT 6-1 *Desalting a Gelatin Hydrolysate with Ion-Retardation Resin* [4]

Slurry 30 g of Dowex AG 11-A-8, 50 to 100 mesh, in water and pour into a column 1.2 $cm^2 \times 40$ cm.

Prepare a partially hydrolyzed gelatin solution by boiling for 2 hr 10 g

[3] G. Semenza, *Chimia* **14,** 325 (1960).
[4] C. Rollins, L. Jensen, and A. N. Schwartz, *Anal. Chem.* **34,** 711 (1962).

of gelatin, 50 g of NaCl, and 10 ml concentrated HCl in 1 liter of water. Neutralize this solution with NaOH before proceeding with the separation.

Drain the excess water from the resin column until the level reaches the resin surface. Carefully pipet 5 ml of the hydrolysate solution onto the resin surface and begin collecting the effluent. Continue the elution with water and collect 2-ml aliquot fractions of effluent.

Test the effluent for amino acids by heating 1 ml of effluent with 0.5 ml of 0.1% ninhydrin in a boiling water bath for 15 min. Dilute to 10 ml and measure the absorbance at 570 mμ. Test for chloride by adding 1 ml of 0.1*M* sodium chromate to 1 ml of effluent and titrate in a porcelain crucible with 0.1*M* silver nitrate.

Plot the results as absorbance (or mg/ml) vs effluent volume (in ml). A typical set of data is shown in Fig. 6-5.

BIBLIOGRAPHY

Amphlett, C. B., *Inorganic Ion Exchangers,* Elsevier, Amsterdam, 1964.

Knight, C. S., *Some Fundamentals of Ion-Exchange-Cellulose Design and Usage in Biochemistry,* in *Advances in Chromatography* (J. C. Giddings and R. A. Keller, Eds.), Vol. 4, Dekker, New York, 1967.

7 Adsorption Chromatography

Liquid chromatographic column methods have begun to show their full potentialities with the development of detectors of very high sensitivity. Adsorption chromatography, the oldest of the various chromatographic techniques, is based on the retention of solute by surface adsorption. Although overshadowed to some extent in recent years by other chromatographic techniques, it remains one of the simplest and most effective ways of separating mixtures of rather nonpolar substances such as lipophilic (fat-soluble) substances and constituents of low volatility. It has also remained one of the least well understood chromatographic methods. Liquid–solid, as opposed to gas–solid (Chapter 11), chromatographic systems will be discussed in the present chapter. Only non-ionic organic compounds will be considered. The adsorption of ionic species and potentially ionic species, such as acids and bases, is frequently accompanied by ion-exchange effects. These are discussed in Chapters 5 and 6.

General Aspects of Adsorption

In liquid–solid chromatography a solid substrate serves as the stationary phase. Separation depends on the equilibrium established at the interface between the grains of the stationary phase and the mobile liquid phase, and on the relative solubility of the solute in the mobile phase. This interplay of forces is diagrammed in Fig. 7-1. Competition between the solute and solvent molecules for adsorption sites establishes a dynamic process in which solute and solvent molecules are continuously coming in contact with the surface, residing there momentarily, and then leaving to re-enter the mobile phase. Which is most strongly bound to the surface depends on the adsorption affinity for the two molecules. While desorbed, the solutes are forced to migrate by the forward flow of the mobile phase. This force acts

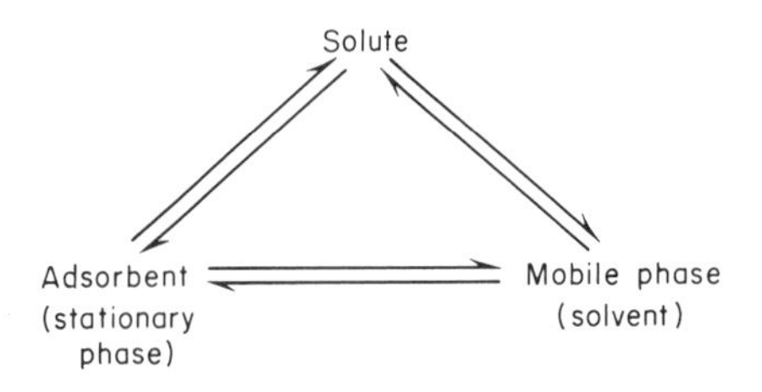

FIG. 7-1. Interplay of forces in adsorption chromatography.

equally upon all the substances while they are in the mobile phase. Only those molecules with the greater affinity for the adsorbent will be selectively retarded.

Tswett's original paper will illustrate the process.[1] A few milliliters of petroleum ether extract of dried leaves were placed on the top of a column of magnesium oxide. When this had penetrated into the upper layer of the column, the chromatogram was developed by more solvent percolating down the column. Several colored zones were formed: Migrating downwards most rapidly is an orange carotene band; above this, a narrow yellow band, a dark green chlorophyll zone, and a light green band are to be seen. Elution was continued until the zones left the column at its lower end—the forerunner of elution chromatography.

Intermolecular forces involving surfaces are by no means negligible. These arise because, by its nature, a surface introduces a discontinuity into the system. Whereas a molecule or atom in the interior of a liquid or a solid is, on the average, exposed to equal forces in all directions, this is no longer the case at a surface. There is, in consequence, a surface-energy effect at any interface. Relatively weak London forces exist between all surfaces and adsorbed molecules. Electrostatic forces exist between polar surfaces and any adsorbed molecule or between nonpolar surfaces and polar adsorbed molecules; they induce dipoles in large nonpolar molecules and increase the existing dipole moments of polar compounds. Between strong electron donors and acceptors there exist charge transfer forces which culminate in the special case of hydrogen bonding. Very strong forces arise between covalent and ionic bonded atoms. These intermolecular forces were discussed in considerable detail in Chapter 3. The topography of a surface is also important. The number of available adsorption sites decides the proportion of the adsorbent surface which can be covered by adsorbed molecules.

Although most normal chromatographic adsorption is of the physical type, chemisorption may be presumed to occur frequently, involving perhaps only a fraction of the solute molecules in any one run. It leads to anomalies, such as irreversible adsorption, tailing, and streaking.

[1] M. Tswett, *J. Chem. Educ.* **44,** 238 (1967); English translation of original paper appearing in *Ber. Deut. Botan. Ges.* **24,** 316, 384 (1906).

Adsorption Isotherms In adsorption chromatography the partition coefficient (K_d) is equal to the concentration of solute in the adsorbed phase (in g/g) divided by its concentration in the solution phase (in g/ml). An adsorption isotherm is a graphical representation of the uptake of any substance by an adsorbent from a given solvent at a fixed temperature. The isotherm portrays the amount of solute adsorbed per unit weight of adsorbent (x/w) from a solution of concentration, C, at equilibrium. Thus

$$(x/w)/C = K_d \tag{7-1}$$

The three principal shapes of adsorption isotherms were discussed in Chapter 4 and were illustrated in Fig. 4-1.

When the partition coefficient is constant, i.e., K_d does not change with the concentration of the solute, a linear adsorption isotherm prevails. The amount of solute adsorbed varies linearly with the concentration of the contacting solution. This condition is met in adsorption chromatography generally only in rather dilute solutions. The resulting elution curves are gaussian in shape, symmetrical about the midpoint, and with trailing and leading edges. This indicates that, as adsorption proceeds, the ease with which it takes place remains constant, suggesting that fresh sites are becoming available. In practice, an adsorption isotherm can never be completely linear. Each unit amount of adsorbent can only take up limited amounts of solute. The adsorptive power of the surface generally falls off markedly even when only a small proportion of the adsorption sites is occupied, since the most active sites are occupied first. The ease with which adsorption takes place decreases until the (adsorbed) monolayer on the surface is complete, i.e., all the adsorption sites are occupied (the plateau on the adsorption isotherm).

Adsorption isotherms usually show a strong curvature towards the saturation concentration. This means that at higher concentrations a larger proportion of the solute will remain in solution. Substances with convex adsorption isotherms exhibit an asymmetric band or zone having a sharp leading front and a diffuse trailing boundary. The "tailing" arises because the portion in the trailing edge, where the concentration of solute is lower, is always more strongly adsorbed. Unsymmetrical bands with diffuse boundaries are a disadvantage and present difficulty in distinguishing the end of the faster band from the beginning of the adjacent slower band.

The isotherm that exhibits upward concavity represents a situation in which, as adsorption proceeds, it becomes easier for solute molecules to be adsorbed. Those already adsorbed on the surface at the most active sites assist further adsorption by intermolecular bonding. Such isotherms are generally given by flat molecules standing edge-on to the substrate surface. An example is phenol adsorbed on alumina, where the hydroxyl group probably forms a hydrogen bond to surface oxygen atoms on the alumina, and the aromatic nucleus associates with other solute molecules.

Adsorbent Linear Capacity Interest in adsorption chromatography centers around the adsorbent linear capacity,[2] defined as the maximum column loading possible without loss in linearity. A satisfactory quantitative measure of linear capacity is afforded by the column loading at which V_R drops to 0.9 times the linear V_R value. Adsorbent linear capacity is high for adsorbents of high surface area and for solids with energetically uniform surfaces, that is, no adsorption sites of widely differing energies, and low for most activated adsorbents (usually less than 10^{-4} g/g). Destruction of the strongest adsorption sites, or their irreversible coverage by some strong selective adsorbate (surface deactivation), increases the range of linear adsorption capacity 50- to 100-fold. Water is frequently used to deactivate the polar adsorbents such as alumina and silica; also glycol and glycerine have been employed. In the nonpolar adsorbents, such as charcoal, high molecular-weight fatty acids or alcohols are used to saturate the active sites. Any change in variables which tends to decrease the partition coefficient of the solutes generally increases isotherm linearity. Thus, higher column temperatures and stronger eluents generally favor linearity of isotherms. The effect of other separation variables, such as eluent type, eluent flow rate, and column geometry, is reflected primarily in their relationship to solute band width.

Plate Height The minimum plate height occurs at low velocities in adsorption chromatography as a natural consequence of the small diffusion coefficients in the liquid phase. Although only limited data are reported, for 1-nitronaphthalene eluted by 15% methylene chloride–isooctane from 1% water-deactivated SiO_2, the optimum eluent velocity was approximately 20 mm/min.

Chromatographic Media

The role of the adsorbent in shaping the adsorption isotherm is determined by numerous factors, including the adsorbent surface area and the chemical composition and geometrical arrangement of the atoms or groups that comprise the surface. The stronger adsorbents possess a more polar structure. However, in addition to the polarity of the adsorbent, whether the adsorbent can donate or accept hydrogen bonds can be significant with certain solutes and solvents. Subtle differences in surface shape may also be significant at times.

In selecting a suitable adsorbent, one has to consider the polarity and lability of the solutes being separated. Hydrophilic adsorbents retain a molecule more strongly the more polar groups the molecule contains and

[2] L. R. Snyder, *Advances in Analytical Chemistry and Instrumentation*, Vol. 3, C. N. Reilley (Ed.), Interscience, New York, 1964.

the more unsaturated it is, especially when the molecule contains conjugated double bonds. Weakly polar compounds are therefore chromatographed on an active adsorbent and strongly polar compounds on weak adsorbents. For the particular solutes in a given sample, the adsorbent should be very active and adsorb quickly, yet without acting chemically on any labile adsorbed substance. Adsorption and desorption should occur with relative ease in order that a large number of theoretical plates are developed. In general, if an adsorption–solvent system involves both a strong adsorbent and a powerful eluent, an analogous separation might also occur on a weak adsorbent and a correspondingly weaker eluent. If so, the latter adsorbent–solvent pair would be preferred. If the compounds to be separated have only a slight affinity for the usual adsorbents, it is best to work with an "active" adsorbent and with "weak" eluents. On the other hand, "weak" adsorbents and "strong" eluents must be used if the adsorption affinity of the compounds is great.

Certain adsorbents are considered "acidic," for example, silica and Florisil, while others are regarded as "basic," such as alumina. Since chemisorption of acidic samples on basic adsorbents frequently occurs, acidic adsorbents are preferred for the separation of acids, and basic adsorbents for the separation of bases.

Silica The method of preparation determines the type of silica or silica gel obtained. Porous glass, ground quartz, and sand are nonporous and give powders of low specific surface, whereas acid precipitation from metal silicates produces silica gels of very high specific surface and therefore high adsorptive capacity. For chromatographic work, silica gel (or silicic acid) is the form most used. It is white, so that bands of colored materials can be seen clearly.

Among the chemical groups present on the surface of silica gel it is possible to distinguish several types. Predominant in the usual commercial gel is the silanol (—Si—OH) group. These surface hydroxyl groups interact with polar or unsaturated molecules by hydrogen bonding. These fundamental types of surface hydroxyls are designated as "bound" (I), "reactive" (II), and "free" (III):[3]

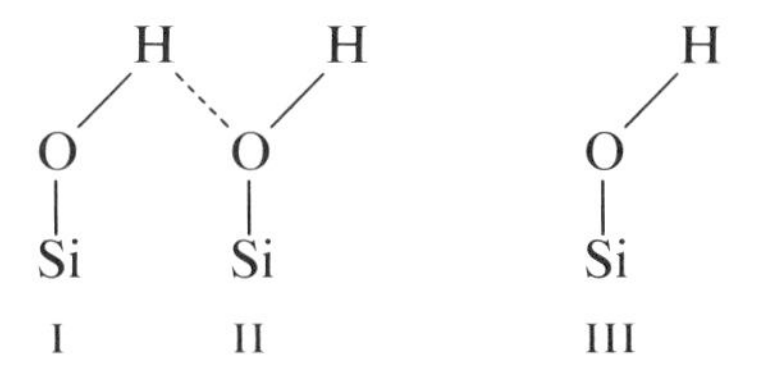

[3] L. R. Snyder and J. W. Ward, *J. Phys. Chem.* **70**, 3941 (1966).

The relative strength of these surface hydroxyls as adsorption sites is in the order I < III < II. Some silanol groups exist also as the hydrated silanol (—Si—OH·OH_2). During the gelation process and subsequent drying of the gel, a certain proportion of these hydroxyl groups will undergo dehydration to form siloxane (—Si—O—Si—) linkages between neighboring silicon atoms. Siloxane groups, being nonspecific (but relatively weak) in their adsorptive properties, adsorb polar, nonpolar, saturated, and unsaturated molecules to the same relative extent.

Alumina Most chromatographic aluminas are a mixture of γ-alumina with a small percentage of alumina monohydrate (and perhaps a little sodium carbonate left during the manufacture). It is not yet possible to specify the precise nature of adsorption sites on alumina. Some workers [4] have attributed the adsorptive properties of alumina to surface hydroxyl groups. Distortion of the hexagonal close-packed lattice of γ-alumina through damage from grinding and general handling leaves "active sites" which consist of positively charged aluminum atoms and negatively charged oxygen atoms exposed on the surface. From the surrounding water, hydroxyl ions will be adsorbed on the positively charged sites and hydrogen ions by the negatively charged sites, both actions producing surface hydroxyl groups which may in part be covalently bound and in part ionized, depending on the extent of damage to the surface. Another viewpoint suggests that adsorption sites arise from exposed aluminum atoms or strained Al—O bonds.[5] In addition to the acidic sites responsible for the adsorption of most polar and unsaturated molecules, the alumina surface also contains basic sites, most likely surface oxide ions.

Three types of alumina are commercially available. Aluminum oxide, neutral (pH 6.9 to 7.1), is a suitable adsorbent for hydrocarbons, esters, aldehydes, ketones, lactones, quinones, alcohols, some nitro compounds, and weak organic acids and bases. It is generally used only with nonaqueous solutions. Basic alumina (pH 10 to 10.5) corresponds to the "technical" oxide which contains sodium carbonate. It is employed primarily for acid-sensitive samples, such as acetals and glycosides. In aqueous or aqueous–alcoholic solutions the basic alumina exhibits strong cation-exchange properties and will adsorb basic amino acids, amines, and other basic substances, as well as inorganic cations. Acid alumina (pH 3.5 to 4.5) is an acid-washed

[4] D. J. O'Connor, P. G. Johansen, and A. S. Buchanan, *Trans. Faraday Soc.* **52,** 229 (1956).

[5] J. B. Peri, *J. Phys. Chem.* **69,** 211, 220 (1965).

preparation which acts as an anion exchanger. It is used for the separation of both inorganic anions and acidic organic molecules in so far as these are not acid-sensitive.

Other Adsorbents Hydroxylapatite, a form of calcium phosphate, finds use primarily for the fractionation and purification of medium and high molecular weight substances encountered in preparative biochemistry. The adsorbent does not retain low molecular weight materials.

Florisil, a magnesium silicate, is an acidic adsorbent which finds some use as a substitute for alumina when handling base-sensitive samples. Its reputation for chemisorption places it among the last choices of the analytical chromatographers.

The adsorptive properties of diatomaceous earth are extremely weak. It is employed only for very polar samples. Frequently it is used in admixture with active adsorbents to promote faster column flow rates. In these cases, it may be regarded as an inert diluent.

Activated carbon exists in two forms: nonpolar charcoal prepared by activation at 1000°C or higher, and polar charcoal prepared by low-temperature oxidation. Nonpolar carbon exhibits a predominantly graphitic surface on which adsorption is the result primarily of London forces; the amount of any one compound adsorbed is governed principally by its volatility or boiling point. The higher the boiling point, the more strongly it is adsorbed. The surface of polar charcoal is covered with various oxygenated groups on which adsorption involves electrostatic forces and hydrogen bonds, similar to the mechanism for metal oxide adsorbents.

Activity Grades of Adsorbents Various activity grades of silica and alumina can be prepared by the controlled deactivation of the fully activated adsorbent with known amounts of water. The term "fully activated" is used for an adsorbent whose adsorptive strength has attained a maximum and cannot be increased by further heating to any temperature. Activity grades according to the Brockmann scale in dependence on water content (as added water) are listed in Table 7-1. It can be seen that more water must be added to the silicic acid for each grade attainment than when alumina is used. The silicic acid therefore has potentially greater selectivity, and a wider range of activities in between the grades listed. The Brockmann activity scale is defined in terms of the action of a series of test dyes on columns or along thin layers of alumina or silica gel. Specific details for this test procedure are given in the laboratory section of this chapter. The R_f values of the test dyes are listed in the laboratory section (Table 7-3). A more rigorous method of denoting adsorbent activity is discussed in the next section.

TABLE 7-1. Activity of Silicic Acid and Alumina in Dependence on Water Content (as Added Water)

% Water added to silicic acid	*Activity grade*	*% Water added to alumina*
0	I	0
5	II	3
15	III	6
25	IV	10
38	V	15

Fundamental Basis of Retention Volume in Adsorption Chromatography

One factor in the separability of a pair of compounds is their difference in retention volume. The major theoretical problem in adsorption chromatography is, therefore, the dependence of solute retention volume on solute molecular structure and the experimental separation conditions.

The Role of the Adsorbent Equilibrium within the chromatographic column may be envisaged in two ways: the adsorbed phase as a continuous medium of adsorbent plus adsorbed material, or the adsorbed phase as a liquid coating (exclusive of adsorbent). There is an abundance of experimental evidence which indicates that the adsorbed phase consists of a monolayer covering the entire adsorbent surface to an average thickness of 3.5Å. With this value, the adsorbent surface volume V_a, in milliliters per gram, is given by $V_a = (3.5 \times 10^{-8}\ \text{cm}) \times (\text{surface area, in cm}^2/\text{g})$, for a calcined adsorbent, and by $V_a = (3.5 \times 10^{-8})(\text{surface area}) - 0.01(\%H_2O)$ when some quantity of water is added to the calcined adsorbent to produce a water-deactivated adsorbent.

For column elution, the retention volume, V_R, is given by

$$V_R = V_M + A_S K_d \qquad (7\text{-}2)$$

where V_M is the volume of the solution phase and A_S is the surface area of the adsorbent (which is proportional to the total weight of the adsorbent). Rearranging Eq. (7-2)

$$(V_R - V_M)/A_S = K_d \qquad (7\text{-}3)$$

Snyder[6] has derived a relationship between K_d and the adsorbent activity

[6] L. R. Snyder, in *Advances in Analytical Chemistry and Instrumentation*, Vol. 3, C. N. Reilley (Ed.), Interscience, New York, 1964.

or surface-energy function ϕ, and a solute-eluent parameter f:

$$\log K_d = \log V_a + \phi f \tag{7-4}$$

Values of V_a may be calculated for a particular adsorbent as a function of its water content. With ϕ set equal to 1.00 for the calcined adsorbent, the eluent–solute parameter f can be evaluated from a standard reference solute–eluent combination and the experimental value of K_d for the calcined adsorbent. Adsorbent activity for the adsorbent deactivated by any quantity of water may then be determined from an experimental value of K_d for elution of the same reference solute by the same eluent.

Example 7-1

Using naphthalene-pentane as the standard solute–eluent system on alumina whose surface area, as measured by nitrogen (BET) adsorption, is 155 meter^2/g, the experimental value of K_d was determined to be 68. For calcined alumina,

$$V_a = (3.5 \times 10^{-8} \text{ cm})(155 \times 10^4 \text{ cm}^2/\text{g}) = 0.0542 \text{ cm}^3/\text{g}$$

With $\phi = 1.00$ for the calcined alumina, the solute-eluent parameter is evaluated from Eq. (7-4):

$$\log 68 = \log 0.0542 + 1.00f$$

and $f = 3.10$.

Now for a deactivated alumina of 3% water content,

$$V_a = (3.5 \times 10^{-8})(155 \times 10^4) - 0.01(3.0) = 0.0242 \text{ cm}^3/\text{g}$$

and, from the experimental value of $K_d = 3.3$ for the deactivated adsorbent,

$$\log 3.3 = \log 0.0242 + 3.10\phi$$

Thus, the adsorbent activity ϕ is 0.69 of the fully activated value.

The Role of the Eluent The choice of the mobile phase in adsorption chromatography is intertwined with the competition between solute and eluent molecules for a place on the adsorbent surface. The adsorption of a solute molecule requires concomitant desorption of one or more eluent molecules, the exact number determined by the relative areas required by solute and eluent on the adsorbent surface. The solute–eluent parameter f of Eq. (7-4) reflects the free energy for the adsorption–desorption equilibrium. It can be related to three parameters. The parameter $S°$ is the dimensionless adsorption energy of the solute from pentane solution onto adsorbent of standard activity ($\phi = 1.00$). The parameter $\epsilon°$ is the eluent strength of the solvent and equal to the adsorption energy of solvent per unit area of adsorbent surface. A_i is the effective molecular area of the solute. These parameters are related as follows:

$$f = S^\circ - \epsilon^\circ A_i \tag{7-5}$$

Substitution of Eq. (7-5) into Eq. (7-4) gives an expression that relates K_d to solute structure and to the conditions of separation:

$$\log K_d = \log V_a + \phi(S^\circ - \epsilon^\circ A_i) \tag{7-6}$$

The quantitative application of Eq. (7-6) requires the evaluation of A_i and ϵ° for solutes and eluents of interest. This has been done for many substances by Snyder. Some eluent parameters are listed in Table 7-2 which gives the classical "eluotropic series" of solvents.

One prediction, based on Eq. (7-6), is that for a given solute and absorbent activity, upon elution by two eluents, the change in log K_d is given by

$$\log (K_d)_1/(K_d)_2 = \phi A_i(\epsilon_2^\circ - \epsilon_1^\circ) \tag{7-7}$$

TABLE 7-2. Eluent Parameters ("Eluotropic Series")

Increasing elutive power ↓		ϵ°	
		Al_2O_3	SiO_2
	n-Pentane	0.00	0.00
	Cyclohexane	0.04	−0.05
	Carbon tetrachloride	0.18	0.14
	Carbon disulfide	0.26	0.14
	Benzene	0.32	0.25
	Ethyl ether	0.38	a
	Chloroform	0.40	a
	Methyl isobutyl ketone	0.43	a
	Acetone	0.56	a
	Dioxane	0.56	a
	Ethyl acetate	0.58	a
	Amyl alcohol	0.61	a
	Diethyl amine	0.64	a
	Acetonitrile	0.65	a
	Pyridine	0.71	a
	Butyl cellosolve	0.74	a
	Isopropanol	0.82	a
	Ethanol	0.88	a
	Methanol	0.95	a
	Ethylene glycol	1.11	a

[a] $\epsilon^\circ > 0.25$.

From L. R. Snyder, *J. Chromatog.* **8,** 178 (1962); *ibid.* **11,** 195 (1963).

and will be proportional to the effective molecular area of the solute and to the change in eluent strength of the solvent.

In general, adsorption effects are strongest in nonpolar solvents. Therefore, one often applies the sample as a solution in a nonpolar solvent of relatively weak eluting power. Development of the chromatogram may be carried out with the same solvent, or with a slightly stronger eluent. It should be possible to elute bands successively by increasing gradually the eluent strength when a wide range of solute adsorptivities are encountered. Mixtures of two solvents of differing eluting power may give better separation than a single solvent. The stronger eluent is added to the developing solvent in increasing amounts (see "gradient elution" technique in Chapter 4).

The eluents must be of high purity and free of moisture, since water is adsorbed strongly by most adsorbents. Traces of alcohol (in ethyl acetate, ether, or chloroform), peroxides (in ethers and hydrocarbons), olefins (in commercial hydrocarbons), heterocyclic compounds and phenols (in aromatic hydrocarbons), or free acids (in halogenated hydrocarbons) can all hamper proper separations.

Solute Adsorption Affinity The molecular structure of the solute is more important than either the adsorbent or the eluent in determining the sample separation order and retention volume. The role of solute structure in determining retention volume is expressed in Eq. (7-6) by the solute parameter $S°$. In addition to measuring $S°$ from an experimental value of K_d using any adsorbent–eluent combination whose parameters are known, $S°$ can be also estimated from adsorption energies of various sample groups. These adsorption energies have been tabulated for alumina and silica gel by Snyder.[7]

To a first approximation, the adsorption energy of a molecule can be assumed equal to the sum of the adsorption energies of its individual constituent atoms or groups. Substitution onto aromatic versus aliphatic skeletons must be distinguished. Alkyl groups are adsorbed weakly on polar adsorbents. The introduction of double bonds raises the adsorption affinity to an extent that increases with their number and degree of conjugation and coplanarity. In aromatic ring systems, the adsorption affinity increases with the number of rings and the degree of resonance.

The introduction of a single functional group into a hydrocarbon skeleton generally raises the overall adsorption affinity, without altering the order of adsorption in terms of double bonds, ring systems, and degree of conjugation already present in the skeleton. On alumina the order of increasing adsorption affinity for an aromatic skeleton is: methyl (methyl-

[7] L. R. Snyder in *Chromatography*, 2nd ed., E. Heftmann (Ed.), Reinhold, New York, 1967, p. 64.

ene), halogeno, sulfide, ether, nitro, nitrile, aldehyde, ester, keto, amino, amide, hydroxyl, thiol, carboxyl. Hydroxy and amino groups are in the reverse order for saturated compounds. Except for thiol and carboxyl groups, the same general order prevails for silica. On silica, thiol falls between sulfide and ether, and carboxyl is much weaker in adsorption affinity and lies under the values for amino and amide groups.

The rigidity of surfaces, as contrasted with the fluidity of solutions, affects the adsorbability when two or more functional groups are present in the same molecule. When the adsorption energy of the stronger group exceeds some minimum value, the stronger group will localize on adsorption sites to the exclusion of the more weakly adsorbing groups in the sample molecule. On alumina, linear molecules are more strongly adsorbed than angular molecules. No such effect is observable on silica.

The presence of a second functional group in a molecule can also change the adsorption energy of the original group(s) as a result of electronic interaction between the groups. This effect is most conspicuous in aromatic molecules where the functional groups are part of, or substituted on, the aromatic ring. For silica and alumina this normally leads to increasing sample adsorption energies when the original group is electron donating and to decreasing energies when it is electron accepting. Steric interaction between adjacent functional groups has to be taken into account. Planar molecules are preferentially adsorbed. Internal hydrogen bonding generally tends to weaken bond formation with the adsorbent.

A general model for correlating sample separation order with structure of solute, type of adsorbent, and eluent strength has been advanced by Snyder.[8]

Liquid Chromatograph

The availability of high-sensitivity detector systems for liquid chromatography (partition as well as adsorption) enables certain inherent advantages of liquid chromatography over gas chromatography to be fully exploited. Materials with adverse vapor pressures can easily be analyzed, while materials which previously had to be pyrolyzed prior to gas–liquid chromatography and those which decompose at the high temperatures required for gas–liquid separations can now be analyzed directly at or near ambient temperatures by liquid chromatography.

In the Barber–Coleman unit (Fig. 7-2) the liquid effluent from the chromatographic column is collected on a continuously circulating platinum chain which revolves (1–22 cm/min) in a clockwise direction. The chain conveys the effluent through an evaporation chamber consisting of a

[8] L. R. Snyder, *J. Chromatog.* **20,** 463 (1965); *Advances in Chromatography,* Vol. 4, J. C. Giddings and R. A. Keller (Eds.), Dekker, New York, 1967; *Advances in Analytical Chemistry and Instrumentation,* Vol. 3, C. N. Reilley (Ed.), Interscience, New York, 1964.

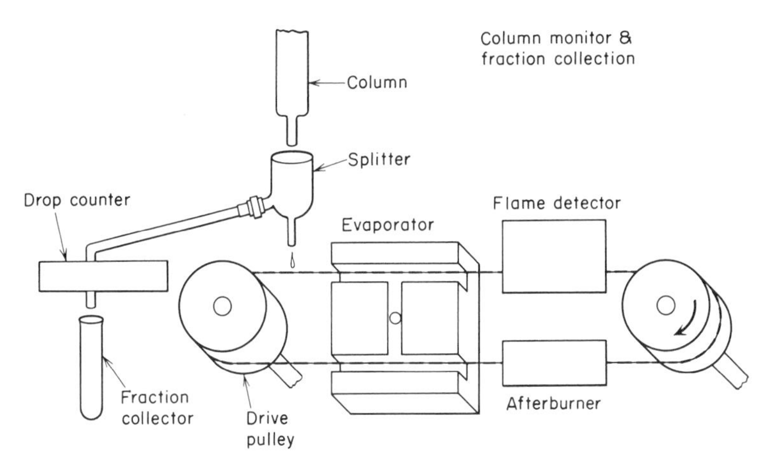

FIG. 7-2. Schematic diagram of Barber–Coleman liquid chromatograph. Column monitoring and fraction collection would be omitted with micro columns. (Courtesy of Barber–Coleman Co.)

grooved stainless steel block incorporating both cartridge heaters and a nitrogen gas jet stream to remove the liquid mobile phase prior to sample detection. The relatively non-volatile sample left on the chain after the eluting solvent has evaporated is transported through a hydrogen flame detector where the sample is burned. Sample combustion results in an ionization current which is proportional to the amount of sample passing through the detector (see Chapter 11). Any residual material left on the chain after sample combustion in the flame detector is removed by passing the chain over a hydrogen afterburner.

In the Pye Liquid Chromatograph (Fig. 7-3) a travelling stainless steel wire, 0.13 mm in diameter, is led from a spool through a glass tube (3.2 mm i.d.) surrounded by a cleaning furnace which is electrically heated and maintained at 650–700°C. A slow stream of argon removes any organic matter initially present on the wire. The cleaned wire then passes round a pulley and through a metal block fitted with a hypodermic needle, to which the column is attached. A slot is cut in the end of the needle and the wire passes through this slot so that it continuously moves through the eluate leaving the column. The wire then passes through the evaporator tube to remove the solvent, and then to the pyrolysis tube which is swept by argon from both ends. The solute deposited on the wire is pyrolyzed and swept by the argon through the exit at the center of the pyrolysis tube to the detector—an argon or flame ionization detector. The wire then passes out of the tube and is wound onto a take-up spool. The wire travels at a fixed, but selectable speed, usually 76 mm/sec. A spool contains 9 miles of wire, enough for 90 hr of continuous running. Wire speed and pyrolysis tempera-

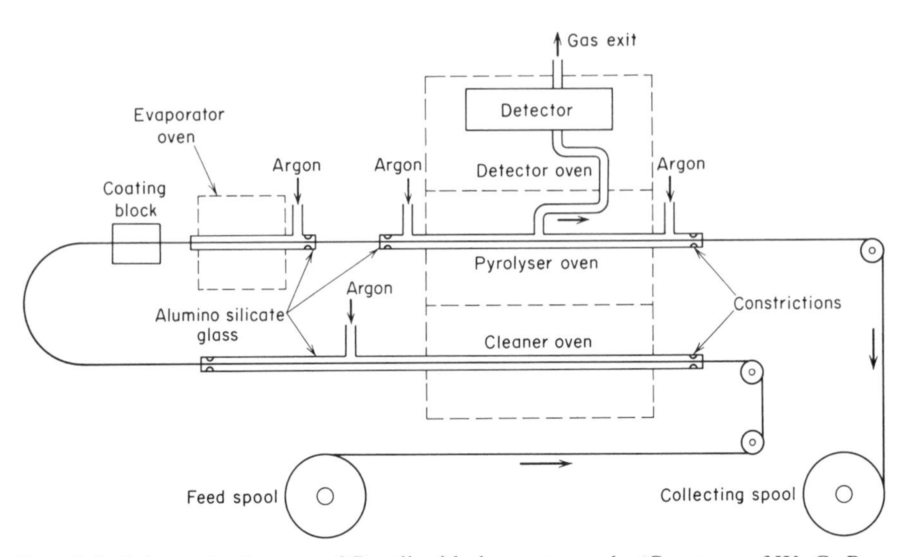

FIG. 7-3. Schematic diagram of Pye liquid chromatograph. (Courtesy of W. G. Pye & Co., Ltd.)

ture are interrelated with the particular compound in the effluent. For high wire speeds the mass of solute transported per unit time is increased and results in high sensitivities for low-boiling or low-concentration samples. For the higher-boiling compounds the response may pass through a maximum when too low a pyrolysis temperature causes inefficient pyrolysis to occur at the higher wire speeds.[9]

With either commercial chromatograph only a very small fraction of the effluent (about 4 μl/min) is required for the detector. The effluent from an analytical or preparative column is split with the major portion directed to the collector and the minor part collected on the wire or chain. This provides a convenient monitoring chromatogram of the column effluent material. An event marker on the recorder actuated by the drop counter or volumetric dispenser of the fraction collector (Chapter 4) indicates the indexing of the collection tubes. For micro columns the total effluent may be deposited on the wire or chain. Quantification of the resolved peaks is made by conventional techniques (Chapter 11). The solvent must have a lower boiling point than the solute and generally a lower surface tension than water. A mixture of water with 5% or more alcohol or any other water-miscible organic solvent is suitable. The type and polarity of the solvent have no effect on the operation of the liquid chromatograph. This obviates the necessity for comparative columns and enables changes of solvent and solvent flow to be made during an analysis. The very short time needed to

[9] T. E. Young and R. J. Maggs, *Anal. Chim. Acta* **38,** 105 (1967).

detect the sample and the small dead volume between the column and the detector prevent excessive peak broadening by diffusion. With present detectors, analysis is restricted to vaporizable or pyrolyzable organic compounds.

Applications

Chromatography on adsorbents such as silica gel and alumina is used primarily for the fractionation of mixtures of rather nonpolar substances into classes of compounds that migrate at different rates by virtue of their respective molecular structures. Separation can be achieved if sample components exhibit differences in (1) number and polarity of constituents and their steric positions, (2) number and location of carbon–carbon double bonds, and (3) spatial configuration. As a rule, adsorption methods are first used for a rough fractionation of the mixture. In crude extracts especially, the mutual competition of substances for active sites on the adsorbent may be an advantage.

Silica gel will remove selectively aromatic compounds from paraffinic and naphthenic materials. It is the best adsorbent for unsaturated hydrocarbons where isomerization and polymerization must be avoided. Alumina gives a better separation among aromatics. With saturated mixtures, molecular sieves (Chapter 12) will be more useful.

Adsorbents impregnated with silver nitrate separate lipid classes according to the degree and geometry of their unsaturation. The clear separation of *cis*- and *trans*-monoenoic esters has been achieved, as well as their separation from saturated and polyunsaturated esters. In addition, saturated, ethylenic, and acetylenic epoxy and hydroxy esters have been resolved. By double impregnation with silver nitrate and boric acid, one can achieve simultaneous separation of *threo* and *erythro,* saturated, and unsaturated dihydroxy esters.

Adsorption chromatography has met with a relatively large amount of success for the separation of optical, geometrical, and diastereoisomers. Separation is accomplished either by an optically active adsorbent or by transforming the racemic compound into a mixture of two diastereoisomers by combination with a *d*- or *l*-compound, followed by passing the mixture through a column of an optically inactive adsorbent.

Adsorption chromatography is the method of choice for the separation and isolation of various unsaponifiable components of lipophilic substances (fat-soluble), such as lipids and steroids. Separation takes place according to the type and number of substituent groups. Adsorption chromatography is the ideal tool when oxygenated functions are present. On alumina and silicic acid columns the order of increasing adsorbability is hydrocarbons, cholesterol esters, triglycerides, cholesterol, free acids, and phosphatides.

Specially treated silicic acid is often employed for these separations and these preparations are available commercially. A favorite solvent has been petroleum ether with increasing proportions of diethyl ether, and final elution with methanol. A column conditioned by washing with pure benzene, followed by 5% benzene in hexane, will separate the cholesterol esters of palmitic, oleic, linoleic, and arachidonic acids.

Magnesium silicate is very suitable for the separation of acetylated sugars as well as steroids and essential oils. It has also been used as a pesticide cleanup prior to gas–liquid chromatography.

Considerable difficulty is encountered when adsorption chromatography is applied to hydrophilic compounds. Severe adsorption effects occur between the solute molecules and the adsorbent. This results in very broad and unsymmetrical elution peaks which limit the degree of separation. Partition chromatography is indicated for these materials.

LABORATORY WORK

Preparation of an Adsorbent Column

A glass or Teflon tube, tapered at the bottom, is really all that is needed in the way of apparatus. The performance of a column, however, can be improved by using tubes with a glass frit or Teflon filter disk. To avoid the possibility of zone remixing the column should have a minimum amount of dead space at the outlet. A needle-valve or screw clamp on Tygon tubing regulates the solvent flow. Stopcocks and their lubricants must be avoided.

To load the column, fill it about half full with the solvent chosen. Push a wad of glass wool to the bottom with a dowel. Dust in a little sea sand through a funnel and level the surface by tapping. These steps can be omitted if the tube has a frit or filter disk.

Columns may be prepared by dry- or wet-filling. The latter method is recommended because of its simplicity and the ease with which homogeneously packed columns can be obtained. A thin, free-flowing slurry prepared from the adsorbent (150–200 mesh) and the solvent is poured in several portions into the chromatographic tube. The tube should be gently tapped and rotated while the adsorbent is settling. When the solvent rises too high in the tube, part of it is drained off, but during the settling period the flow of solvent from the column should be discontinued. Addition of slurry is continued until the desired quantity of adsorbent has been poured into the column. As the column builds up the flow rate decreases and pressure may have to be applied from a cylinder (lecture-size) of compressed gas via a suitable pressure-reducing valve. Suction on the lower end of the column should never be employed since this can lead to cracks and gas bubbles in the presence of volatile solvents.

While the adsorbent packing is settling, run out excess solvent. The column is repeatedly refilled with solvent until the adsorbent no longer shows any tendency to settle. During this process the adsorbent bed must ALWAYS REMAIN COVERED WITH SOLVENT. Finally, a disk of filter paper and a layer of glass beads are placed on top of the adsorbent bed to maintain a level surface and prevent agitation of the adsorbent during further addition of solvent and samples. The solvent is then allowed to run down to the top of the glass beads.

The ratio of diameter of adsorbent packing to its height is usually between 1:4 and 1:8. The ratio of adsorbable ingredients to adsorbent can range around 1:30 for alumina to 1:60 for silica gel.

The purity of solvents is extremely important. All organic solvents should be dry and freshly distilled. A number of solvents are available commercially in a "chromatographic grade."

To avoid "wall effects" the column tube should be coated with a solution of 1% dichlorodimethylsilane in benzene. The silane solution is heated to 60°C, poured into a clean glass column, allowed to stand for several minutes, the solution decanted and the benzene evaporated in a drying oven. The entire procedure is then repeated to double the coating which will last for several months without further treatment.

Solvent is fed into the top of the column from a reservoir. If gravity flow proves insufficient to maintain a suitable flow of mobile phase, solvent can be fed into the column by a constant-volume flow pump or an overpressure from a small cylinder of inert gas can be kept above the stationary phase. Arrangements for gradient elution techniques are described in Chapter 4.

Preparation of Alumina and Silica Gel in Different Activity Grades

The fully activated grade of alumina is prepared by heating in a porcelain dish small portions of alumina (chromatographic grade such as Alcoa F-20) at 500°C for at least 3 hr with frequent stirring, then cooling to room temperature in a vacuum desiccator. An appropriate amount of water is added to the fully activated (grade I) adsorbent to prepare less-active grades.

The fully activated grade of silica gel is prepared by heating the chromatographic grade of silica gel (such as Davison Code 12 or Code 62) at 400°C for 16 hr, cooling to room temperature in a vacuum desiccator, and then storing in a tightly sealed container. An appropriate amount of water is added to this fully activated (grade I) gel to prepare less-active grades.

To prepare intermediate activity grades, appropriate amounts of water (Table 7-1) are added to known weights of fully activated adsorbent. For the preparation of an adsorbent containing $X\%$ free water, X parts of distilled water are placed in a bottle and $100 - X$ parts of fully activated ad-

TABLE 7-3. Activity Scale of Alumina and Silica Gel

Position on column	*Activity grade*								
	I	*IIa*	*IIb*	*IIIa*	*IIIb*	*IVa*	*IVb*	*IVc*	*V*
Top	2	...	3	...	4	...	5	...	6
Middle	...	2	...	3	...	4	...	5	...
Bottom	1	...	2	...	3	...	4	...	5
In effluent	...	1	...	2	...	3	...	4	...

Numbers refer to dyes: 1, azobenzene; 2, *p*-methoxyazobenzene; 3, Sudan yellow; 4, Sudan red III; 5, *p*-phenylazoaniline; 6, *p*-phenylazophenol.

sorbent are added at once. The bottle is closed tightly, shaken until the lumps formed upon mixing disappear, then permitted to equilibrate for 24 hr on a mechanical shaker.

To determine the Brockmann activity grade of alumina and silica gel, six standard dyes, used in pairs, are applied to a column of adsorbent and washed through by a series of solvents. A chromatographic tube, 10 cm long by 1.5 cm i.d., is packed to a height of 5 cm with the dry adsorbent and covered with a disk of filter paper. The test mixture, 2.0 mg of each dye in 2 ml of benzene and 8 ml petroleum ether, is applied to the column and developed with 20 ml of solvent consisting of 16 ml benzene mixed with 4 ml petroleum ether. The flow rate should be 1.0 to 1.5 ml per min. Table 7-3 shows the position of the dyes at each activity grade. These test kits are available commercially. Table 7-4 lists the R_f values of these same test dyes.

TABLE 7-4. R_f Values of Test Dyes on Alumina or Silica Adsorbents

Dye	*Activity on the Brockmann-Schodder scale*			
	II	*III*	*IV*	*V*
Azobenzene	0.59	0.74	0.85	0.95
p-Methoxyazobenzene	0.16	0.49	0.69	0.89
Sudan yellow	0.01	0.25	0.57	0.78
Sudan red	0.00	0.10	0.33	0.56
p-Aminoazobenzene	0.00	0.03	0.08	0.19

From S. Hermanek, V. Schwarz, and Z. Cekan, *Coll. Czech. Chem. Commun.* **26,** 3170 (1961).

Elution Curves

A column of suitable dimensions is prepared from the adsorbent selected by wet-filling with the eluting solvent to be used. Sufficient pressure is applied to maintain a flow rate of 1–2 ml per min in the empty part of the tube. As soon as the solvent is about to disappear in the layer of glass beads, the pressure is discontinued by opening the needle valve in the pressure line. The sample is now pipetted onto the column. The same pressure as before is again applied and the collection of filtrate fractions is started. Eluting solvent is added, first in very small portions until the liquid above the adsorbent no longer contains any sample, then in larger amounts. A number of filtrate fractions are collected by suitable means and their volumes are exactly measured. Determine the concentration of the solutes in each fraction by appropriate analytical methods. Elution curves are established by plotting the total amount of solute eluted as well as the solute concentration in each fraction against the amount of filtrate collected.

The developing solvents can either be added manually to the column or a separating funnel, filled with solvent, can be placed on top of the column. By closing the head of the funnel with a stopper and opening the stopcock at the bottom, the solvent level in the column will adjust itself automatically to the desired height. Then, the liquid can flow out only if the solvent level falls below a prescribed point, which causes an air bubble to enter the tip and migrate into the funnel. This results in the automatic discharge of an appropriate amount of solvent. Details are shown in Fig. 5-7.

In Fig. 10-4 is depicted a nomograph by means of which the required solvent may be found if one knows the activity of the adsorbent and the class of compound being separated.

Suggested Systems

(1) *Adsorptive:* 100 μg of benzo[α]pyrene in 200 μl of eluent. *Adsorbent:* 2.4 g of silica gel containing 3, 6, or 12% of water. *Eluent:* Cyclohexane containing 8% (*v*/*v*) of benzene. Determine the absorbance at the 296-mμ peak for which $\epsilon_{max} = 7.67 \times 10^4$. For details see *Anal. Chem.* **29,** 1309 (1957).

(2) *Adsorbent:* Alumina, grade I; column 75 cm long by 1 cm in diameter. *Eluent:* Petroleum ether (60°–80°C) containing 10% (*v*/*v*) of chloroform. *Adsorptives:* 2.5 mg load consisting of fluorene (300 mμ), phenanthrene (293 mμ), pyrene (334 mμ), fluoranthene (287 mμ), anthracene (375 mμ), and carbazole (291 mμ). The order of elution is as listed and the most characteristic absorption maxima are given in parentheses. *Procedure:* Examine the ultraviolet absorption spectra of successive 50-ml fractions of eluate between 280 and 380 mμ. On irradiation with an UV lamp, fluoranthene shows a pale blue fluorescent zone and anthracene gives a well-defined deep blue fluorescent zone.

(3) Purification of crude 2-aminoanthraquinone can be accomplished on activated alumina with aqueous pyridine as the solvent. 1-Aminoanthraquinone appears as a fast-travelling pale yellow band which is followed by 2-aminoanthraquinone and 1,2- and 1,4-diaminoanthraquinones, the last two giving tightly held pale violet bands.

(4) Separation of mixtures containing products of a reaction or just two colorless solids. The effluent can be examined by change in refractive index or through UV absorption between 220 and 380 mμ. Adsorbents can be alumina or silica gel (grade II or III) and the eluent benzene. After the initial run, the student may wish to modify the procedure to achieve a more complete and perfect separation. However, for the first chromatogram, use 25 g of adsorbent, 0.5 g of the unknown mixture, and collected two 10-ml fractions with benzene as eluent, then two 10-ml fractions with methanol as eluent. Each eluate is evaporated to dryness in the hood by warming on a steam bath. Determine the melting range of the residue and compare with the melting points of the pure materials. Suggested binary mixtures: biphenyl (69°–71°C) and acetanilide (114°C); biphenyl and benzanilide (161°C); *p*-dimethoxybenzene and acetanilide; benzophenone (49°–50°C) and biphenyl; acetanilide and *p*-dibromobenzene (87°C).

(5) Separation of some dyestuffs on a column prepared from 28 g alumina (60–80 mesh), activity grade II or III. Column is washed with 95% ethanol. Sample containing 5 mg each of fluorescein and methylene blue, dissolved in 3–4 ml ethanol, is added to the top of the column. Ethanol is used as developer until the effluent becomes colorless following elution of the methylene blue band. The receiver is changed and water is used as eluent until the fluorescein has completely eluted. Methyl orange (1 mg) and 5 mg of methylene blue, dissolved in 2.2 ml of ethanol, may be separated as described above; also a mixture of methyl orange and Victoria Blue B (1 mg).

Estimation of Adsorption Isotherms

In practice the degree of adsorption is estimated by shaking a known amount of the adsorbent with the solution containing the substance in question. Fifteen minutes is usually sufficient for the attainment of equilibrium. After this period the adsorbent is removed by decantation and centrifugation, and the decrease in concentration of the solute is estimated. Data obtained at varying concentrations of solute are plotted as adsorption isotherms.

Problems

1. Two systems have been found to give a constant adsorption isotherm: (a) the adsorption of disperse dyes onto hydrophobic fibers, such as cellulose acetate,

and (b) the adsorption of amino acids and peptides on the calcium (but not on the hydrogen) form of montmorillonite. Suggest reasons for the behavior observed.

2. Predict the chromatographic separability order of these halogen derivatives of fluorescein on alumina: Eosin (4Br), erythrosin (4I), phloxin (4Br, 4Cl) and rose bengale (4I, 2Cl).
3. Suggest a reason for the stronger adsorption of 2-aminoanthraquinones on silica as compared to the 1-isomer.
4. Suggest a reason for the preferential adsorption of such compounds as the *o*-dihydroxybenzenes and *o*-hydroxy- and *o*-aminoazobenzenes on alumina.
5. Suggest a reason for the observation that *o*-alkylphenol derivatives are less strongly adsorbed on silica than *p*-isomers.
6. In the adsorption of pyrrole derivatives on alumina, bulky groups adjacent to the nitrogen actually increase adsorption. Suggest a reason.
7. Which are preferentially adsorbed on silica and alumina, the 2-vinyl-, 2-phenyl-, and 2-naphthylnaphthalenes or their 1-isomers?
8. Calculate the partition coefficient (K_d) for the system naphthalene(solute)–*n*-pentane(eluent) water deactivated silica ($V_a = 0.342$ ml/g) from these equilibrium data for the solute concentrations:

Adsorbed phase, (g/g) $\times 10^6$:	459	153	38.0	15.3	3.82
Nonadsorbed phase, (g/ml) $\times 10^6$:	161	52.5	13.4	5.36	1.32

9. What are the partition coefficients for the adsorbents (a) D-12 4.6% H_2O–SiO_2, (b) D-12 15.0% H_2O–SiO_2, and (c) D-62 1% H_2O–SiO_2 (see Problem 14). The experimental value of K_d for elution of ethoxybenzene by pentane from D-12 6.9% H_2O–SiO_2 is 45.6. For this adsorbent, $V_a = 0.23$ ml/g, and $\phi = 0.69$.
10. Predict the effect of eluent and adsorbent activity change on K_d for the elution of methyl benzoate by these adsorbent–eluent combinations (see Problem 11 for other data). $K_d = 174$ for elution from 16% H_2O–SiO_2 and $A_i = 14.9$ for methyl benzoate.

System	Adsorbent	Eluent	$\epsilon°$
	D-12 16% $H_2O—SiO_2$	*n*-Pentane	0.00
I	D-12 1.6%	50% Benzene–pentane	0.207
II	D-12 16%	10% Benzene–pentane	0.077
III	D-12 16%	25% Benzene–pentane	0.137
IV	D-12 16%	Benzene	0.250
V	D-62 1%	*n*-Pentane (see Problem 14)	0.00

11. The partition coefficient is 12.6 for acenaphthylene eluted by pentane from D-12 grade 4.6% water-deactivated silica. Predict the K_d values for elution of acenaphthylene with pentane from three other code 12 adsorbent samples: (a) 1.0% water-deactivated silica, (b) 8.0% water-deactivated silica, and (c) 12% water-deactivated silica. For code 12 silica these activity functions and adsorbed volume were found from experimental measurements of naphthalene eluted by pentane:

% H_2O-SiO_2	Activity function, ϕ	Adsorbed volume, V_a	K_d (ml/g)
0.0	1.00	0.30	32
1.0	0.94	0.29	21
2.0	0.85	0.28	14
4.0	0.75	0.26	9.0
7.5	0.68	0.22	5.1
10.0	0.64	0.20	4.0
15.0	0.61	0.15	2.6

12. In Problem 11, calculate solute–eluent function (f) for naphthalene–pentane.
13. In Problem 11, estimate the surface area of the calcined code 12 adsorbent.
14. For code 62 silica, which has had 1.0% water added to the calcined adsorbent whose surface area is 287 m^2/g, K_d for the naphthalene–pentane elution was found equal to 2.6. Calculate K_d for acenaphthylene eluted by pentane.
15. Estimate the limits which must be imposed on sample size to maintain linearity on uniform surfaces. Assume a column with radius 0.25 cm and length 100 cm, generating 1000 plates, with sites spaced 3×10^{-8} cm apart, an adsorbent with 10^2 meters/g, and 10% coverage without nonlinearity.
16. Suggest an explanation for these two observations: (a) Very weakly adsorbing solutes show a high linear capacity; (b) pure compounds show a lower linear capacity than do mixtures of compounds.
17. On fully activated silica gel, of surface area 155 m^2/g, $K_d = 68$ for naphthalene as solute and pentane as eluent. (The solute–eluent parameter, f, was found to be 3.10.) These values of K_d were obtained on water-deactivated alumina: 30 (0.5% water); 17 (1.0%); 7.1 (2.0%); 3.3 (3.0%); 1.3 (4.0%). (a) Calculate the adsorbent surface volume and the adsorbent activity for each grade of alumina. (b) Plot the variation of adsorbent surface activity against the coverage of surface by water.
18. What is the ratio of retention volumes of the solute coronene, $C_{24}H_{12}$ from 3.7% $H_2O-Al_2O_3$ by the eluents CCl_4 and benzene(B). $A_i = 15$.
19. Deduce the configuration of the adsorbed solute (C_6H_5—S—$C_{10}H_{20}$—S—C_6H_5). It is eluted from 3.7% water-deactivated alumina by the eluents pentane and CCl_4. With both ends anchored, $A_i = 21$; but with only one end anchored to the adsorbent surface, $A_i = 8.5$. An experimental ratio of 230 was observed for $(K_d)_{pentane}/(K_d)_{CCl_4}$.

BIBLIOGRAPHY

Cassidy, H. G., *Fundamentals of Chromatography,* Vol. X in A. Weissberger (Ed.), *Technique of Organic Chemistry,* Interscience, New York, 1957.

Mair, B. J., *Chromatography: Columnar Liquid–Solid Adsorption Processes,* Chap. 34 in Part I, Vol. 3 of I. M. Kolthoff and P. J. Elving (Eds.), *Treatise on Analytical Chemistry,* Interscience, New York, 1961.

Snyder, L. R., *Adsorption,* Chap. 4 in E. Heftmann (Ed.), *Chromatography,* Reinhold, New York, 1967.

8 *Liquid–Liquid Partition Chromatography*

In liquid–liquid partition chromatography a separation is effected by distribution of the sample ingredients between a stationary liquid phase, usually held in place by an inert scaffold of solid particles, and a mobile liquid phase, the two chosen to be of limited miscibility. Liquid column methods, as opposed to sheet methods, are the subject matter of this chapter.

The Chromatographic Column

The partition chromatographic column consists of a finely divided solid (the support) on which a solvent is fixed with such tenacity that it will not migrate. A second liquid phase (mobile phase), immiscible with the stationary liquid substrate, flows over the latter. Packed granular material serving as support provides contact over a very large interface. The components of a sample mixture participate in a partition between the stationary phase, where they are held in a fixed position, and the mobile phase, where they migrate. Those components which partition more readily into the stationary phase are retarded in their passage through the column with respect to those which partition more into the mobile phase.

The Support The essential requisites of a support are its ability to hold a certain amount of stationary liquid phase, as well as insolubility in both the mobile and stationary liquids. Preferably the support should be a porous granular solid, of particle size lying between 100 to 300 mesh, which can hold over 50% of its dry weight of the stationary liquid. As supports, kieselguhr, diatomaceous earth, cellulose powder, and silica gel have proved especially suitable. Silica gel will adsorb up to 70% of its weight of water, for example, without becoming wet in the ordinary sense. The product is still a dry powder which can be packed into a chromatographic tube in exactly the same way as solid media were handled in adsorption chromatography. Cellulose powder may be envisioned as consisting of connected "puddles" of swollen fibrous material when treated with a liquid. By contrast, kieselguhr, or diatomaceous earth, holds the stationary liquid in droplet form between the spines and holes of the diatom skeletons.

The Partitioning Liquids In normal partition columns, selection of solvents involves a hydrophilic solvent held stationary by the support, and usually a hydrophobic solvent serving as the mobile phase. Solvents may be classified on the basis of their ability to form hydrogen bonds. Arranged according to this property, solvents form a series similar in some respects to the eluotropic series for adsorption chromatography. At the head of this series (Table 8-1) are solvents which are either donors or acceptors of electron pairs and have the ability to form intermolecular hydrogen bridges, whereas at the end of the series the solvents are lacking this property.

The solvent pair selected should possess a low mutual solubility. Both phases should exhibit a high absolute solubility towards the sample ingredients; if they do not, the column loading capacity will be low. Each phase must be saturated with the other at the operating temperature.

In many applications, water is bound to the support and a water-immiscible organic solvent forms the mobile phase. Such a system is applicable to the separation of hydrophilic substances and substances of medium polarity. It is possible to add buffers to the stationary water phase for ad-

TABLE 8-1. Solvent Series for Liquid–Liquid Partition Chromatography

Decreasing ability to form hydrogen bonds ↓		
	Water	*n*-Amyl alcohol
	Formamide	Ethyl acetate
	Methanol	Ether
	Acetic acid	*n*-Butyl acetate
	Ethanol	Chloroform
	Isopropanol	Benzene
	Acetone	Toluene
	n-Propanol	Cyclohexane
	t-Butanol [a]	Petroleum ether
	Phenol	Petroleum
	n-Butanol ↓	Paraffin oil ↓

[a] The solvents up to *t*-butanol are miscible with water in all proportions; other solvents form two layers with water.

justing the pH value of the system or masking agents to alter partitioning characteristics. Hydrophilic organic solvents may be employed in place of water. Isomeric hexachlorocyclohexanes have been resolved with nitromethane as the stationary liquid and *n*-hexane as the mobile phase. Excellent partitions of the fatty acids have been carried out with methanol and liquid paraffin mixtures. The method has also been extended to longer chain (C_{11} to C_{19}) fatty acids by using a mixture of 2-aminopyridine and furfuryl alcohol as the stationary liquid and *n*-hexane as the mobile phase.

Reversed-Phase Chromatography The scope of column partition chromatography can be extended by modifying the support whereby the hydrophobic organic solvent becomes the stationary phase and the hydrophilic solvent the mobile phase. This is reversed-phase chromatography. By exposing kieselguhr or silica gel in a closed container to the vapors of a chlorosilane, the surface of the support is rendered unwettable by strongly polar solvents and will then retain the less polar phase of numerous solvent systems. The stability of a column depends upon the interfacial tension of the two phases. When this is high, stable columns can be made from silanized silica gel. Typical reversed-phase systems include an olive oil–50% aqueous ethanol system and a liquid paraffin–25% aqueous ethanol system on which fatty acids have been separated. Rubber powder has been impregnated with hydrocarbons or triglycerides to partition the esters of higher fatty acids with methanol as the mobile phase. Polyethylene powder acts both as the migration medium and as the nonpolar stationary phase. This procedure with reversed phases is particularly useful if no common solvent system can be found for the normal partition technique by which a sample component gives a distribution coefficient of about 5 to 10. It is useful for the separation of nonpolar to medium polar hydrophobic substances. Reversed-phase columns with stationary phases of hydrocarbons or silicone oil require precise temperature control. Variations in temperature will result in either stripping of the stationary phase from the column or demixing of the stationary phase in the mobile phase.

Theoretical Basis of Partition Chromatography

The effectiveness of separation in liquid–liquid partition chromatography, as in other chromatographic methods, depends on a disengagement of zone centers and maintenance of narrow and compact zones. The minimum in the plate height–elution velocity curve almost invariably occurs at impractically low flow rates and, consequently, there is normally less interest in determining the "optimum" value as in gas–liquid chromatography. This situation exists as a fundamental consequence of the slower diffusion of solute molecules in liquids relative to gases. Mass transfer between phases determines the column plate number. Decreasing the eluent flow rate will generally improve column efficiency. Increasing the column length will increase the total number of plates in the column proportionately, providing that the linear eluent flow rate is held constant. Conservative estimates place the plate height in a silica gel column at 0.02 cm. Thus, a column 10 cm in height has approximately 500 plates.

By contrast with separations by countercurrent distribution (Chapter 3), most liquid–liquid chromatographic separations are not accurately predictable from the relative solubility of a solute in the two phases because of adsorption by the solid support or other interaction between the support and the solute. When the support is truly inert so that solute–support interaction is absent, the relation of the column retention volume, V_R, to the partition coefficient K_d, is given by the familiar expression

$$V_R = V_M + K_d V_S \tag{8-1}$$

where V_M is the volume of the interstitial mobile phase and V_S is the volume of the stationary phase in the column. Thus, $V_R - V_M$ is the adjusted retention volume, and $(V_R - V_M)/w_L$ is the specific retention volume where w_L is the total weight of solvent comprising the stationary phase.

For convenience, the sample ingredients should have an average partition coefficient such that the product $K_d V_S$ is approximately equal to V_M. In a partition column the relative volumes of the two liquid phases are fixed within narrow limits, the volume ratio of the mobile phase to that of the stationary phase being usually 5 to 10. Hence, the solvent systems should be chosen so that the partition coefficient lies close to 5.

Equations for determining column efficiency and resolution are no different from those set forth in Chapter 4 (*q.v.*).

Apparatus

The apparatus used in liquid–liquid partition chromatography ranges from little more than a column of tubing to a fully automated machine. Equipment has been developed which closely parallel those used with gas–liquid chromatography and gel permeation columns (see, for example, Fig. 12-1). The liquid–liquid column unit described by Lambert and Porter[1] utilizes a pressurized flow scheme incorporating an automatic recording differential refractometer as detector. For preparative applications, the system is coupled to an automatic fraction collector and solenoid-operated funnel arrangement. Samples are added to the column without release of pressure by means of a micrometer syringe assembly. The liquid chromatographs described in Chapter 7 are applicable to liquid–liquid partition methods within the limitations mentioned in regard to solute–solvent volatilities.

[1] S. M. Lambert and P. E. Porter, *Anal. Chem.* **36,** 99 (1964).

Applicability

Separations can be achieved if sample components exhibit differences in regard to their solubility—which might be attributed to (1) number and polarity of constituents and their steric positions; (2) molecular size and shape; (3) chain length of aliphatic groups; and (4) number and location of carbon–carbon double bonds. In normal partition chromatography the sample may range from polar to medium polar, whereas in reversed-phase partition chromatography the sample should be non-polar to medium polar. Normal partition chromatography is most successful in fractionations of short-chain compounds and of oxygen-substituted long-chain compounds, whereas reversed-phase partition is most suitable for the fractionation of long-chain aliphatic compounds according to chain length and/or degree of unsaturation. Liquid column partition chromatography is one of the methods of choice for the preparative separation of fragile biological materials and compounds of very high molecular weight. Some suggested column systems are listed in Table 8-2.

The system involving silica gel with water as stationary phase is valuable to the separation of hydrophilic substances and substances of medium polarity. An extensive use of this system has been in the separation of amino acids. Another type of solvent system involves a hydrophilic organic solvent, such as methanol, formamide, or polyethylene glycol as the stationary

TABLE 8-2. Typical Liquid–Liquid Partition Systems

Compound	*Column packing*	*Solvent system*
Fatty acids, monocarboxylic	Celite impregnated with 0.5*N* sulfuric acid	Chloroform/butanol/ether
Fatty acids, dicarboxylic	Silicic acid impregnated with 1*N* sulfuric acid	*t*-Butanol/chloroform
Ketosteroids	Silicic acid impregnated with methanol/water	Benzene/chloroform, methylene chloride, petroleum ether using gradient elution
Aliphatic alcohols	Silicic acid impregnated with water	CCl_4/$CHCl_3$/acetic acid system using stepwise elution
Glycols (C_2 to C_4)	Silicic acid/Celite impregnated with water	*n*-Butanol/$CHCl_3$ using stepwise elution
Phenols	Silicic acid impregnated with water	iso-Octane or cyclohexane
Aldrin-dieldrin	Silicic acid impregnated with nitromethane	*n*-Hexane

phase and a hydrophobic solvent as the mobile phase. This type of solvent system serves to separate substances of medium to low polarity, such as the more polar steroids and lipids. A reversed-phase system, such as paraffin or silicone oil held stationary and the hydrophilic solvent used as the mobile phase, is used primarily for the separation of hydrophobic substances as, for example, the isolation of epoxy and hydroxy esters and various natural polyunsaturated esters.

Liquid–liquid chromatography is an invaluable analytical and preparative tool for many compounds which decompose on gas–liquid chromatographic systems or are difficult to separate by gas–liquid chromatography because of unfavorable vapor-pressure relationships. Among the first class are compounds which are thermally unstable, those which are subject to catalytic degradation, and those which decompose because of substrate interaction. Among the second class are *cis* and *trans* isomers.

Reversed-phase chromatography can utilize and improve the techniques of liquid–liquid solvent extraction. Separations which would not be quantitative by a batch extraction method become quantitative and highly selective when the organic phase is supported on a column and the aqueous phase is used as the eluent. This technique can be extended to many solvent systems for the separation of organic and inorganic compounds. Chromatography on liquid anion exchangers is, in a sense, reversed-phase partition chromatography. Columns with tributyl phosphate adsorbed on silanized kieselguhr have been used to separate traces of arsenic from germanium and traces of scandium from calcium, also traces of niobium from molybdenum. Using columns with tri-*n*-octylamine, traces of cobalt from nickel and traces of iron(III) from manganese have been separated. Applications of this type should be capable of further extension, especially with the use of relatively highly dissociated agents as the organic phase.

LABORATORY WORK

Experiment 8-1 *Separation of Organic Acids*

In this experiment a variety of aliphatic acids will be separated on a column packed with chloroform-butanol mobile phase and $0.5N$ sulfuric acid on Celite as stationary phase.[2]

Apparatus Use a glass column 45 cm in length and 1 cm in diameter, equipped with coarse glass frit and screw clamp on the exit tubing.

Column Packing The stationary phase, 9.6 ml of $0.5N$ sulfuric acid, is mixed thoroughly with 12 g of Celite 545 (100–200 mesh). The moist

[2] E. F. Phares *et al.*, *Anal. Chem.* **24**, 660 (1952).

Celite is then slurried with 10% butanol in chloroform (previously the butanol–chloroform mixture should have been equilibrated with 0.5N sulfuric acid) and poured into the column by tamping with a close-fitting perforated disk. Excess mobile phase is allowed to drain from the column until the level of the solvent is only a few millimeters above the top of the packing. The free column volume is 20 ml.

Chromatographic Separation The mixture of acids is soaked up on 0.5 g of Celite, acidified with 0.1 to 0.2 ml of 25% sulfuric acid, then transferred quantitatively with 1–2 ml of solvent onto the top of the column, and tamped down. The liquid is drawn down almost to the top of the packing, and the rinsing repeated with 1–2 ml of organic phase. Continue development with the organic liquid at flow rates of 0.5 to 1 ml per min. Collect the effluent in 5-ml portions. The organic acids are determined by titration with standard 0.1N alkali using a microsyringe. Plot the volume of titrant against volume of effluent.

Individual acids must be identified by comparison with knowns run on the same column. Suggested acids include acetic, fumaric, kojic, and lactic, all of which should be reasonably separable from each other and will elute in the order listed. A few drops of phenol red indicator gives a moving colored band that provides visual evidence of the degree of success in packing the column.

EXPERIMENT 8-2 *Separation of Phenols by Reversed-Phase Chromatography*

In this experiment monohydroxy phenols will be separated on columns packed with Teflon-6 or microporous polyethylene.[3]

Apparatus Use glass columns 15 cm in length and 1 cm in diameter, equipped with coarse glass frits and stopcocks of Teflon.

Column Packing Columns are filled with the stationary phase, 6% 1-hexanol in cyclohexane. The dried support (Teflon-6 or microporous polyethylene), previously ground in a mortar with dry ice and ethyl ether, then sieved to 80 mesh, is added to the column until the desired bed height is obtained. The excess stationary liquid is then replaced by the mobile liquid phase (0.5M aqueous sodium chloride) using gentle suction.

Elution Curves Standard phenol solutions are prepared by dissolving an accurately weighed 100-mg sample in 5 ml of isopropanol and diluting to 100 ml with 0.5M aqueous sodium chloride. Samples are added to

[3] J. S. Fritz and C. E. Hedrick, *Anal. Chem.* **37,** 1015 (1965).

the column in 1.00-ml aliquots from the standard solutions. The flow rate is 2 ml per min.

Scan the ultraviolet spectrum of each phenol to ascertain the proper wavelength for scanning (in continuous flow method) or determining the phenol of collected fractions.

A suggested system consists of phenol and *o*-cresol. Usually the cresol will have eluted with 80 ml of mobile phase.

High-molecular-weight phenols are best separated using cyclohexane and aqueous methanol as mobile phase. A suggested system is *o*-isopropyl phenol and 2,6-di-isopropyl phenol. Elute with 50% aqueous methanol until about 45 ml of effluent is collected, then continue the elution with 75% methanol until an additional 35 ml of effluent has been collected.

Experiment 8-3 *Separation of Aromatic Amines by Reversed-Phase Chromatography*

In the reversed-phase procedure, a 1.1- × 60-cm column of 80-mesh Teflon-6, with cyclohexane as stationary phase, and water as mobile phase, is employed to separate 1.5 mg of *o*-methoxyaniline from an equal weight of *m*-methoxyaniline. See Experiment 8-2 for specific directions on preparing the column. Use a flow rate of 1.3 ml per min. Collect 90 ml of effluent in 1-ml increments up to 45 ml, and 2-ml increments thereafter. Determine the fraction of solute in each fraction of effluent from its absorption spectrum or by potentiometric acid–base titration.[4]

BIBLIOGRAPHY

Cassidy, H. G., *Fundamentals of Chromatography,* in A. Weissberger (Ed.) *Technique of Organic Chemistry,* Vol. X, Interscience, New York, 1957.

[4] C. E. Hedrick, *Anal. Chem.* **37,** 1044 (1965).

9 *Paper Chromatography*

In paper chromatography, introduced in 1944 by Consden, Gordon, and Martin,[1] the substrate is a sheet of filter paper, or other special type of paper, and the mobile phase is a liquid which percolates within the porous structure of the paper. A small amount of sample is placed on a limited area of the paper as a spot or streak, and then irrigated by the solvent system. Usually development of the chromatogram is stopped before the mobile phase reaches the farther edge of the paper, so that solute zones are distributed in space instead of time.

Paper chromatography is especially useful when the quantitative or qualitative analysis of a sample only necessitates a separation of constituents, and not their recovery in a purified form. Only small volumes of sample are required and many samples can be run at one time. The major limitations of paper chromatography are the relatively long development times and less sharply defined zones, as compared to thin-layer techniques.

Paper as a Chromatographic Medium

Paper is a somewhat random arrangement of fibers; only a limited degree of orientation is introduced during the manufacturing process. The packing of the fibers forms a porous medium for the retention of the stationary phase, and the spaces between the fibers provide channels for the flow of the mobile phase. The capillary nature of these channels provides both a resistance to solvent flow and a capillary driving mechanism. As the solvent front advances, it draws behind it a column of liquid in the bundle of capillaries that constitute the pore structure of paper. These are irregular-sized, tortuous, yet for the most part, interconnected channels.

[1] R. Consden, A. H. Gordon, and A. J. P. Martin, *Biochem. J.* **38,** 224 (1944).

TABLE 9-1. Papers for Chromatography

Type	*Commercial brands*
Standard papers	Whatman 1 and 2 Schleicher & Schüll 2043b and 602h:P Ederol 202 Macheray–Nagel 260
Rapid flow	Whatman 31 ET, 54, and 4 Schleicher & Schüll 2040a
Preparative cardboards	Whatman 31 ET and 3 MM Schleicher & Schüll 2071
Carboxyl papers	Schleicher & Schüll I and II
Acetylated papers	Schleicher & Schüll 2043b/6, 2043b/21, and 2043b/45 Macheray–Nagel 214 Ac, 261 Ac, and 263 Ac
Kieselguhr paper	Schleicher and Schüll 287
Silicone-treated papers	Whatman 1, 4, and 20 (silicone-oil impregnated) Schleicher & Schüll 2043b/hy Macheray–Nagel 212 WAA, 261 WAA, and 263 WAA
Ion-exchange papers	Erdol 208/IK and 208A Schleicher & Schüll cation- and anion-exchange papers Reeve–Angel SA, SB, WA, and WB Whatman AE, DEAE, CM, P, and citrate

The dead-end pores that do exist trap the mobile phase to render it stagnant (analogous to active adsorption sites that contribute to "tailing" in column chromatography). Perhaps the best integrated picture of paper as a chromatographic medium is that of Stewart.[2]

In addition to the standard types of paper, many special papers have been introduced in recent years. Table 9-1 lists these categories and correlates them with the manufacturer's designation. All these papers are available in the form of rectangular sheets; many are also sold in rolls or strips, in circular sheets, and in other special shapes.

Papers of Pure Cellulose Ordinary filter paper is a somewhat random pile of cellulose fibers. Cellulose may be pictured as a fiber network of polymeric carbohydrate chains of molecular weight up to 500,000, hydrophilic in nature, and cross-linked by a strong hydrogen-bonding system. Each fiber

[2] G. H. Stewart, in Vol. I of *Advances in Chromatography*, J. C. Giddings and R. A. Keller (Eds.), Dekker, New York, 1966, p. 93.

is a bundle of oriented smaller units called fibrils. These fibrils have regions of a high degree of order called crystallites and regions of a low degree of order termed amorphous regions. The stationary liquid phase is not uniformly distributed over these fibrils but appears to be concentrated in the amorphous regions. Water or other hydrophilic solvents in these amorphous regions may be viewed as a different kind of water, possessing a high degree of organization and having properties different from those of bulk water. It may be envisioned as being present as "connected puddles" of swollen fibrous material, organized and dense near the cellulose chains, but becoming more like bulk water the farther it is removed from the chains. These puddles resemble a concentrated solution of a polysaccharide, restrained from physically dissolving by the polymer network to which the polysaccharide groups are attached. The surface and the amorphous regions that hold the imbibed water are mutually responsible for retention.

Modified Cellulose Papers Hydrophobic substances can be separated on kieselguhr filter paper or analogous adsorbent-loaded papers. Pure cellulose paper is impregnated with highly purified silica gel, alumina, or diatomaceous earth. In these loaded papers, adsorptive characteristics of the silica, alumina, or diatomaceous earth dominate over those normally associated with the supporting cellulose matrix.

Hydrophobic substances can also be separated on acetylated papers or on silicone-treated papers. By these treatments the paper is rendered hydrophobic and can be used as the support for reversed-phase chromatography of lipophilic materials.

Additional possibilities are offered by papers containing ion-exchange resins and the various cellulose ion-exchange papers. In the series of papers loaded with ion-exchange resins, micropulverized resins are loaded into a slurry of high-quality alpha-cellulose pulp and the mixture made into paper sheets that contain up to 55% resin by weight. Papers with both strong and weak cation- and anion-exchange groups are available. Inorganic ion exchangers, particularly zirconyl phosphates and ammonium 12-molybdophosphate, are formed within the paper by precipitation. Chapter 5 should be consulted for properties of ion-exchange resins; inorganic ion exchangers and the cellulose ion exchangers are discussed in Chapter 6.

Glass Fiber Papers Papers manufactured from glass fibers are useful with reagents that are too corrosive for cellulose papers. Adsorption effects can be avoided in some instances.

Solvent Systems

Chromatography on paper is essentially liquid–liquid partition in which the paper serves as carrier for the solvent systems. Many combinations of solvent systems may be used, provided that they are not infinitely miscible with one another. In terms of the stationary phase, these systems may be divided into three groups: an aqueous stationary phase, a stationary hydrophilic organic solvent, and a stationary hydrophobic solvent.

Aqueous Stationary Phase Aqueous systems are used for strongly polar or ionic solutes. Water is held stationary on the paper and the mobile phase passes through. The aqueous stationary phase is attained by exposing strips or sheets of suspended paper to a saturated atmosphere in a closed chamber. If an aqueous buffer solution or a salt solution is to be used as the stationary phase, the paper is drawn through the solution, allowed to dry, and then exposed to an atmosphere saturated with water vapor. For aqueous stationary phases, the mobile phase might be *n*-butanol and higher alcohols for a neutral system, *n*-butanol/acetic acid/water (40:10:50 *v*/*v*) for an acidic system, or *n*-butanol/ammonia/water (75:8:17 *v*/*v*) for a basic system. The latter two systems are prepared by shaking all components in a separatory funnel; the less-polar phase serves for development.

Water itself can be used as the developer. Partitioning occurs between the "water–cellulose complex" or puddles of water, organized and dense near the amorphous regions of the cellulose chains, and the free water. The bound water has a thermodynamic activity radically altered by the cellulose; the energetics indicate a hydrogen bond.

Stationary Hydrophilic Organic Solvent When the stationary phase is a hydrophilic organic solvent, either of two methods may be used, depending on the volatility of the stationary phase. When the stationary phase is sufficiently volatile, the paper is suspended in a saturated atmosphere until equilibrated. If the stationary liquid is not sufficiently volatile, the paper is drawn through a solution of the liquid in a volatile diluent, then suspended in air until the diluent has evaporated.

Formamide is a suitable hydrophilic organic solvent. Dissolved in ethanol, a 40% solution is used to impregnate the paper, after which the ethanol is removed by evaporation. Progressively less polar mobile phases used for development might be chloroform, various mixtures of chloroform with benzene, benzene, and mixtures of benzene with cyclohexane. A list of other hydrophilic stationary liquids would include the glycols, the Cellosolves, Carbitol, glycerol, and benzyl alcohol.

Stationary Hydrophobic Solvent Specially treated papers are used for reversed-phase chromatography. These are available commercially or may be prepared in the laboratory. Two methods render the paper hydrophobic.

The hydrophilic character of the paper may be modified chemically through treatment with vapors of dimethyldichlorosilane or by acetylation of the cellulose. In the second method, the paper is impregnated with the hydrophobic solvent dissolved in a volatile diluent, then the diluent allowed to evaporate in the air.

Silanized or acetylated papers will take up the less-polar solvent from the atmosphere in a chamber. Impregnated papers already contain the stationary hydrophobic phase, and require no further preparation.

Systems for reversed-phase chromatography, in order of decreasing polarity, include dimethylformamide (as the stationary phase; applied as a 50% solution in ethanol) with cyclohexane; kerosine (applied as 10% solution in petroleum ether) with 70% isopropanol or glacial acetic acid; and paraffin oil (applied as a 10% solution in benzene) with dimethylformamide/methanol/water (10:10:1) as the mobile phase.

Mechanism of Paper Chromatography

Solute migration is initiated from a small compact spot or narrow streak where the sample is applied to the paper prior to development. Differential migration of solute molecules begins when the initial zone is engulfed by the solvent front. The main driving force for flow of the mobile phase is capillary in nature. It originates from the tendency of wetting liquids to flow into empty capillary or pore spaces. Surface tension is the driving force for capillarity. For dynamic flow, liquid transport is hindered mainly by viscosity, while in the ascending technique surface tension is balanced against gravity.

Solute Migration The migration of a solute in paper chromatography depends, in a complex manner, on the nature of the forces acting on the solute. As described earlier, cellulose is an adsorptive and partitioning medium, consisting of pools of liquid in the amorphous regions connected by crystallite bridges. Cellulose also has ion-exchange properties; the exchange capacity ranges between 5 and 64 μEq/g. Partition will be limited to solutes that can diffuse freely to different depths in the swollen cellulose structure. Large molecules may be excluded from portions of the labyrinth of glucoside chains. Thus, different mechanisms may exert a greater or lesser role in the partition process, although liquid–liquid partition seems to be the dominant mechanism. Impregnation of cellulose to produce a reversed-phase system avoids many of the vagaries of the hydrophilic stationary phase of cellulose.

For partition on paper, the mathematical relationship between the R_f value and the partition coefficient K_d is given by

$$R_f = A_M/(A_M + K_d A_S) \quad \text{or} \quad K_d = (A_M/A_S)[(1/R_f) - 1] \tag{9-1}$$

Due to difficulty encountered in measuring the cross-sectional areas of the stationary and mobile phases, and the variability of the ratio A_M/A_S from one type of paper with one set of solvents to the next, it is more convenient to develop a chromatogram and measure the R_f value directly. Often the development is done concurrently with samples and known materials run on adjacent paths of a single sheet or on separate strips of identical paper. This is necessary because R_f values depend on the geometry of the paper, whether circular, rectangular, tapered, or other special design, the distance of the initial zone from the solvent reservoir, and the duration and direction of development (ascending, descending, or horizontal). Keller and Giddings have discussed these variables.[3]

For a given solvent–paper–solute system at constant temperature, the ratio A_M/A_S has a constant value. Taking logarithms of Eq. (9-1), and substituting the term $R_M = \log [(1/R_f) - 1]$, we obtain

$$\log K_d = \log (A_M/A_S) + R_M \tag{9-2}$$

When R_M is plotted against the number of structural units for a homologous series, a smooth curve is usually obtained. Interpolation, and limited extrapolation, enables members of a homologous series to be identified when present together in samples. Such a plot is also useful in studying the effect of phase composition and in correlation of chromatographic behavior in different solvent systems. Plots of R_M values in one solvent system versus R_M values in a second solvent system, with suitably chosen pairs of partition liquids, often provide information concerning the molecular interactions of solutes and solvents, and also their molecular structure. This, of course, is analogous to two-dimensional chromatography; also the treatment in gas–liquid chromatography shown in Fig. 11-21. Attention is drawn to analogies between Eqs. (9-2) and (7-4); Snyder's theoretical treatment of similar variables in adsorption chromatography could prove rewarding if applied to paper chromatography.

Zone Spreading The dimensions of the individual spots increase in all directions (as compared to the initial zone) as development continues. Both longitudinal spreading of each solute in the flow direction of the mobile phase, and lateral spreading occurs. The increase is greater in the flow direction since all three spreading factors—nonequilibrium caused by the partition kinetics, ordinary diffusion, and eddy diffusion—are operative, while only the latter two contribute to lateral spreading. Consequently, the zones tend to acquire an elliptical shape with the major axis in the flow direction. In the early stages of chromatographic development, solvent velocity is high and the spreading rate is dominated by the nonequilibrium factor. As the mobile phase penetrates farther into the paper, the velocity decreases in inverse proportion to the distance traveled; the molecular diffusion becomes

[3] R. A. Keller and J. C. Giddings in *Chromatography*, 2nd ed., E. Heftmann (Ed.), Reinhold, New York, 1967, Chap. 6.

increasingly more important until it dominates at large distances from the solvent reservoir. Thus, the plate height is not constant, but increases with increasing development time. Separation power is highest at the beginning. Theory suggests that the best flow rate corresponds to a length/width ratio of about $\sqrt{2}$ for zones.

Experimental Technique

In the following sections the operations involved in the spotting of the sample on the paper and the subsequent development of the chromatogram, including the various arrangements for flow of mobile phase, will be discussed.

Placing the Sample on the Paper The quantity of substance used depends upon the purpose of the chromatographic separation, the solubility of the sample ingredients in the solvent system selected, and the ease of detection of the material on the chromatogram. A line is generally drawn about 2.5 cm from the edge of the paper strip or sheet. The sample is applied along this line as spots (diameter 5–15 mm) about 2–3 cm apart or as a streak. For the usual paper chromatographic analysis it is best to use 5–20 μg of material, dissolved in 5–10 μl of solvent which is applied to the starting point with a micropipette. If volumes larger than 10 μl are to be applied, the spots must be dried before a second application to the same spot is made. Commercial applicators in a variety of designs are available for spotting and streaking.

Chambers The development tank or chamber is usually a glass container with a tightly fitted cover. The cover keeps the atmosphere of the tank confined. Solvent is placed in the bottom of the chamber to saturate the chamber atmosphere. For highly volatile solvents, it is advisable to place sheets of filter paper on the chamber walls with one end dipping into the solvent reservoir. This aids in rapidly saturating the chamber atmosphere and avoiding evaporation losses from the developing chromatogram.

For ascending development, only the paper sheets or strips need support; usually inert clips on the upper end. A sheet rolled into a cylinder, held together by clips, will stand up on the bottom of the tank.

For descending development, a support or shelf around the tank is required to hold the solvent trough. An exploded view is shown in Fig. 9-1. The anchor rods keep the paper from slipping out of the trough owing to the increase in weight during development. The anti-siphon rods serve to prevent siphoning of the solvent down the paper due to capillary attraction, which would occur if the paper were simply hung over the side of the trough. By making the solvent rise a few centimeters, this effect is eliminated.

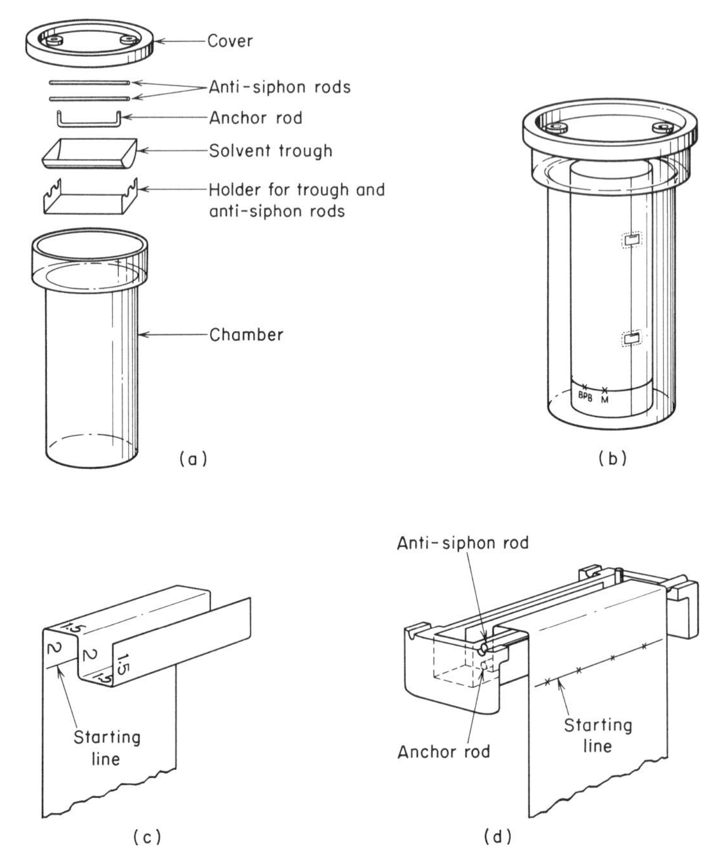

FIG. 9-1. (a) Exploded view of a cylindrical glass chamber for ascending or descending chromatography; (b) ascending with sheet in form of a cylinder; (c) method of folding strip for descending technique (distances in cm); (d) strip in position for descending development.

Development Methods The individual development techniques differ by the fact that the mobile phase either flows downward (descending technique), rises in the paper (ascending technique), or flows horizontally (horizontal or radial technique). R_f values obtained by the different methods differ somewhat.

In descending development the paper, to which the sample has been applied, is inserted with its upper end in the trough and the mobile phase is allowed to run down the paper assisted by gravity. Development is rapid. The chromatogram may be developed continuously if the bottom of the paper is cut to a serrated edge to facilitate disengagement of solvent from the lower edge of the paper.

In ascending development the paper, spotted with the sample, is immersed in the solvent mixture in the bottom of the development chamber. The mobile phase ascends the paper by capillary action. Only simple apparatus is required, but the rate of development is slow.

Horizontal development utilizes compact equipment. The chamber consists of a shallow container with the paper supported horizontally on glass rods. Development is solely by capillary action and is not aided or hindered by gravity.

The radial technique uses a circular piece of filter paper with a wick cut parallel to the radius from edge to center. The sample is deposited either at the upper end of the wick or streaked in a circle a short distance from the center of the paper. Small tanks are made from two petri dishes, one inverted over the other. The mobile phase ascends the wick and flows radially through the paper disk to form concentric rings of components.

Two-Dimensional Chromatography In two-dimensional separations the sample is placed as a spot near one corner of a sheet. One edge of the sheet is immersed in the first solvent and separation is conducted in one direction until the solvent front approaches the opposite edge of the paper. After thoroughly drying the sheet to remove the first solvent, the other edge adjacent to the original spot is placed in the second solvent and the chromatogram is developed at right angles to the preceding development. Usually different types of solvent systems are used in each direction to bring about more effective development of certain components, lying too closely together after a one-dimensional separation using either solvent alone. A limiting factor in two-dimensional development is the small amount of sample that can be run. Streaks cannot be separated by this procedure.

Multiple Development The stepwise or multiple-development method is useful where a mixture contains a group of substances with low R_f values and other material with relatively large R_f values. If, initially the developer is the more polar solvent, compounds with the lower R_f values will be separated while those with higher R_f values travel with the front. After intermediate drying, development with the less-polar second solvent serves to separate the compounds with high R_f values farther up the paper. Usually the compounds originally separated are not appreciably moved by the second less-polar solvent.

In an alternative procedure, the less-polar components are first separated with a solvent of low eluting strength. This separates the compounds with lower R_f values while those with higher R_f values travel reasonable distances from the origin. After intermediate drying, the chromatogram is developed with a more strongly eluting solvent, but to a lesser distance than with the first solvent.

Continuous Development When two substances have almost identical R_f values, a solvent system is sought in which they exhibit low R_f values but show signs of separation. The separation is then carried out by continuous flow of solvent through the paper. The solvent is allowed to drip off the ends of the paper strip or edge of the sheet. The bottom of the paper is cut in saw-tooth fashion, each tooth the width equal to the distance apart of the individual spots (or 2 cm with streaks) and descending development is employed.

Wedge Technique To achieve separation in a mixture of two substances with relatively similar R_f values, the paper is cut in the form of a wedge. As the solvent flows over the initial spot, located near the apex of the wedge, it spreads out at an angle. The developing zones take on a flattened appearance with sharper front and rear edges, and less tendency to overlap.

When the amount of one sample ingredient greatly exceeds that of the others, the sample should be applied to the tip of the wedge if the minor components have the smaller R_f value, but to the wide part of the wedge when they have the larger R_f value.

Centrifugal Chromatography In principle, centrifugal chromatography[4] is classical paper chromatography in which the flow of solvent is accelerated by centrifugal force at speeds of 300–1500 rpm. A constant solvent velocity and thus better resolution can be obtained in this way. The apparatus consists of a closed centrifuge with a clear cover. Circular sheets of filter paper are attached (in a horizontal plane) to the shaft of a variable speed rotor. Compounds to be chromatographed are placed on the circular sheet just slightly outward from the point at which the stream of developing solvent, delivered as a fine jet stream through a capillary from an external reservoir, will hit the paper. The reservoir of mobile phase is kept under pressure with an inert gas. Development takes only a few minutes.

Visualization and Evaluation of Chromatograms

The methods of detection and estimation of individual components following development can be carried out directly on the paper chromatogram.

Colored fractions are observed visually. A substantial body of knowledge has grown up regarding reagents to produce colored reaction products with originally colorless fractions. The reagent is usually applied as an aerosol spray or by drawing the chromatogram through a solution of the reagent. Duplicate chromatograms, treated with different selective reagents, will often reveal overlapping components. In addition to viewing the

[4] Z. Deyl, J. Rosmus, and M. Pavlicek, *Chromatog. Rev.* **6**, 19 (1964).

developed spots in visible light, ultraviolet radiation is employed to detect a large number of organic chromophores either through absorption at characteristic wavelengths or through quenching on the faintly fluorescent background of the filter paper.

Spots can be evaluated photometrically after color development with a suitable reagent. Special attachments for commercial photometers provide a means for drawing the chromatogram across a window in front of a photocell. The output signals are plotted on a recorder or manually (Fig. 9-2). An assessment of the area under the photometric curve completes the measurement (see Chapter 11 for a discussion of the several methods used to evaluate areas).

Spot area is proportional to the logarithm of the amount of substance. To compensate for substrate differences, known amounts of the pure material are run concurrently on the same chromatogram.

The most accurate results are obtained by excising the spot area and then eluting the compound to be analyzed from the adsorbent with suitable solvents. In the extractant the solute is then determined colorimetrically or spectrophotometrically.

If suitable radioactive isotopes can be incorporated into the compound, the location of the spot and its quantitative evaluation is possible with a radiation detector mounted on a scanning device. An autoradiogram can also be prepared.

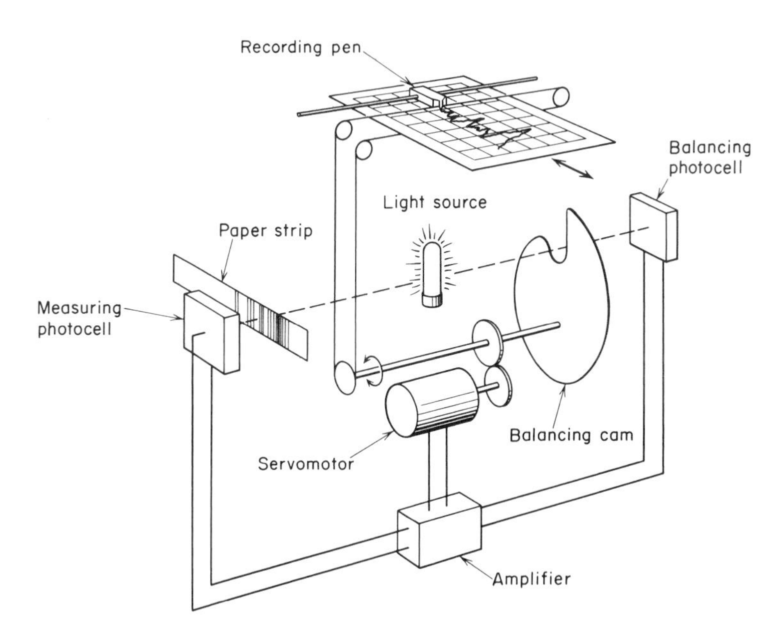

FIG. 9-2. Photometric scanning of paper chromatograms (schematic).

LABORATORY WORK

EXPERIMENT 9-1 *Separation of Indicators*

Common acid–base indicators provide solutes whose location is easily ascertained after development. Spot 10 μl of each stock indicator solution (0.05% *w*/*v*) and develop with the organic phase [obtained upon shaking together *n*-butanol–aqueous ammonia–water (75:8:17)] until the solvent front has progressed at least 10 cm from the starting line.

Measure the R_f value for each indicator. By comparison with the standards identify the components in the unknown issued to you.

These are approximate R_f values in the basic system: Congo red and indigo carmine, 0.0; chlorophenol red and phenol red, 0.18; cresol red and bromocresol purple, 0.42; bromophenol blue, methyl orange and methyl red, 0.55; neutral red, 0.66; bromothymol blue, 0.80; methyl violet, thymol blue, phenolphthalein, and thymolphthalein, 0.9.

EXPERIMENT 9-2 *Inorganic Paper Chromatography*

The chlorides of copper, cobalt, and nickel migrate in paper with isobutyl methyl ketone or methyl ethyl ketone. The R_f values can be measured as a function of the HCl–water ratio in the solvent system.

Spot Whatman No. 1 paper with varying concentrations of solutions of the mixture (200 μg/ml of each metal) and the individual cations. After air drying, place the spotted paper in a humidity desiccator [52% relative humidity; obtained with moist $Ca(NO_3)_2$ or $Mg(NO_3)_2$ crystals as desiccant] for 0.5 hr or more. Immerse the paper in the developer until the solvent front has risen at least 10 cm from the starting line. Remove, drain, and air dry the paper for 2–3 min. Then suspend the paper above 10 ml of concentrated aqueous ammonia in a large open vessel until fuming has stopped (use a hood) and the free acid is neutralized. Spot locations are revealed by spraying with 0.1% (*w*/*v*) dithiooxamide in 60% ethanol. Colors: copper, green; cobalt, yellow; and nickel, blue.

The developer is prepared just before using by mixing HCl–water–ketone (5:3:25). With a higher proportion of water the separation between copper and cobalt is sharply decreased. The solvent system—chloroform–acetone–isoamyl alcohol [HCl (2:1:1:0.2)] provides an improved separation of cobalt and copper.

EXPERIMENT 9-3 *Chromatography of Organic Acids*

A series of known organic acids, plus an unknown containing one or more of these acids, is run on a sheet of Whatman No. 1 paper.

Spot 10 μl of each organic acid solution (0.5% *w/v* in 10% isopropanol–water) along a line 5 cm from one end of the sheet, each spot located 2–3 cm apart. Immerse the lower 1 cm of paper in the developer and allow the chromatogram about 4 hr to develop. The developing solvent is *n*-butyl formate–formic acid–water (10:4:1) and containing 0.05% *w/v* sodium formate and 0.02% *w/v* bromophenol blue.

After development is complete, remove the paper from the chamber and dry in a hood for 8–10 hr. On the dry chromatogram, the organic acids will appear as yellow spots on a blue background. Measure the R_f values. If members of a homologous series were used, plot R_M values against the carbon number.

EXPERIMENT 9-4 *Reversed-Phase Separation of Fatty Acids*

In paper impregnated with liquid paraffin, the longer-chain fatty acids will be retarded more than the short-chain homologs.

Immerse strips or sheets of Whatman No. 3 paper in a 5% *v/v* solution of liquid paraffin in ether for 5 min. Remove the paper and hang in a hood until the ether has evaporated completely (about 12 hr).

Spot 10 μl of fatty acid solution (2 mg/ml dissolved in ether) and allow the ether to evaporate. Suspend the paper in the atmosphere of the chamber for 30 min, then immerse in the developing solvent and allow the chromatogram to develop until the solvent front has traveled 10–12 cm from the starting line (about 3 hr). The developer is acetone–water (4:1).

Remove the paper from the tank and allow the solvent to evaporate. Mark the position of the solvent front with a pencil. Spray with 0.4% *w/v* solution of Nile blue. Acids are revealed as blue spots on a pink background. The spray reagent is prepared from 50 ml of 0.4% Nile blue sulfate in ethanol to which is added triethanolamine until the blue color changes to red without any blue tinge, then an additional 50 ml of ethanol is added.

Suggested solutes: myristic, palmitic, and stearic acids. Mixtures for analysis can be obtained from ether extraction of an acidified soap solution, or from the hydrolysis of fats.

Problems

1. For these changes in solute composition, predict the effect on R_f values: (a) the addition of an amino group on the partition between phenol and water and (b) on the partition between collidine and water, (c) the addition of a hydroxyl group on the partition between phenol and water and (d) on the partition between collidine and water.
2. Explain these observations. Using phenol as mobile phase and water as the stationary phase, proline migrates faster than valine; but with *n*-butanol, valine is faster than proline.
3. In the developing solvent *n*-butanol–1.5*M* NH_3 (1:1) the lower fatty acids are separated rather well (R_f values follow each compound): formic, 0.10; acetic,

0.11; propionic, 0.19; *n*-butyric, 0.29; *n*-valeric, 0.41; *n*-caproic, 0.53; *n*-heptanoic, 0.62; and *n*-octanoic, 0.65. (a) Plot R_M values against the number of —CH_2— units. Note the limits for separability. (b) Suggest a method for separating the higher fatty acids.

4. On paper impregnated with liquid paraffin and developing with acetone–water, (a) predict the elution order of the C_{12}–C_{18} fatty acids. (b) By increasing the concentration of liquid paraffin used to impregnate the paper, what will be the effect on R_f values of the fatty acids? (c) Increasing the concentration of acetone in the developer will have what effect on R_f values?
5. These R_f values were obtained on paper impregnated with buffer solution of specified pH for the individual solutes. (a) Plot R_f value vs pH for each substance. (b) Determine the optimum pH for liquid–liquid extraction. (c) Estimate the pK_a value(s).

	R_f value		
pH	Phenobarbital	Perphenazine	Sulfadiazine
0.0	0.85	0.0	0.0
2.0	0.84	0.02	0.2
2.5		0.13	0.35
3.0		0.49	0.46
3.5		0.87	0.48
4.0	0.83	0.97	0.49
4.5		0.99	0.49
5.0		1.00	0.48
6.0	0.82	1.00	0.42
7.0	0.78	1.00	0.17
8.0	0.65		0.05
8.5	0.32		
9.0	0.20		0.01
10.0	0.11		0.0
11.0	0.04		0.0

6. The R_f values of selected amino acids are given in three different solvent systems:

Amino Acid	1	2	3
β-Alanine	0.10	0.22	0.66
Arginine	0.02	0.17	0.89
Aspartic acid	0.01	0.21	0.19
Cystine	0.02		
Glutamic acid	0.02	0.20	0.31
Glycine	0.07	0.24	0.41
Histidine	0.07	0.27	0.69
Hydroxyproline	0.10	0.28	0.63

continued

Amino Acid	1	2	3
Isoleucine	0.31	0.45	0.84
Leucine	0.36	0.45	0.84
Lysine	0.02	0.11	0.81
Norleucine	0.42	0.45	0.84
Norvaline	0.23	0.38	0.80
Phenylalanine	0.36	0.48	0.85
Proline		0.28	0.88
Serine	0.08	0.28	0.36
Tyrosine	0.24	0.51	0.51
Valine	0.18	0.36	0.78

Solvent 1: *t*-Amyl alcohol saturated with water.
Solvent 2: 2,6-Lutidine–collidine–water (1:1:1).
Solvent 3: Phenol–water (10:2), used in presence of 3% NH_3 or 1:1 acetic acid.

(a) Plot a two-dimensional chromatogram with the initial spot in the lower left corner and using solvent 3 (rising development) and solvent 2 (development to the right). (b) Plot a two-dimensional chromatogram using solvent 3 and solvent 1. (c) For the chromatograms obtained in Parts (a) and (b), draw a smooth curve through the monofunctional amino acids from glycine through valine. Correlate the location of each solute with its structure. Notice in particular the location of the dicarboxylic acids, the diamino acids, and the hydroxyl derivatives.

BIBLIOGRAPHY

Block, R. J., E. L. Durrum, and G. Zweig, *A Manual of Paper Chromatography and Paper Electrophoresis,* Academic Press, New York, 1958.

Hais, I. M., and K. Macek (Eds.), *Paper Chromatography,* 3rd ed., Academic Press, New York, 1964.

Lederer, E., and M. Lederer, *Chromatography. A Review of Principles and Applications,* Elsevier, Amsterdam, 1959.

Macek, K., and I. M. Hais (Eds.), *Stationary Phase in Paper and Thin-Layer Chromatography,* Elsevier, Amsterdam, 1965.

Macek, K., and I. M. Hais, *Bibliography of Paper Chromatography 1944–1956,* Publishing House of Czechoslovak Academy of Sciences, Prague, 1960.

Macek, K., I. M. Hais, J. Gasparic, J. Kopecky, and V. Rabek, *Bibliography of Paper Chromatography 1957–1960,* Publishing House of Czechoslovak Academy of Sciences, Prague, 1962.

Macek, K., I. M. Hais, J. Kopecki, and J. Gasparic, *Bibliography of Paper and Thin-Layer Chromatography 1961–1965,* Academic Press, New York, 1967.

10 *Thin-Layer Chromatography*

Thin-layer chromatography has, within the last few years, leaped into prominence as one of the simplest, most useful, and most widely applicable forms of chromatography yet devised. The first stage of its development was the work of Ismailof and Schraiber[1] in 1938. An adsorbent coated on a glass plate serves as the stationary phase. Development of the chromatogram takes place as the mobile phase percolates through the stationary phase. In fact, thin-layer chromatography may be regarded as an "open column" method of chromatography that is much faster and more convenient than analogous column methods. Separation may be effected by adsorption, partition, exclusion, or ion-exchange processes, or a combination of these. Because of its convenience and simplicity, sharpness of separations, high sensitivity, speed of separation, and ease of recovery of the separated compounds, thin-layer chromatography has found wide acceptance.

Chromatographic Media

In thin-layer chromatography a variety of coating materials and solvent systems are available. By selecting the correct combination of these basic variables, one can carry out very specific separations.

Coating Materials As a general rule, any of the stationary phases used in column chromatography can be used in thin-layer chromatography provided that they are available in a uniformly fine particle size (between 1

[1] N. A. Ismailof and M. S. Schraiber, *Farmatsiya,* No. 3 (1938); *Chem. Abstr.* **34,** 855 (1940).

and 5 μ) and will adhere to a glass plate or other support. Special grades of coating materials for thin-layer chromatography are available from laboratory suppliers.

Silica gel is used more often than any other coating material. Penetration of solute into the interior of the silica gel particles is limited; the gel functions largely as a surface adsorbent. Zone formations are rapid, sharp, and regular. Silica gel is used for the resolution of acidic and neutral substances and, being more hydrophobic than cellulose, is useful for separating mixtures of substances with relatively low water solubility. It is effective for inducing molecular polarization in nonpolar materials.

Cellulose powder can be considered as a substitute for paper chromatography. It is recommended for the separation of water-soluble materials. Available capacity of a cellulose thin layer is higher than that of silica gel layer due to the cellulose structure which permits access well into the macro-particle. This factor may result, however, in separations which are slower than those obtained with silica gel.

Plates coated with alumina are used often for the resolution of basic mixtures. After coating, the plates are dried under specified conditions to obtain the required activity grade. Kieselguhr (diatomaceous earth) is neutral and is used for the separation of strongly hydrophilic compounds.

Normally, adsorbents do not adhere very well to glass plates and, therefore, it is often customary to add various types of binders to achieve a more adherent coating. Binder-free materials do not provide good adhesion and satisfactory mechanical stability on handling. On the other hand, a binder may modify the adsorbent properties adversely or give side-effects which frequently are bothersome. Calcium sulfate (plaster of Paris) is a common binder; starch (preferred for chromatographing inorganic ions), collodion, plastic dispersions, and hydrated silicon dioxide are others.

Most of the coating materials are also available in a composition which incorporates a chemically inert phosphor such as zinc silicate, which does not affect the separating characteristics of the coating. The preparations fluoresce when exposed to ultraviolet radiation at 2540 Å (and a few also at 3660 Å).

In order to increase resolution or improve the shape of spots the adsorbent layers may be modified. By incorporation of boric acid a separation of sugar isomers is achieved since the borate forms complexes with the isomers to varying degrees. Layers impregnated with chelate-forming reagents are used for the separation of inorganic cations and phenolic carboxylic acids. Silver nitrate impregnation is useful in the separation of unsaturated compounds. Alkaloids can be separated on basic silica gel layers obtained by mixing silica gel with dilute sodium hydroxide instead of water. Acidic silica gel layers are obtained by preparing the layers with nonvolatile acids or buffers instead of water.

Applicators Slurries of finely divided adsorbents and other solid materials are applied to glass plates or plastic sheets in a thin, uniform layer by a spreading procedure, usually with the aid of a commercial applicator. Two types are on the market. The design shown in Fig. 10-1 consists of a hollow metal block with exit gates on each of two opposite long faces. An internal rotating reservoir chamber with a wide longitudinal slot fits into this block. Flow of the suspension does not begin until the chamber is rotated to its open position. Simultaneously, the applicator is pulled across a series of plates laid out on a mounting board. Up to 5 (20 × 20 cm) plates or 20 (20 × 5 cm) plates can be coated in one operation. The applicator is self-adjusting to the surface of the plates because it rides on the plate during

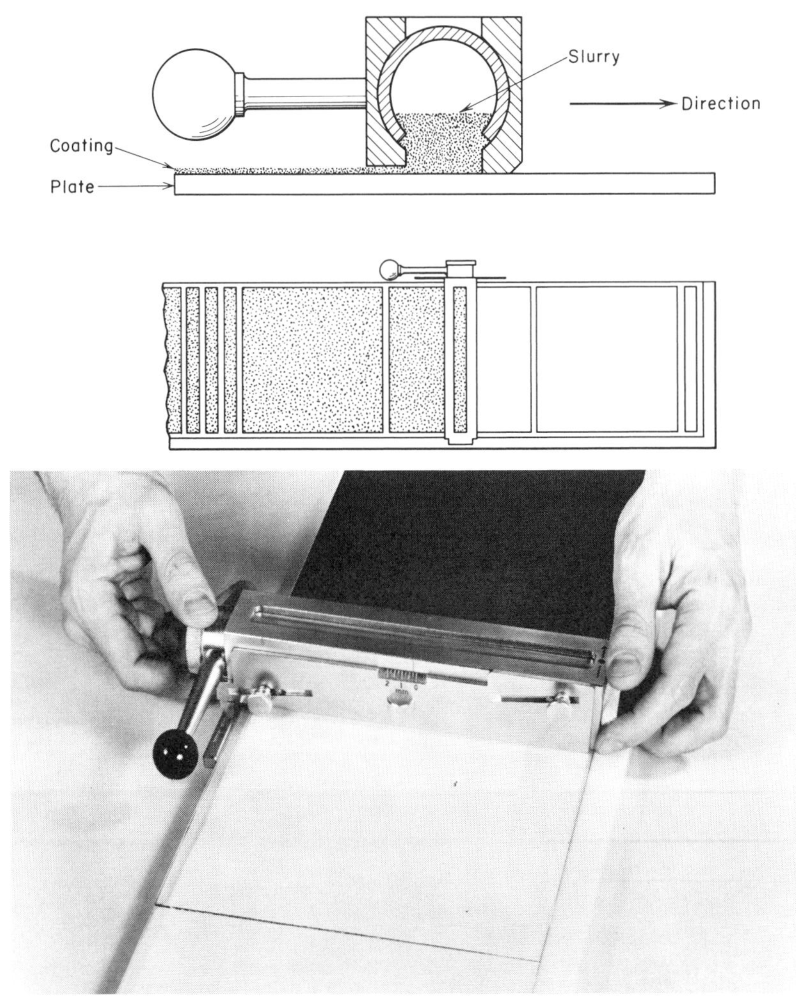

FIG. 10-1. Mounting board with plates and applicator.

the coating procedure. The height of the exit gate, adjustable from 0 to 2000 μ, controls the thickness of the layer. Applicators of the Miller–Kirchner type are stationary. The glass plates are pushed under a reservoir containing the coating slurry. Inexpensive applicators with a fixed exit gate can be constructed or purchased. Plates should be carefully beveled for safe handling. For heating above 130°C (but below 200°C) plates should be fabricated from borosilicate glass.

Precoated Systems Ready-to-use thin layers, prepared with the most widely used adsorbents, are available as precoated glass plates, plastic foils, and chromatotubes. Although somewhat more expensive, precoated layers avoid the difficulty of producing uniform coatings, handling messy slurries, and the investment in the complicated equipment needed to deposit the slurries. Coating procedures, however, will still be used to make special layers, for which there is limited demand commercially.

Activation of Adsorbent In partition chromatography unactivated plates are used, the chromatoplate being dried overnight at room temperature. Cellulose ion-exchange adsorbents need no activation and may be dried at room temperature.

In adsorption chromatography, the activity is directly proportional to the duration and temperature at which the chromatoplate is dried and the dryness of the environment in which it is stored. Silicic acid is activated at 100–250°C. Drying time depends on activity grade desired. The R_f values vary according to the degree of activation of the layer. R_f values in thin-layer chromatography are much less reproducible than in paper chromatography, unless great pains are taken to standardize chromatographic conditions.

After activation, plates should normally be protected against moisture, especially if the relative humidity in the laboratory is high. For this purpose, either a desiccator, which is large enough to hold a drying rack with plates, or a storage cabinet with shelf inserts can be used.

Adsorbent Standardization Relative R_f values are determined by the extent of adsorption of the migrating compound. The simplest expression for R_f as a function of the partition coefficient assumes that K_d is constant throughout the separation. Three factors are implicit in this assumption: (1) K_d does not change with the concentration of solute or its position on the sheet; (2) the ratio of quantities of adsorbed (w_a, in g) and solution (V_M, in ml) phases are constant at any point behind the solvent front at any time during separation; and (3) all of the solution phase on the sheet arises from capillary migration of solvent up the sheet. The R_f value is then given by

$$R_f = 1/[1 + (w_a/V_M)K_d] \tag{10-1}$$

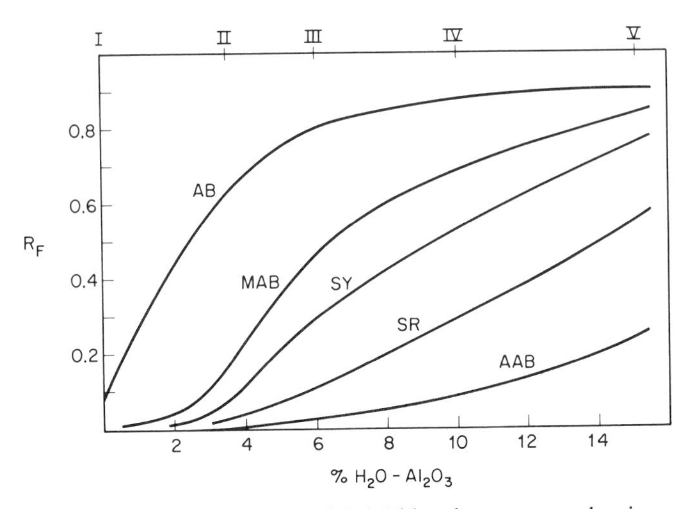

FIG. 10-2. R_f values of Brockmann and Schödder dyes versus alumina–water content. See Table 10-1 for dye abbreviations. [From L. R. Snyder, in *Advances in Chromatography,* Vol. 4 (Eds. J. C. Giddings and R. A. Keller), Dekker, New York, 1967, p. 32.]

From Chapter 7 the fundamental equation [Eq. (7-6)] that relates K_d to solute structure and to the conditions of separation is

$$\log K_d = \log V_a + \phi(S^\circ - \epsilon^\circ A_i) \tag{10-2}$$

By analogy with the R_M function, an R'_M function can be defined for systems obeying Eq. (10-1):

$$R'_M = \log K_d + \log (w_a/V_M) \tag{10-3}$$

Combination of Eqs. (10-2) and (10-3) with elimination of K_d gives a general relationship for the prediction and correlation of R_f values in thin-layer chromatography:

$$R'_M = \log (V_a w_a/V_M) + \phi(S^\circ - \epsilon^\circ A_i) \tag{10-4}$$

When the various parameters on the right side of Eq. (10-4) are known for a particular solute in a given thin-layer system, the R_f value of that solute can be predicted directly. For a given batch of adsorbent, or for two adsorbent samples that are sufficiently similar, both adsorbent parameters decrease together as the water content of the adsorbent is increased. That is, the water content of the adsorbent simultaneously defines the values of both adsorbent parameters. It is possible, according to Snyder,[2] to define adsorbent activity on a single scale using the R_f values of the solute dyes employed by Brockmann and Schödder. For alumina, the curves of Fig. 10-2 can be

[2] L. R. Snyder, in *Advances in Chromatography,* Vol. 4, J. C. Giddings and R. A. Keller (Eds.), Dekker, New York, 1967, pp. 3–46.

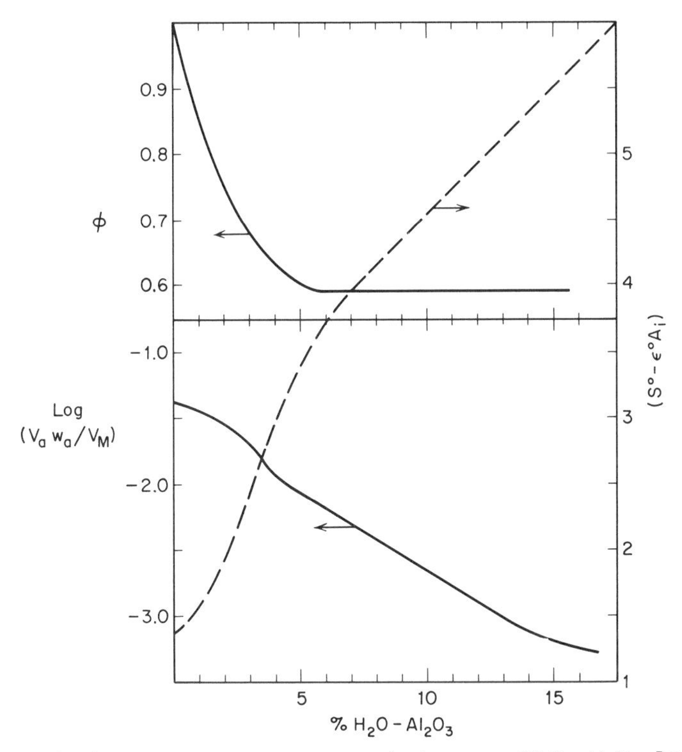

FIG. 10-3. Adsorbent parameters versus nominal percent H_2O–Al_2O_3. [From L. R. Snyder, in *Advances in Chromatography,* Vol. 4 (Eds. J. C. Giddings and R. A. Keller), Dekker, New York, 1967, p. 31.]

used for standardization. The curves of Fig. 10-3 then provide values of the adsorbent parameters versus nominal percent H_2O–Al_2O_3 for all five activity grades of alumina. A similar approach is probably applicable to other commercial adsorbents. However, whenever adsorbent surface area differs significantly from that typical of commonly used adsorbents (i.e., 200 to 400×10^4 cm²/g), a single activity scale is not expected to apply.

Whether a single activity scale holds or not, the adsorbent can be rapidly and quantitatively standardized by development of the standard dyes with CCl_4. If preliminary comparison of the dye R_f values with Fig. 10-2 shows that apparent adsorbent activity to vary significantly depending on which dye is used to calculate activity, the single activity scale must be abandoned and the adsorbent parameters determined by plotting R'_M for each dye against its value of $(S° - \epsilon° A_i)$ from Table 10-1.

Sample Development

The choice of solvent system and the composition of the thin layer is determined by the principle of chromatography to be employed.

TABLE 10-1. Standardization of Alumina on Chromatoplates by the Brockmann-Schödder Procedure with CCl_4 Development

		Adsorbent parameters			
Solute	$(S° - \epsilon° A_i)$	*Activity grade*	$\%H_2O-Al_2O_3$	$\log (w_a V_a / V_M)$	ϕ
Azobenzene (AB)	2.26	I	0.0	−1.37	1.00
p-Methoxyazobenzene (MAB)	3.70	II	3.4	−1.80	0.66
Sudan yellow (SY)	4.24	III	6.0	−2.18	0.59
Sudan red (SR)	5.07	IV	10.0	−2.66	0.59
p-Aminoazobenzene (AAB)	6.2	V	15.0	−3.20	0.59

From L. R. Snyder, in *Advances in Chromatography,* Vol. 4, J. C. Giddings and R. A. Keller (Eds.), Dekker, New York, 1967, p. 31.

Solvent Systems In adsorption chromatography on alumina or silica gel it is relatively easy to select the required solvent system when the substances to be separated belong to a known class of compound. In Fig. 10-4 is depicted a nomograph showing the relationship among the activity grade of the adsorbent, the compound class, and the solvent system. For unknown mixtures the proper developing solvent is ascertained by spotting the mixture onto microslides and developing the microslides with individual solvents spaced throughout the eluotropic series (see Table 7-2).

Separations on cellulose resemble those achieved in paper chromatography (*q.v.*) with the advantage that thin-layer methods always give superior

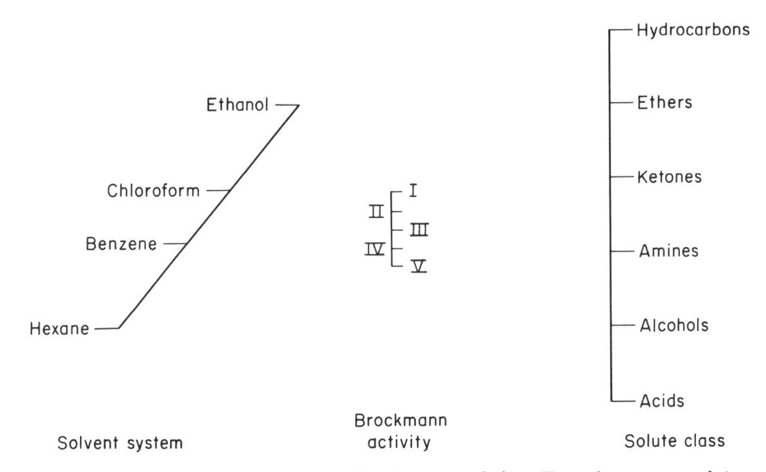

FIG. 10-4. Nomograph interrelating adsorbent activity (Brockmann scale), solvent system, and solute classification.

resolution. Thin layers of cellulose or silica gel incorporating buffers and special reagents have been mentioned earlier. An enumeration of typical solvents for liquid partitition chromatography was given in Chapter 9.

Sample Application The sample to be chromatographed is applied with a microsyringe near one end of the chromatoplate. Samples may range from 1 μl to 1 ml, containing from 0.01 μg to 10 mg of material. The solvent used to dissolve the sample should be as nonpolar as possible and reasonably volatile.

A spotting template aids in giving reproducible placement of samples. Spots can be located 1 cm apart, enabling up to 18 locations on a single plate, 20 cm in width. If desired, a layer can be divided into individual columns by scoring the layer either with a pointed tool or a mechanical scriber.

A wide-band streaking pipet is useful for applying larger amounts of sample as a streak. The pipet consists of two glass plates clamped together with a stainless steel holder. One edge of each plate is beveled so that the point of contact is a thin line. The smooth plate contains the sample reservoir notch; the other plate is indented with numerous channels through which the sample is drawn in and later expelled.

Development of Chromatoplate Chromatoplates are usually developed by the ascending technique at room temperature to a height of 15–18 cm. Tanks (Fig. 10-5), lined with filter paper to saturate the atmosphere, will accommodate two plates. A single-plate sandwich-type chamber (Fig. 10-6) provides very narrow chamber dimensions. It requires only limited quantities (15–30 ml) of solvent and the reduced space means better saturation of the atmosphere. In 20–40 minutes the chromatograms usually will be complete. The plates are then removed and allowed to dry.

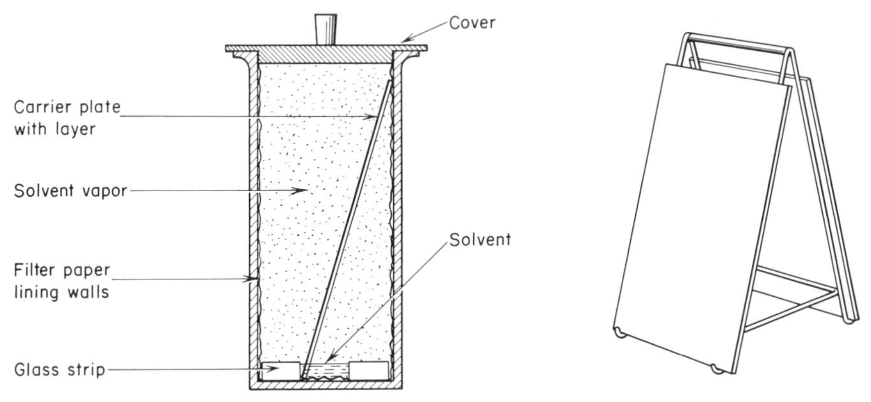

FIG. 10-5. Developing chamber for chromatoplates.

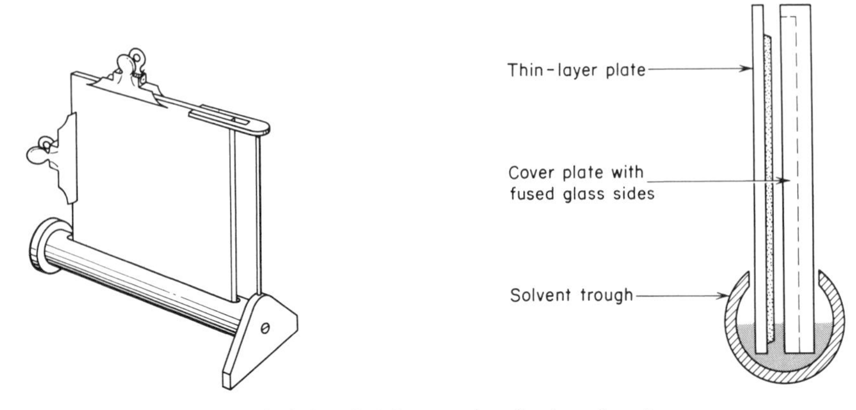

FIG. 10-6. Sandwich-type developing chamber.

Types of Development All the development techniques discussed in the chapter on paper chromatography are applicable to thin layers. The superior resolution obtained in thin-layer chromatography is due to the extremely small diffusion rate in the adsorbent layer. Reactions such as oxidation, reduction, dehydration, hydrolysis, and the formation of derivatives can be carried out directly on the chromatoplate.

Substances not resolved at room temperature may be separated at elevated temperatures. Continuous development in one direction may also improve separations. The solvent is simply allowed to run over the top of the plate where it is sucked up by an adsorbent pad. Where a mixture contains groups of compounds which do not separate satisfactorily using one solvent system, step-wise development often yields better separations. Generally two systems are used that vary in polarity and/or pH. Another method of separating mixtures having a very large polarity range is gradient elution, that is, by developing with a solvent of continuously changing composition. The same effect can be achieved by thin-layer chromatography on a gradient layer[3] using a solvent of constant composition. In this procedure two different adsorbent slurries are applied simultaneously in such a way that the final layer is a gradient mixture of both.

Reversed-phase development is done on plates which have been impregnated with paraffin, silicone oil, undecane, and *n*-tetradecane. They are generally applied by immersing the plate very slowly in a 5–10% solution of the impregnating agent in petroleum ether or diethyl ether. After evaporating the solvent the layer is ready for chromatography. Paraffin and silicone oil provide permanent impregnation whereas undecane can be removed after development by heating the plate for 45 min at 120°C. Polyethylene glycol, another impregnant, can be incorporated in the original slurry before

[3] E. Stahl, *Angew. Chem.* **3,** 784 (1964) (English Edition).

the layers are cast. The solvents used in reversed-phase development generally contain acetic acid or acetonitrile. When using the reversed-phase technique, longer development times are required and the capacity of the layer is decreased.

All of these variants are applicable to separations on a preparative scale. In contrast, two-dimensional, two-solvent development can be used only for analytical purposes. A sample is placed in a spot on the lower left-hand corner of a chromatoplate. A mixture of standards is placed in a similar position in the lower right-hand corner and also in the upper left-hand corner. The plate is positioned in a chromatographic tank containing the first solvent. When the sample and reference compounds have ascended to a distance of 16 cm, the plate is removed from the tank and permitted to air dry for 10 min. It is then turned 90° to the other side adjacent to the origin and developed in a tank containing the second solvent. This method provides separation of standards, as well as sample, in both directions, so that the right-angle migrations of the two reference mixtures form the coordinates of a graph which permits the unknowns to be located in two dimensions.

Visualization and Documentation

Colored substances can be viewed directly against the color of the stationary phase. Some compounds fluoresce in ultraviolet light. Colorless substances may be detected by spraying over the surface reagents which produce colored areas in the positions occupied by the solutes. Among the widely used reagents are indicators for acidic and basic substances, ninhydrin for amino acids, antimony trichloride for steroids and essential oils, and bromofluorescein for unsaturated substances. In fact, all reagents normally used in paper chromatography, and which do not involve a mechanical washing step, can also be used in thin-layer chromatography. In addition, corrosive reagents, such as chromic–sulfuric acid, can be sprayed onto inorganic layers and any organic compounds charred by heating.

Fluorescing material can be incorporated in the adsorbent or in the solvent used to prepare the slurry, or can be sprayed on the plate after development. In such cases the solute's location will appear as a dark spot on a fluorescent background when viewed under shortwave ultraviolet light. After inspection of the developed plates, the fluorescence of fluorescein compounds can be extinguished by exposure to bromine vapor and, if olefins are present, their bromo derivatives then fluoresce bright yellow.

The surface of the layer or sheet can be scanned with a radiation detector when radiotracers are incorporated in the sample components. For tracers, an autoradiograph is exposed long enough to bring up the impurity and this is determined on a densitometer. A direct photoprint can be obtained for substances which absorb ultraviolet light. The plate is placed directly onto copy paper and the whole exposed to ultraviolet light for several seconds, and then the photocopy is developed to reveal the spots.

In preparative thin-layer chromatography, destructive spray reagents cannot be used to visualize materials that they modify or destroy. However, reagents can be streaked or brushed along one edge of the plate to locate band positions.

After development, desired zones can be removed from the plate by scraping or by cutting apart when flexible sheets are utilized. A "vacuum cleaner" collects samples directly in an extraction thimble held in a vacuum flask; this is ideal for preparative work.

A permanent record of the developed chromatogram can be cheaply prepared by tracing the revealed spots onto semi-transparent paper laid on the plate. Alternatively the plate can be photographed in black and white or color (with a Polaroid camera) or an ultraviolet contact photoprint can be made. After visualization of the spots with a color reagent the adsorbent can be treated with paraffin to make it translucent and the plate then used in an office copying machine. A plastic dispersion (e.g., Neatan from Merck) can be sprayed on the plate and allowed to set. The whole layer is then peeled from the glass plate and stored as a flexible film.

Quantitative Methods

Quantitative analysis of separated components on thin layers is most commonly performed by some measurement of the density and area of the spot by photodensitometry of plates or photographs in transmitted or reflected light and direct fluorimetry of spots on the plate. All of these procedures require comparison with known amounts of model mixtures that, for reasonable accuracy, must be chromatographically examined alongside the sample on the same plate.

Radiochemical analysis can be effected on the intact layer with a proportional flow counter or indirectly by photodensitometry of autoradiographs.

Analysis of separated components after removal from the plate is generally carried out on the materials eluted from the scraped-off adsorbent. Compounds are normally determined by ultraviolet, infrared, visible, or fluorescence spectrophotometry or, if no chromophoric groups are present, by colorimetry after reaction with a suitable chromogenic reagent. Bioassay of biologically active materials is another possibility.

Applications

Thin-layer chromatography is particularly effective for rapidly analyzing large numbers of samples. It has proved to be especially useful for separations of classes of compounds which are either insufficiently volatile or too labile for analysis by gas–liquid chromatography. Reactions can be studied *in situ* in the adsorbent layer. The products of a synthesis or purification procedure can be followed. It can check on impurities in solvents and reagents, and can be used routinely both for isolation of material and determination of final purity.

Forensic chemists in crime and toxicology laboratories find thin-layer chromatography perfectly suited for analysis of evidence obtained in narcotics, forgery, arson, and a wide range of other legal matters. It can be used to establish that a bit of ink or a fleck of a signature came from a particular writing device or proprietary formulation. It can even identify the constituents of the paper on which the writing is done.

Production formulas can be checked for omission of ingredients. Other uses include the analysis of competitive products, separation of plasticizers and antioxidants, separation of ink and dye formulations, and quality control of products.

LABORATORY WORK

One may prepare his own chromatoplates, or prepared plates and "instant" thin-layer sheets may be purchased. A number of commercial houses market TLC kits for student use and as an introduction to the technique.

Preparation of Chromatoplates

The glass plates must be thoroughly cleaned and completely free from grease. Scrub the plates with a scouring powder, or immerse for several hours in sulfuric–chromic acid mixture or a concentrated solution of sodium carbonate. This is followed by brushing under tap water and rinsing clean with distilled water. The cleaned plates are left to dry on a rack at room temperature.

For coating, the cleaned plates are placed in a row, with edges in alignment. For *silica gel* plates, 25 g of silica gel (Merck silica gel G, 200 mesh, containing 13% purified plaster of Paris as binder, or the equivalent types from other suppliers; silica gel GF_{254} contains a fluorescence indicator upon exposure to ultraviolet light of 254 mμ) is shaken thoroughly with 50 ml of distilled water in a 150-ml flask fitted with a plastic stopper for 30 sec. The thin paste is transferred immediately to the spreader which is then pulled across the row of plates with a uniform motion (about 3 sec per plate), not stopping until all the plates are covered. If wrinkles appear in the thin layers, take each plate while still wet and gently bump the edge of the plate on a hard vertical surface. Clean the spreader immediately, brushing under tap water and rinsing with distilled water.

The thickness of the layer is usually 0.250 mm, except in preparative work where it may be as large as 2 mm. Twenty-five grams of silica gel suffices for three 20 × 20 cm plates of 0.250-mm thickness. With some applicators the thickness can be controlled by setting the exit blade at the inscribed thickness; less expensive models provide only the standard thickness. Plates can also be prepared without a spreader. Tape or spacers of desired thickness are attached to the edges of each plate. The slurry is spread or poured on the plates, then drawn or rolled out with a rod or knife edge into a uniform layer. For 0.250 mm, 32-gauge metal strips provide the correct depth.

Leave the prepared chromatoplates on a level surface or on the template for superficial drying (about 15 min or until the surface of the layer becomes dull), then separate and place the individual plates in a rack with guide rails and dry them in an oven for 1½ hr at 110°C for activation. They are then ready for use or they may be transferred to a storage cabinet which contains "blue" silica gel as desiccant.

For *alumina* plates, 25 g of aluminum oxide G is shaken with 50 ml of distilled water and handled as described for silica gel plates. The final activation depends upon the compounds to be separated. For highest activity, alumina plates require 4 hr at 135°C followed by storage over blue silica gel or alumina of the required Brockmann activity.

Basic silica gel layers are prepared by mixing silica gel with 0.1*N* NaOH, Sorensen buffer (pH 6.8), or 0.15*M* sodium acetate solution instead of water. *Acidic silica gel* layers are prepared by using 0.2*M* oxalic acid or 2.5% (*v*/*v*) sulfuric acid solution instead of water. *Impregnation* with 0.1*M* boric acid solution is used for the separation of sugars, and impregnation with a 10% silver nitrate solution for the separation of unsaturated compounds. These modified silica gel layers are best dried at room temperature overnight.

For layers containing no binder (such as silica gel H, Merck), great care must be taken not to shake the plate which is of necessity developed in a near horizontal position. Suspensions without a binder can be kept for quite a long time after preparation. They are not recommended for beginners.

Fluorescent plates can be prepared by using a 0.04% solution of fluorescein instead of water to prepare the slurry. Silica gel is available commercially with zinc silicate or other phosphors added.

Reversed-phase chromatoplates are prepared by immersing the adsorbent layer (on the plate) very slowly in a 5 or 10% solution of the impregnating agent (paraffin, silicone oil, undecane, or *n*-tetradecane) in petroleum ether or diethyl ether. After removing the plate and evaporating the solvent, the layer is ready for chromatography. Paraffin and silicone oil provide permanent impregnation whereas undecane can be removed after development by heating the plate for 45 min at 120°C.

Cellulose powder and *ion-exchange celluloses* for thin-layer chromatography (designated DEAE, ECTEOLA, and CM; see Chapter 6) adhere well to glass without a binder. Shake 10 g of powder with 60 ml of distilled water for 30–45 sec; homogenizing in a mixer is best. The suspension is spread onto the plates in the usual manner. Plates may be dried for 1 hr at 50°C or overnight at room temperature. They need not be stored in a desiccator.

For *gel filtration,* plates which are clean and dry are coated with the swollen superfine gel (10- to 40-μ particle size). The gel adheres instantly to the plates without the need for added binder. Any commercial spreading device may be used. The plates are stored in a moist chamber or dried very gently, if not immediately required. To obtain a gel of suitable consistency, the different Sephadex types must first be swollen for varying lengths of time in the amount of buffer shown below.

Type of Sephadex	Amount of buffer (ml) to be added to 1 g of dry Sephadex	Swelling time (hr)
G-25 Superfine	5	5
G-50 Superfine	11	5
G-75 Superfine	15	24
G-100 Superfine	19	48
G-150 Superfine	22	60
G-200 Superfine	25	72

A slurry of suitable consistency for making thin layers is obtained by allowing a suspension of the particles in buffer to settle for ½ hr, then removing as much excess of buffer as possible through a narrow tube with gentle suction. Layers 500 μ in thickness are generally more uniform than thinner layers.

Spotting the Sample

A solution of the sample, 0.1 or 1.0% (*w*/*v*) in concentration, and 5–10 μl in volume, is applied from a micropipet on a straight line about 1.5–2.0 cm from the edge of the chromatoplate. Spots are located 1 cm apart and at least 1 cm from the edge parallel to the direction of development. Care should be taken when spotting onto the absorbent layer to prevent the pipet digging into the layer. A plastic template is very useful to place over the plate while applying the sample.

The area of the spot should be kept as small as possible. Larger sized samples are repeatedly spotted, allowing the solvent to evaporate in between applications. A hot-air blower speeds drying with solvents of low volatility.

Development of the Chromatoplate

To saturate the atmosphere in the chamber with the developing solvent: it is advisable to line the sides of the chamber with sheets of filter paper which dip into the solvent in the base of the chamber. Approximately 200 ml of solvent is required for a rectangular chamber (25 × 30 × 10 cm); less is needed for sandwich-style chambers.

After evaporation of the solvent used to spot the sample onto the plate, the chromatoplate is immersed in the developing solvent to a depth of 0.5 cm (for ascending chromatography) and leaned against the side of the chamber. Close the chamber and allow development to proceed until the solvent front has advanced the desired distance, usually 10–15 cm from the starting line. Elapsed time will vary from 15–45 min. Remove the plate from the chamber promptly as soon as the solvent front has reached the finishing line, and allow the solvent to evaporate. Record the liquid front with a pencil line, made either before the start of the development (as a guide for the desired distance of travel) or immediately after removing the plate from the chamber.

Redistilled or chromatographic grade solvents should be used. The quality of both solvent and adsorbent coating may be checked by two-dimensional thin-layer chromatography using the same solvent in both directions. Discolored solvent fronts in both directions are an indication of an impure solvent, whereas concentration of a discolored substance at one edge of the plate points to an impure coating material.

Visualization of Solutes Colored substances can be detected directly and their distance from the starting line measured. Fluorescent plates are inspected with ultraviolet radiation (CAUTION: Wear protective goggles to avoid damage to the eyes from scattered ultraviolet radiation) by means of a viewing box. Spots can be conveniently marked by dotting with a needle. Exposure of the plate to bromine vapor will extinguish the fluorescence of fluorescein and cause certain unsaturated compounds to fluoresce.

Many materials are rendered visible by spraying with selective reagents. Examples of suitable spray reagents are listed in the books of Bobbitt, Randerath, Stahl, and Truter listed in the Bibliography at the end of this chapter and in the paper by Wollich, Schmall, and Hawrylyshyl.[4] A sprayer with a changeable glass flask and disposable aerosol propellant can is convenient. In any case, the spray should be very fine and as little of it used as possible. The sprayer is used in a horizontal motion, beginning in the upper left of the plate, proceeding to the right, down and across the paper or plate to the left, in an alternating motion to right and then left, finally moving up and down the plate until the entire area has been covered twice.

[4] E. G. Wollich, M. Schmall, and M. Hawrylyshyl, *Anal. Chem.* **33,** 1138 (1961).

Suggested Systems

Concentrations are roughly 1 mg of a substance in 1 ml of a suitable solvent.

1. Amino Acids A mixture of aspartic acid, glycine, α-alanine, isoleucine, and tyrosine is developed for 45 min on silica gel G with water–ethanol–acetic acid (1:5:0.1) containing 10 mg ninhydrin. After development, air dry and heat in an oven at 110°C, whereupon the colored spots appear. In the order of listing, the R_f values are 0.11, 0.31, 0.45, 0.70, 0.73. Spots are pink in color except for the violet color of aspartic acid.

2. Carbohydrates A mixture of glucose, fructose, and sucrose is developed for 40 min on silica gel G with *n*-butanol–acetic acid–ethyl ether–water (9:6:3:1). After development, air dry and spray with ethanol–sulfuric acid–anisaldehyde (18:1:1) solution then heat at 100°C for 5–10 min. R_f values are 0.55 (light blue), 0.45 (violet), and 0.31 (violet), respectively.

3. Dyes A mixture of *p*-dimethylaminoazobenzene (butter yellow), Sudan red G, and indophenol blue is developed for 12 min with benzene. The yellow, red, and blue colors are self-evident at R_f values of 0.58, 0.22, and 0.11.

Using a developing solvent consisting of methyl ethyl ketone–acetic acid–isopropyl alcohol (2:2:1) and a developing time of 20 min on chromatoplates of silica gel G, these dye mixtures (color, R_f values) are separable: I, fluorescein (yellow, 0.85), rhodamine B (red, 0.43), and malachite green (green, 0.12); II, Sudan yellow (orange, 0.83), crystal violet (violet, 0.20), and methylene blue (blue, 0.02); III, Sudan IV (red-brown, 0.83), *o*-cresolsulfonephthalein (yellow, 0.65), and Victoria blue (blue-grey, 0.30, 0.00).

4. Fatty Acids A mixture of caproic, caprylic, capric, lauric, myristic, palmitic, and stearic acids (spotted from an ether solution) are developed with *n*-butanol (saturated with water) at 20°C until the solvent front has ascended 15 cm (about 12 hr). Dry the plate and stain with concentrated sulfuric acid containing 0.1% potassium dichromate. Average R_f values for the acids are 1.00, 0.85, 0.72, 0.59, 0.45, 0.33, and 0.19, respectively.

5. Humectants in Tobacco Break apart and soak one cigarette in 5 ml of distilled water for 30 min, centrifuge, and apply 4 μl of the solution to a silica gel G plate. On the same plate, spot pure samples of glycerol (0.30), ethylene glycol (0.49), 1,2-propylene glycol (0.61), and butane-2,3-diol (0.68) [and also fructose (0.09), glucose (0.15), and sucrose (0.01)]. Develop with acetone for 45 min. Dry and spray the chromatoplate with 1% lead tetraacetate in dry benzene, then heat at 110°C for 5 min. White spots appear on a brown background. R_f values are given in parentheses after each compound.

6. Two-Dimensional Separation of Amino Acids Spot the unknown mixture in the lower left corner of the chromatoplate of silica gel G. Spot a known mixture of amino acids in both the lower right and upper left corners. Develop in the first direction (ascending relative to sample spot in lower left corner of plate) with water–ethanol–acetic acid (10:50:1) until the solvent front has advanced 10 cm. Air dry, and immerse the plate in the second chamber that contains as developing solvent a mixture consisting of *n*-butanol–methyl ethyl ketone–water (2:2:1) and a beaker of cyclohexylamine (1 ml for every 25 ml of solvent mixture). Develop until the solvent front advances 10 cm. Spray with a 0.25% (*w*/*v*) ninhydrin solution in acetone containing 7% glacial acetic acid; dry and heat for 15 min at 45°C. If the amino acids enumerated under System No. 1 are employed, the R_f values in the second direction are 0.11, 0.31, 0.35, 0.70, and 0.73, respectively.

7. Dicarboxylic Acids Plates are coated with a mixture consisting of 30 g kieselguhr G, 0.05 g sodium diethyldithiocarbamate, 45 ml water, and 15 g polyethylene glycol *M* 1000. The coated plates are dried in air 10 min and activated at 100°C for 30 min. The sample is spotted on the plate and developed with diisopropyl ether–formic acid–water (90:7:3) [previously saturated with polyethylene glycol *M* 1000] until the solvent front has advanced 12 cm. Dry the chromatoplate at exactly 100°C in an acid-free environment for 10 min. Spray with 0.04% (*w*/*v*) solution of bromocresol purple in 50% ethanol (adjusted to pH 10). Light yellow spots on an intense blue background distinguish the location of the dicarboxylic acids (R_f values in parentheses): oxalic (0.14), malonic (0.21), succinic (0.28), glutaric (0.36), adipic (0.43), pimelic (0.55), suberic (0.67), azelaic (0.82), and sebacic (0.92).

Problems

1. On buffered (pH 6.8) silica gel G, with 70% ethanol as solvent, these R_f values were obtained: methylamine, 0.10; ethylamine, 0.20; *n*-propylamine, 0.30; isoamylamine, 0.45; ethanolamine, 0.10; benzylamine, 0.50. (a) Predict the R_f value for *n*-butylamine. (b) Comment on the positions of ethanolamine, isoamylamine, and benzylamine.
2. On silica gel G, activated at 105°C, these R_f values were obtained for Rhodamine B in the several solvent mixtures: Chloroform–acetone–diethylamine (5:4:1), 0.58; chloroform–diethylamine (9:1), 0.49; and cyclohexane–chloroform–diethylamine (5:4:1), 0.20. Comment on the observed R_f values in terms of the solvent eluting strength and the structure of Rhodamine B.
3. Graph the relationship between R_f value and carbon number for (a) the fatty

acids enumerated under "Laboratory Work, Suggested System 4"; (b) the dicarboxylic acids enumerated in "Suggested System 7"; and (c) the amino acids enumerated in "Suggested Systems 1 and 6."

BIBLIOGRAPHY

Bobbitt, J. M., *Thin Layer Chromatography,* Reinhold, New York, 1963.
Kirchner, J. G., *Thin-layer Chromatography,* Interscience, New York, 1967.
Randerath, K., *Thin Layer Chromatography,* Academic Press, New York, 1963.
Stahl, E. (Ed.), *Thin Layer Chromatography—A Laboratory Handbook,* Academic Press, New York, 1964.
Truter, E. V., *Thin Film Chromatography,* Cleaver-Hume Press, London, 1963.

11 *Gas Chromatography*

Gas chromatography differs from other forms of chromatography only in that the mobile phase is a gas and the solutes are separated as vapors. Gas–liquid chromatography accomplishes a separation by partitioning a sample between a mobile gas phase and a thin layer of nonvolatile liquid held on a solid support. Gas–solid chromatography employs a solid sorbent as the stationary phase.

The sequence of a gas chromatographic separation is as follows. A sample containing the solutes is injected onto a heating block where it is immediately vaporized and swept as a plug of vapor by the carrier gas stream into the column inlet. The solutes are adsorbed at the head of the column by the stationary phase and then desorbed by fresh carrier gas. This partitioning process occurs repeatedly as the sample is moved toward the outlet by the carrier gas. Each solute will travel at its own rate through the column. Their bands will separate to a degree that is determined by their partition coefficients and the extent of band spreading. The solutes are eluted, one after another, in the increasing order of their partition coefficients, and enter a detector attached to the exit end of the column. Here they register a series of signals, resulting from concentration changes and rates of elution. If a recorder is used, the signals appear on the chart as a plot of time versus the composition of the carrier gas stream (Fig. 11-1).

The appearance time, height, width, and area of these peaks on the chromatogram can be measured to yield qualitative and quantitative analytical data. The main advantage of gas chromatography as compared with other column chromatographic methods is that longer columns can be used and higher separation efficiencies obtained. Speed results from the low viscosity of gases and vapors and the fast partition equilibrium between thin films of gas and liquid. Sensitivity is indicated by the detection of 10^{-12} g per sec in optimum instances. The gaseous phase reacts less with the stationary phase and the solutes than most liquid mobile phases. Although the gas chromato-

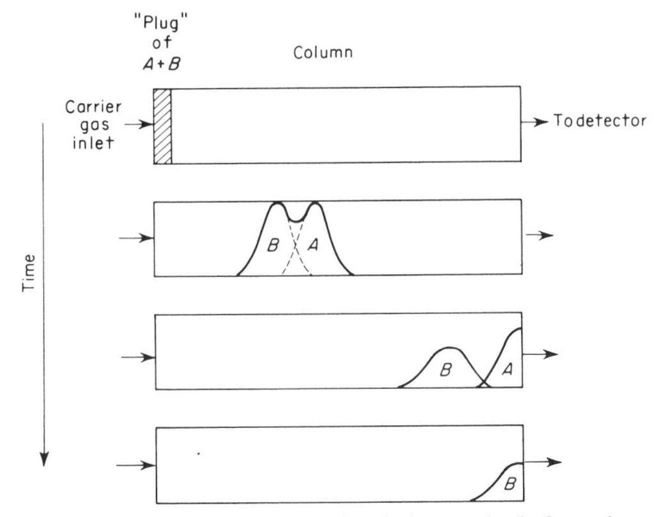

FIG. 11-1. Schematic diagram illustrating the elution method of gas chromatography.

graphic technique is limited to volatile materials, the availability of gas chromatographs working at temperatures up to 500°C and the possibility of converting almost any material into a volatile compound extend the applicability of the method.

GAS CHROMATOGRAPHS

Basically a gas chromatograph consists of six parts: (1) a supply of carrier gas in a high-pressure cylinder with attendant pressure regulators and flow meters, (2) a sample injection system, (3) the separation column, (4) the detector, (5) an electrometer and strip-chart recorder, and (6) separate thermostated compartments for housing the column and the detector so as to regulate their temperature. The components are shown schematically in Fig. 11-2. Each of these components will now be discussed in some detail.

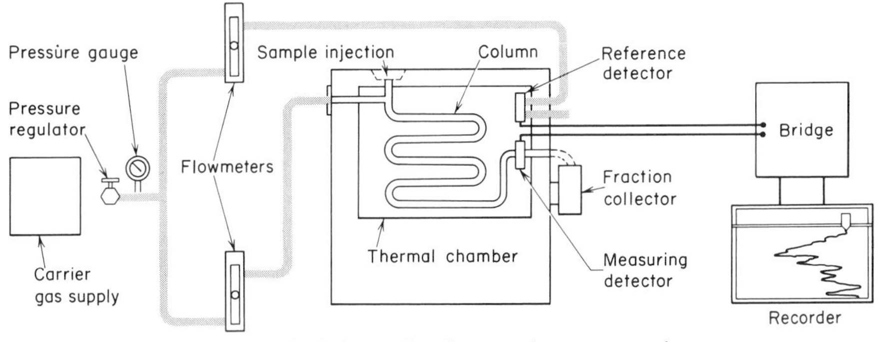

FIG. 11-2. Schematic of a gas chromatograph.

Pressure Regulator and Flow Meter for Carrier Gas

Operating efficiency of a chromatograph is directly dependent on the maintenance of a highly constant carrier gas flow. Carrier gas from the tank passes through a toggle valve, a flow meter, a few feet of metal capillary restrictors, and a pressure gauge (1–4 atm). The flow meter, with a range of 1–1000 liters per min, indicates flow rate. Flow rate is adjustable at this point by means of a needle valve mounted on the base of the flow meter and is controlled by the capillary restrictors.

Contaminants in the carrier gas may affect column performance and detector response when ionization detectors are used. Inclusion of a trap containing a molecular sieve 5Å is usually sufficient for removal of hydrocarbon gases and water vapor; for measurements requiring extreme sensitivity a trap at −180°C in a bath of liquid nitrogen may be used.

Carrier gas is either helium, nitrogen, hydrogen, or argon. Availability, purity, consumption, and the type of detector employed determine the choice. Helium is preferred for thermal conductivity detectors because of its high thermal conductivity relative to that of most organic vapors, but it is expensive and not readily obtainable in some countries. Nitrogen is used in preparative gas chromatographs because of the large consumption of carrier gas involved.

Sample Injection System

The most exacting problem in gas chromatography is presented by the sample injection system. This deceptively simple little device must introduce the sample in a reproducible manner. It must vaporize a liquid rapidly without either decomposing or fractionating the sample.

Liquid samples are injected by a microsyringe with the needle inserted through a self-sealing, silicone-rubber septum into a heated metal or glass-lined block—the flash evaporator. A typical arrangement is shown in Fig. 11-3. The metal block is heated by a controlled resistance heater. Here the sample is vaporized as a "plug" and carried into the column inlet by the carrier gas. Every effort should be made to get the needle tip close to the column packing or well down into the heated block ahead of the column. The manipulation of the syringe is virtually an art developed with practice. Insertion, injection, and withdrawal of the needle should be performed quickly but smoothly. For maximum efficiency one should use the smallest possible sample size (0.5 to 10 μl) consistent with detector sensitivity. Repeatability is about 2%.

Gaseous samples are injected by a gas-tight syringe or by means of a calibrated bypass loop. In the simplest form the latter device is merely a glass system of three stopcocks between two of which there is a standard volume in which gas is trapped. The sample loop is switched in and out of the carrier gas stream by means of valves.

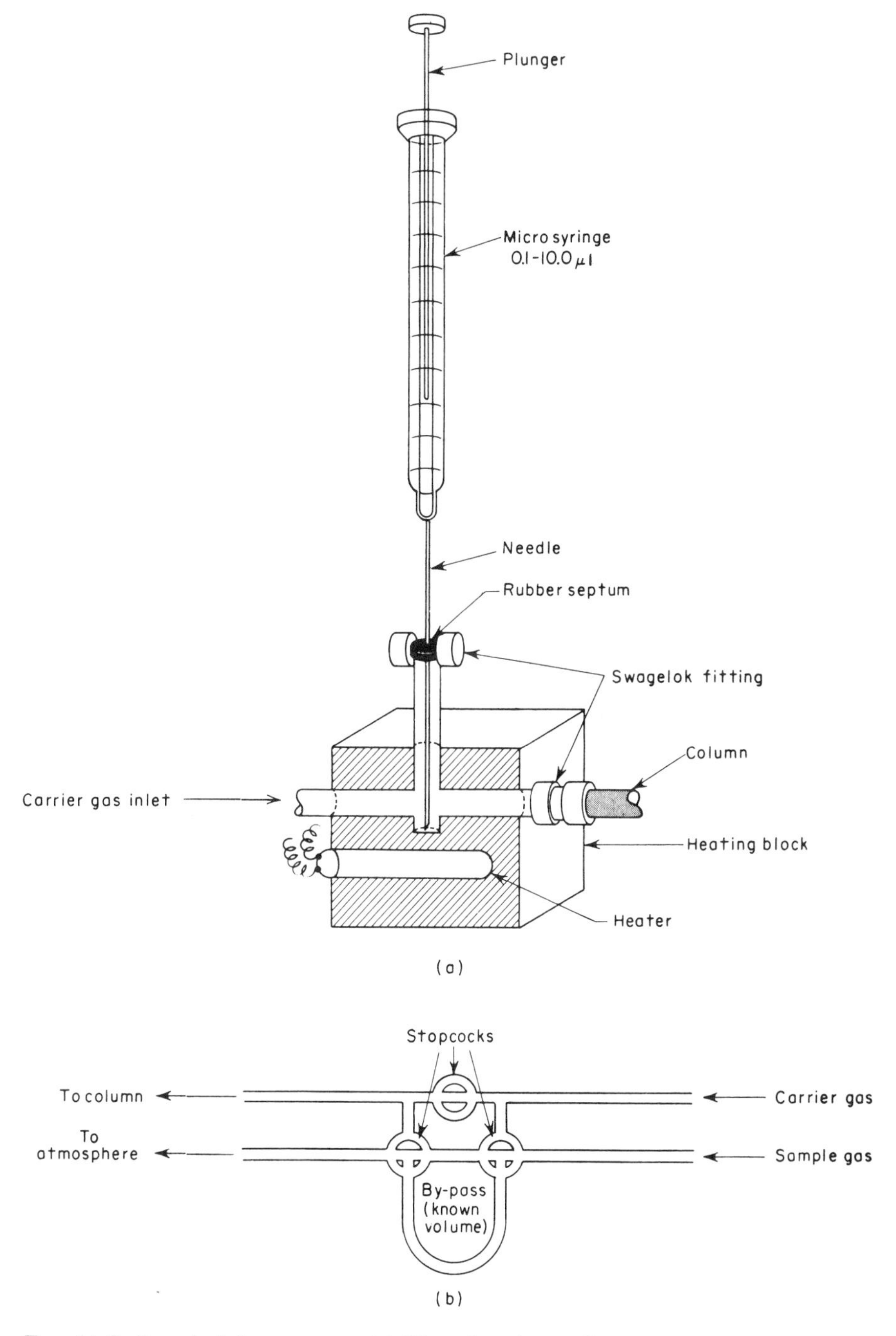

FIG. 11-3. Sample inlet systems: (a) Hypodermic needle syringe and heater block for liquids; (b) gas sample introduction.

A stream splitter serves to obtain minute liquid samples for capillary columns. When placed between the column and detector, the splitter allows any portion of the sample leaving the column to be diverted to a second detector or to an outlet to which a fraction collector may be attached. The stream splitter must be highly non-discriminatory and separate each component of the sample mixture in exactly the same ratio.

Chromatographic Column

The heart of the chromatograph is the column in which the separation takes place. Columns are made of tubing either bent in a U-shape or coiled into an open spiral or a flat, pancake shape. Copper is recommended for temperatures up to about 250°C, and stainless steel for higher temperatures. Glass columns must be used if materials are decomposed by metallic contact. Column diameter varies from capillary dimensions to columns whose outside diameter may be 1/16, 1/8, 1/4, 3/8 in., or even larger for preparative work. Swagelok fittings make installation and change of columns easy and rapid. When separation problems are not critical a very short column (2–5 cm) can be used to advantage. Packed columns normally are used in lengths of 0.7–2.0 meters. Open tubular column lengths run anywhere from 30 to 300 meters. Table 11-1 compares columns of different diameters.

TABLE 11-1. Comparative Parameters of Gas Chromatographic Columns

	Open tubular	*1/16-in. O.D.*	*1/8-in. O.D.*	*1/4-in. O.D.*	*3/8-in. O.D.*
Inside diameter, mm	0.25–0.50	1.2	1.65	3.94	8.0
Practical length, meters	100	20	20	20	30
Maximum plates	300,000	60,000	48,000	30,000	15,000
Maximum plates/meter	3000	3000	2400	1500	750
Liquid phase, %	...	3	5	10	20
Liquid film thickness, μ	1	5	5	10	20
(Average) linear velocity, cm/sec	25	10	10	7	7
Maximum sample size, μl	0.010	1.0	2.0	20	1000

Supports Inasmuch as the column is the heart of the chromatographic system, the support[1] plays a key role in the resultant chromatogram. In choosing a material as support, two factors must be considered: (1) structure and (2) surface characteristics. The structure contributes to the efficiency of the material as a support, and the surface characteristics

[1] J. F. Palframan and E. A. Walker, *Analyst* **92,** 71 (1967); D. M. Ottenstein, *J. Gas Chromatog.* **1** (No. 4), 11 (1963) and in *Advances in Chromatography*, J. C. Giddings and R. A. Keller (Eds.), Vol. 3, pp. 137–196, Dekker, New York, 1966.

govern the degree to which it enters into the separation. Ideally the function of the support is to act as an inert platform, capable of holding immobile a relatively large volume of the liquid phase as a thin film over its surface. The surface area of the material should be reasonably high so as to expose a large surface area to ensure rapid attainment of equilibrium between the stationary and mobile phases. Furthermore, the support material should have good handling characteristics, that is, it must be strong enough to resist breakdown in handling and be capable of being packed into a uniform bed in a column. The ideal chromatographic support has not yet been found.

The most commonly used supports are made from diatomaceous earth. Natural diatomite (kieselguhr, filter-aid), which is somewhat fragile, is treated by calcining with a flux of sodium carbonate above 900°C. The fluxing agent causes incipient fusion of the finer particles forming coarser aggregates. A portion of the microamorphous silica is converted into crystalline cristobalite. This white, flux-calcined material is marketed under trade names such as Chromosorb W, Celite, Diatoport W, Gas-Chrom, and Anakrom. The pink material consists of diatomite that has been crushed, blended, and pressed into a brick, then calcined above 900°C without the addition of a flux. The mineral impurities form complex oxides or silicates which impart a pink color to the material which is marketed as crushed firebrick or Chromosorb P. The pink support has a greater density than the white support, it is less fragile, and is capable of holding a large volume of liquid phase before becoming too sticky to flow freely. Differences in the structure of the two supports are reflected in the void space in the column. A packed column of the uncoated white is approximately 90% void space; the pink is approximately 80% void. If there were no internal void in the particles, the void space in the column would be approximately 67%. Because of the differences in the bulk density of the two supports, with the pink twice as heavy as the white, columns prepared with the same percentage by weight of liquid phase will differ considerably. The pink column will contain twice as much of the liquid phase as the white. The white support has pore sizes of about 9 μ, while the pink support has a smaller pore size of about 2 μ. This explains the higher efficiency of the pink support which holds the liquid phase in smaller and shallower pools which require shorter transit times for the solute.

Both pink and white supports prepared from diatomaceous earths suffer from active sites on their surfaces which act to cause unsatisfactory tailing with the more polar solutes. These active sites are due partly to metallic impurities and partly due to silanol (SiOH) and siloxane groups. Both give rise to hydrogen-bonding effects. Acid washing removes mineral impurities from the support surface and reduces surface activity caused by —OH groups associated with iron and aluminum, but does nothing to mitigate the effect of the silanol and siloxane groups. Silanization (usually preceded by acid washing) converts the surface silanol groups to silyl ethers. Dimethyldichlorosilane or hexamethyldisilazane are frequently used reagents. As a result, surface activity and peak tailing are considerably reduced. At the same time, however, silanization reduces the surface area of the support. Therefore, silanized supports should not be used with more than about 10% stationary phase loading.

Coating of the untreated support with a polar stationary phase sometimes reduces peak tailing because the polar groups of the liquid phase permanently block the active sites of the support particles. In certain cases, special additives are mixed with the liquid phase to enhance this effect; for example, a small amount of stearic acid incorporated in the silicone oil eliminates tailing in the separation of fatty acids. To reduce tailing for amines, the support must be coated with several percent of KOH, then coated with the polar stationary liquid phase.

Two special supports are popular for certain applications. Glass beads have a low surface area and low porosity and can be obtained in very uniform size. Only very low stationary phase loadings can be tolerated; the maximum practical limit is about 3%. In most applications phase loadings as low as 0.05 to 0.2% are employed. Such low loadings permit the analysis of high boiling substances at fairly low temperatures. Porous polymer beads, differing in the degree of cross-linking of styrene with an alkyl vinyl benzene, separate the solute components by direct partition from the gas phase throughout the solid polymer (cf. Chapter 12). As adsorption sites are non-existent, highly polar materials are handled without tailing, and as no liquid phase is normally used, column bleed is not a problem even when temperature programming. The beads (marketed as Porapak P, Q, R, S, and T or Polypak) are thermally stable to about 250°C.

The mesh size of the support determines the average particle diameter in packed columns. Since for these columns the height equivalent to a theoretical plate is directly proportional to the particle diameter, theoretically the smallest possible particles should be preferred. On the other hand, the permeability of a column is proportional to the square of the particle diameter and the pressure drop is inversely proportional to the square of the particle diameter. Thus, decreasing particle size will rapidly increase the necessary pressure drop. One answer to the complex interrelationship between plate height, pressure drop, and mesh size, is to lengthen the column and to use coarser support particles. In practical work, the best choice is 80/100 mesh for a 1/8-in. (3-mm diameter) column and 60/80 mesh for a 1/4-in. (6 mm) column, each with diatomaceous-earth-type supports. In fact, for effective packing of any column, the internal diameter of the tubing should be at least eight times the diameter of the solid support particles. A second possibility involves the use of open tubular columns, coated with a liquid (*q.v.*).

Liquid Phases The versatility of gas–liquid chromatography lies in the availability of an almost infinite variety of liquid partition materials. The range of liquids which can be used is limited only by their volatility, thermal stability, and ability to wet the support. No single liquid phase will serve for all separation problems or be usable at all column temperatures. Indeed it is rarely possible to separate all components of any complex mixture by the use of a single liquid substrate. A selected list of liquid phases is given in Table 11-2.

Liquid phases can be classified into five categories: (1) *Nonpolar.* This class comprises the hydrocarbon-type liquid phases such as paraffin, squalane, silicone greases, Apiezon L grease, and silicone gum rubber (SE-30), but not aromatic materials. Generally, nonpolar liquids separate solutes in order of increasing boiling points. (2) *Intermediate polarity.* These liquid phases possess a polar or polarizable group on a long nonpolar skeleton. Typical members include esters of high-molecular-weight alcohols, such as di-2-ethylhexyl phthalate. Members of this class dissolve both polar and nonpolar solutes with relative ease. (3) *Polar.* Liquid phases containing a large proportion of polar groups are in this class (e.g., the carbowaxes). These materials differentiate between polar and nonpolar solutes, retaining only the polar materials. (4) *Hydrogen bonding.* This special class of polar liquid phases contains materials which possess a large number of hydrogen atoms available for hydrogen bonding. Glycol, for example, is capable of forming a network of multiple hydrogen bonds. (5) *Specific or special-purposes phases* rely on a specific chemical interaction between solute and solvent to perform separations. Silver nitrate dissolved in glycol separates unsaturated hydrocarbons.

By altering the nature of the liquid phase, interaction with the solutes may also be altered. Changes in the relative volatility are effected which make difficult separations very easy. This can be illustrated by the separation of benzene and cyclohexane which have boiling points of 80.1° and 80.8°C, respectively. With a paraffin oil, such as hexadecane as the liquid phase, benzene precedes cyclohexane. However, if the liquid phase is more polar, as in the case of tricresyl phosphate, cyclohexane precedes benzene. The greater the interaction between the solute and the liquid phase, the greater will be the retention time. On a paraffin column *t*-butyl alcohol (b.p. 82.6°C) precedes cyclohexane, but with a highly polar liquid phase containing hydroxyl groups the reverse is true since the alcohol is involved in hydrogen bonding with the liquid phase. In many cases subtle differences can be achieved by selecting a different member of a homologous series or by mixing two liquid phases.

The maximum temperature at which a liquid phase may be used is determined by its volatility. Excessive volatility shortens the column life, contaminates the gas stream in preparative work, and affects the baseline stability of ionization detectors. A column may be programmed for a short time over its limit, but it should not be heated isothermally for an extended time even to the maximum. The lower-temperature limit is determined by the viscosity of the column liquid, which limits the column efficiency, or by the solidification point of the liquid phase.

TABLE 11-2. Typical Liquid Phases

Solvent	*Fields of application*	*Upper temperature limit, °C*
Squalane	Nonpolar; saturated hydrocarbons (paraffins and cycloparaffins) separated from oxygen-bearing compounds close to them in physical properties	150
Apiezon L grease	For separation of high-boiling compounds, particularly aromatic	250–300
Silicone gum rubber (SE-30)	Nonpolar general-purpose column, for highest-temperature work; useful for analysis of steroids, polycyclic aromatics	400
Silicone oil (DC-200)	General-purpose phase used for moderately polar samples	200
Silicone oil (DC-550)	Moderately polar substrate; used for alkylbenzenes and naphthalene homologs	180–200
Polyphenyl ether (5 ring)	Moderately polar; particularly suited for separation of aromatic isomers and mono- and diolefins from paraffins	200–225
Diisodecyl phthalate (and phthalate esters in general)	For all types of compounds; shows no particular selectivity; separates most substances in order of boiling point	175
Diethylene glycol succinate	Mainly used for analysis of fatty acid methyl esters in range C_8 to C_{24}	225
Tricresyl phosphate	General application; shows good selectivity for chlorinated hydrocarbons	150
Carbowax 1540; poly(ethylene glycol)	Polar samples such as alcohols, aldehydes, ketones	150
Carbowax 20M (or Ucon oil LB-550, polypropylene glycol)	High-boiling polar compounds	250

Stationary Phase Loading The relative amount of stationary phase in the column packing is usually expressed in terms of the percent by weight of stationary phase present in the column packing. For example, 15% loading means that in 100 g column packing, there is 15 g of stationary phase coating the surface of 85 g of support.

The amount of liquid phase present in the column influences two important column characteristics: column efficiency and sample capacity. With only a finite surface area available to be coated with a given support, a larger stationary phase volume will result in a thicker film on the support particles. The effect on column performance is discussed later when the factors influencing the plate height and resolution are considered. In terms of the sample capacity of the column, the lower the amount of stationary phase present, the smaller the possible amount of sample. As a guide, an 8–10% loading (and 80/100 mesh support) is appropriate for ⅛-in. packed columns. Since ¼-in. columns are more often used with thermal conductivity detectors, about 15% loading (and 60/80 mesh support) is recommended for these columns. However, two exceptions should be kept in mind. In the analysis of high-molecular-weight or unstable substances, columns with very low stationary phase loading are used. The second exception concerns support materials with low specific surface area. Naturally, in such cases less stationary phase is used or the average film thickness would be too high. Glass beads and polymers belong in this category. Values in the range 0.05–0.2% are common for glass beads.

Open Tubular (Capillary) Columns An open tube coated on the interior with a retentive layer was the invention of Golay (1956). The liquid phase is applied in the form of a very thin film on the inner wall of the tubing. As a result, the carrier gas has an open, unrestricted path along the column. Consequently, with this type of column, the pressure drop is orders of magnitude smaller than that of packed columns of the same length and number of theoretical plates. This permits the use of very long columns (30–100 meters) which, in turn, gives a very high overall column efficiency. While the number of theoretical plates obtainable with standard packed columns of convenient length usually does not exceed 60,000, it is not difficult to obtain theoretical plate numbers well over 100,000 with an open tubular column.

The diameter of an open tubular column can be varied over a relatively wide range. The most frequently used inside diameter is 0.25 mm, but columns with diameters down to 0.07 mm and up to a few millimeters have been reported. Sample capacity is only about 0.1 to 0.3% that available with standard packed columns. A sample splitter must be used ahead of the separation column, and highly sensitive and fast ionization detectors are required.

Support-Coated Open Tubular Columns Colloquially called "fuzzy-wuzzy" columns, a thin layer of porous support is deposited on the inside wall of an open tubular column and then coated with the liquid phase. The characteristics of the support-coated columns differ in two principal aspects from the characteristics of standard open tubular columns.[2] First, the inside surface area is increased without an increase of the internal gas volume. Thus, the β-value—the ratio of the volumes of gas and liquid phases in the column—is decreased. Second, while the volume of the liquid phase is increased by up to an order of magnitude, the film thickness is decreased even more [see Eq. (11-19)].

Practical benefits are derived from these column characteristics. The decrease in β-value means that fewer theoretical plates are necessary than with standard open tubular columns. Thus, either shorter columns can be used, resulting in shorter analysis time or, when using columns of equal length, better resolution can be obtained. The increase in the volume of the liquid phase also means an increased sample capacity. The possibility of injecting larger samples might make unnecessary the need for a sample splitter or, using larger samples, result in a lower detection limit of the whole system. The thinner liquid-phase film reduces the liquid diffusion term and, as a result, the plate height of support-coated columns will generally be less than that of comparable standard open tubular columns.

Detectors

Located at the exit of the separation column, the detector senses the arrival of the separated components as they leave the column and provides a corresponding electrical signal. In principle, the various detectors may be divided into two categories. The first group includes those where at constant temperature and pressure the detector signal depends directly on the concentration of the sample in the carrier gas, and is independent of the mass flow rate of the sample. The sample must not undergo chemical changes by the measurement. The response of the detector is expressed in units of signal per concentration increment.

The second group comprises detectors where the signal is a function solely of the mass flow rate of the sample and will be independent of the sample concentration in the carrier gas. The sample is changed chemically by the measurement. The response is given in units of signal per mass flow rate of the sample.

Detectors representing different categories can be compared only under the condition that concentration and mass flow rate are fixed. In addition, the sensitivity ratio depends on the absolute level of the concentration and the mass flow rate. For example, with increasing mass flow rate and de-

[2] I. Halasz and C. Horvath, *Anal. Chem.* **35,** 499 (1963); L. S. Ettre, J. E. Purcell, and S. D. Norem, *J. Gas Chromatog.* **3,** 181 (1965).

creasing sample concentration, the sensitivity of the flame ionization detector will to an ever-increasing extent exceed that of a thermal (hot-wire) conductivity cell. On the other hand, the sensitivity of a flame ionization detector can be smaller than that of a thermal-conductivity cell at very low mass flow rates.

Linear dynamic range is a term used to characterize detectors. It is the largest signal which a system can handle divided by the smallest signal for which the response is linear with respect to sample size. A wide linear dynamic range makes it possible to detect both trace and major sample components with suitable attenuation. The lowest limit of detection is the minimum detectable amount of sample for which the signal-to-noise ratio is two. These terms are given in Table 11-3 for several detectors.

A very important question concerning all detectors is related to the temperature of the detector compartment. If it is lower than the temperature of the column, condensation can occur and cause poor quantitation; if it is too high, sample decomposition prior to detection is possible. The installation of the detector unit should be as close to the column exit as possible. For some detectors, a maximum temperature limit is imposed due to the component parts.

The type of column is to some extent a determining factor in the choice of detector. For packed columns, this will be a conventional-size thermal-

TABLE 11-3. Characteristics of Gas Chromatographic Detectors

Detector	*Lowest limit of detection*	*Upper limit of linearity*	*Linear dynamic range*	*Time constant*
Hot-wire thermal conductivity	10^{-5} mg/ml	10^{-1} mg/ml	10^4	1 sec
Gas density	10^{-5} mg/ml	10^{-1} mg/ml	10^4	1 sec
Micro thermal conductivity	10^{-10} mg/ml			0.5 sec
Micro cross section	10^{-4} mg/sec	to 100% conc.		
Flame ionization	10^{-9} mg/sec	10^{-3} mg/sec	10^6	1 msec
Argon	10^{-10} mg/sec	10^{-8} mg/sec	100	1 msec
Helium	10^{-7} mg/ml		5×10^4	
Electron capture (H^3)	5×10^{-11} mg/sec [a]	2.5×10^{-8} mg/sec	500	
Ni^{63} high-temperature electron capture	2×10^{-11} mg/sec [b]	10^{-9} mg/sec	50	
Phosphorus	4×10^{-10} mg/sec [c]	4×10^{-7} mg/sec	1000	1 msec

[a] For CCl_4, pulse interval 150 μsec. Concentration range drops to 1×10^{-7} to 1×10^{-4} mg/sec for pulse interval of 5 μsec.
[b] For CCl_4.
[c] For Parathion.

conductivity cell, gas-density cell, or cross-section detector. For open tubular columns, it will be some type of ionization detector. The latter group of detectors can also be used for packed columns, if the effluent is suitably attenuated by stream splitters.

Thermal-Conductivity Cells A thermal-conductivity detector consists of four heat-sensing elements, each situated in a separate cavity in a brass or stainless steel block which serves as a heat sink. The heat-sensing elements are either thermistors or resistance wires. In the hot-wire cell, filaments are straight (or helical) bare wires which are stretched along the axis of the cavity, as shown in Fig. 11-4. The wire is kept under tension by platinum–iridium springs—essential because the length changes with temperature. Filaments are fabricated from gold-sheathed tungsten or teflon-coated tungsten having a high temperature coefficient of resistance. Thermistors are electronic semiconductors of fused metal oxides whose electrical resistance varies with temperature. A thermistor in bead form, encapsulated in glass, may replace the hot wire in thermal-conductivity cells.

The internal design of the cell varies with the intended use. For fast response of less than 1 sec a fraction of the gas stream is passed directly through the cell. When used with the relatively large flows (60–100 ml/min) associated with ¼-in. columns, the cell cavity has a volume of about 2.5 ml. This is decreased to 0.025 ml in the micro cells which are designed for use with capillary and small diameter columns and a flow of 5 ml per min.

The sensitivity of the hot-wire detector may be raised by fabricating two helical wires in series or increasing the current through the wires. High currents increase electrical noise and shorten filament life. Moreover, a temperature limit is imposed by the glowing point of the heated wire and also by the sensitivity to decomposition of the sample components.

The sensitivities of a four-filament, dual helices, tungsten–gold hot-wire assembly and a thermistor cell are comparable when the cell block is near room temperature. The thermistor has the advantage of smaller cell volume and no direct contact with the gas stream, but this is offset by the thermal inertia of the thermistor cell and decreased sensitivity when the cell block is maintained at elevated temperatures.

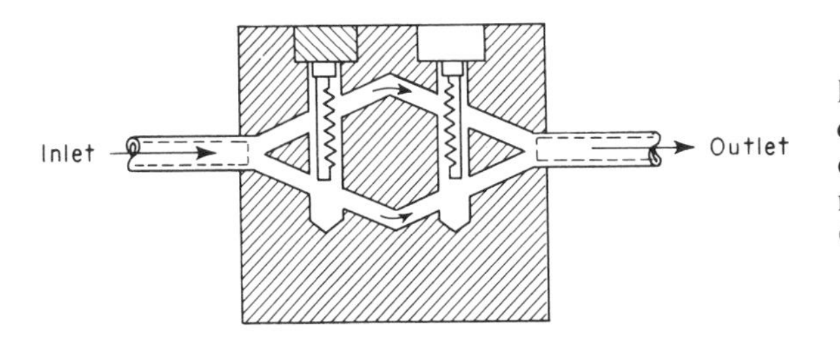

FIG. 11-4. Bypass design of hot-wire thermal conductivity cell for rapid response. (Courtesy of Gow-Mac Instrument Co.)

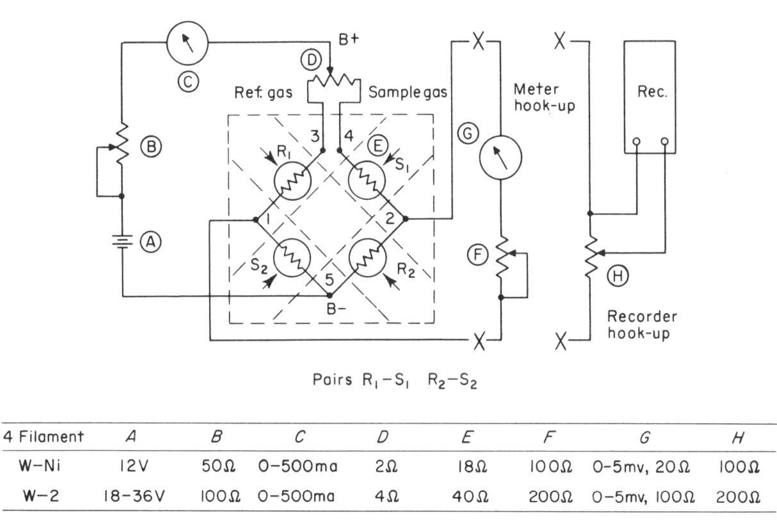

4 Filament	*A*	*B*	*C*	*D*	*E*	*F*	*G*	*H*
W–Ni	12V	50Ω	0–500ma	2Ω	18Ω	100Ω	0–5mv, 20Ω	100Ω
W–2	18–36V	100Ω	0–500ma	4Ω	40Ω	200Ω	0–5mv, 100Ω	200Ω

FIG. 11-5. Circuitry for four-filament, hot-wire, thermal conductivity cell. (Courtesy of Gow-Mac Instrument Co.)

What is measured is the difference in thermal conductivity between the carrier gas and the carrier/sample mixture. The four thermal-conductivity cells, mounted in a cell block, are connected to form the arms of a Wheatstone bridge. The electrical circuitry is shown in Fig. 11-5 for a four-filament, hot-wire, thermal-conductivity cell. The electrical circuit for use with thermistor cells differs only in recognizing the larger resistance (8000 Ω) as compared with a resistance wire (18 to 42 Ω). One lead from a low-voltage d-c power supply is connected to the Wheatstone bridge through a resistance for adjusting the heating current, which is read off the indicating milliammeter. The other lead goes to the contactor of a 2-Ω potentiometer which serves for a precise balancing of the bridge when pure carrier gas is flowing through all cells. Two filaments in opposite arms of the bridge are surrounded by the carrier gas, which conducts away heat generated by the electrical current. The temperature of the filaments will rise until the rate of heat flow away from the filament to the metal block matches the heat generated by the current in the hot-wire resistance element. Similarly, the other two filaments are surrounded by the effluent from the chromatographic column. When pure carrier gas passes both the reference and sample filaments, the heat loss due to cooling is the same in both and the system is in balance. When a solute emerges from the column and passes through the sample side of the detector, the rate of cooling of the sample filaments changes and the Wheatstone bridge is out of balance. This imbalance is registered as a peak on the recorder.

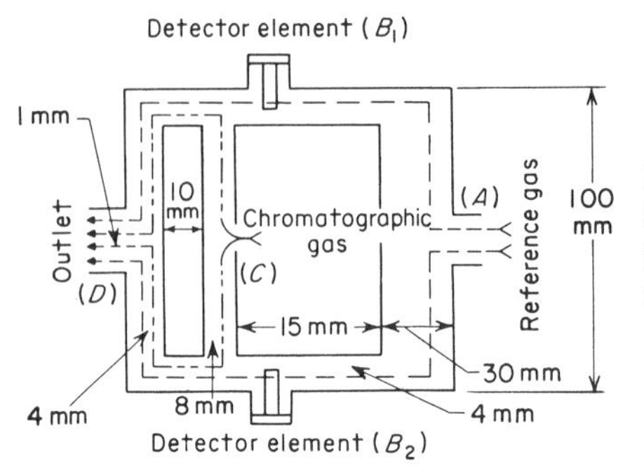

FIG. 11-6. Nerheim gas-density balance. Schematic view as mounted in a vertical plane. (Courtesy of Gow-Mac Instrument Co.)

This type of detector requires that the carrier gas have a thermal conductivity that is considerably different from that of the sample components. Two gases, hydrogen and helium, best meet this requirement. Since helium is inert, it is favored.

Gas-Density Detector The Nerheim configuration,[3] Fig. 11-6, illustrates the principle of operation of the gas-density detector. With the conduit network mounted vertically, the reference (pure carrier) gas enters at A, splits into two streams, and exits at D. Two flow meters, B_1, B_2, are installed, one in each stream, and are wired in a Wheatstone bridge (similar to Fig. 11-5). When the flow is balanced, the detector elements, which are a matched pair of hot wires or thermistors, are equally cooled and the bridge is balanced, thus giving a baseline trace. The effluent from the chromatographic column enters at C, splits into two streams, mixes with the reference gas in the horizontal conduits, and exits at D. The effluent never comes into contact with the detector elements, thus avoiding problems caused by corrosion or carbonization.

When the effluent carries transients which are heavier than the reference gas (virtually always the case when helium is the carrier gas), the density of the heavier molecules will cause a net downward flow, partially obstructing the flow A–B_2–D, with a temperature rise of element B_2, and permitting a corresponding increase in the flow A–B_1–D, with a temperature decrease of element B_1. Bridge unbalance is linear over a broad range and directly related to the gas density difference. The sensitivity of the gas-density detector with a thermistor sensor is comparable to that of a thermistor type of thermal-conductivity cell. The hot-wire sensor, which is less sensitive than thermistors at 25°, may require low-level amplification at 100°C. The low noise level of the detectors permits effective use to 300°C, although the sensitivity decreases rapidly with increasing oven temperature for thermistors. The Nerheim design has an effective sample volume of about 5 ml, making it suitable only for ¼-in. or larger packed columns.

[3] A. G. Nerheim, *Anal. Chem.* **35,** 1640 (1963).

Flame Ionization Detector A schematic view of the flame ionization detector is shown in Fig. 11-7. The column effluent enters the burner base through a millipore filter which removes contaminating particulate matter. Hydrogen, mixed with the carrier gas stream, is burned in air to obtain a plasma (approximately 2100°C) which has sufficient energy to ionize any organic solute passing through it. The ions are collected at the anode and the electrons at the cathode. The resulting ion current is monitored by measuring the voltage drop across a series resistor.

In the design shown, the flame burns at the tip of a stainless steel capillary, which also functions as one electrode. The burner tip is insulated from the body of the burner assembly by a Teflon seal. In an alternate design, the base of the flame head holds a quartz burner tip to minimize thermionic "noise" caused by overheating a metal burner tip. With either design a loop or cylindrical screen of platinum, located just above the tip of the flame (and 6 mm above the burner tip), serves as a collector electrode. An ignition coil is also located above the burner. Voltage is applied to the collection electrodes by means of a 300-V battery. The small electron current is changed to a measurable voltage signal by sending it through a very high input resistance (10^7 to 10^{10} Ω) in series with the flame gases. To work satisfactorily, the flame ionization detector must have a good signal-to-noise ratio and the device must be coupled to an amplifier capable of responding to currents as low as 10^{-14} A.

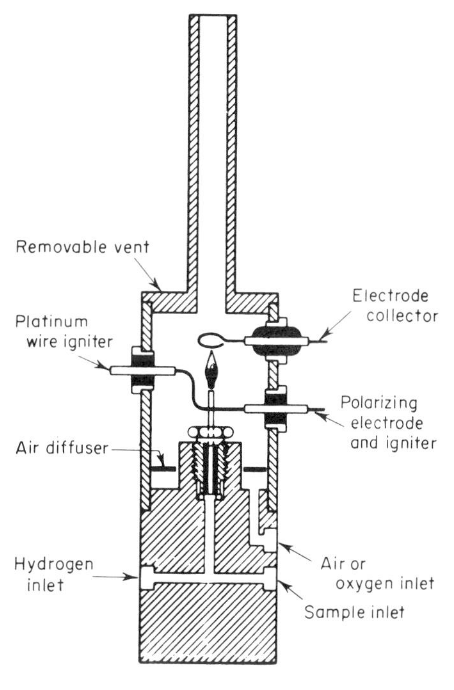

FIG. 11-7. Schematic diagram of a flame ionization detector. (Courtesy of Beckman Instruments, Inc.)

With only the carrier gas and hydrogen burning, a constant signal will be obtained. Addition of organic compounds to the flame results in a large increase in its electrical conductivity due to ionization. The response of a flame ionization detector is directly proportional to the number of carbons in the solute molecule bound only to hydrogen or to other carbon atoms, and diminishes with increasing substitution of halogens, amines, and hydroxyl groups. There is no contribution from fully oxidized carbons such as carbonyl or carboxyl groups (and thio analogs). Apart from vapors of elements in groups I and II of the periodic classification, these being elements easily ionized in flames, it does not respond to inorganic compounds. Insensitivity to water, the permanent gases, CO, CO_2, and most inorganic compounds is advantageous in analysis of aqueous extracts and in air-pollution studies.

The linear dynamic range is 10^6. A sample splitter is necessary; 10 mg per sec is the largest amount of solute that the flame can handle without being swamped. With stationary phases possessing a slight vapor pressure, 1 ng is the smallest amount of sample that can be measured. Precise temperature control is not a requirement for this detector, an obvious advantage in programmed temperature applications.

Phosphorus Detector A modified flame ionization detector with an emission tip of alkali salt exhibits a large increase in the electrical conductivity when compounds containing phosphorus are burned. In one commercial burner, a cesium bromide salt reservoir in the form of a pressed salt tip, and containing a filler pressed under high pressure to form a rugged ceramic-like material, is located directly on top of the quartz tip of the hydrogen burner (Fig. 11-8). The remaining functional components are the same as those employed with the ordinary flame ionization detector. Precise flow control is required for H_2 and air. Lifetime of the salt tip is in excess of 400 hr; equilibration time after initial flame ignition is about 6 min.

When a compound containing phosphorus is burned in the flame, the rate of release of alkali metal vapor is increased. The alkali metal vapor ionizes in the flame quite readily and increases the current flow. Picogram quantities of phosphorus compounds give a detectable response; nanogram responses are given by halogenated compounds and normal hydrocarbons. Linear dynamic range is 1000 above the minimum detectable quantity. The ability of the detector to respond to one class of compounds while ignoring all others reduces the separation requirement. It also permits the operation of a column where "bleed" would otherwise interfere with the analysis. The phosphorus detector finds application to phosphate additives in gasoline and in pesticide residue analysis.

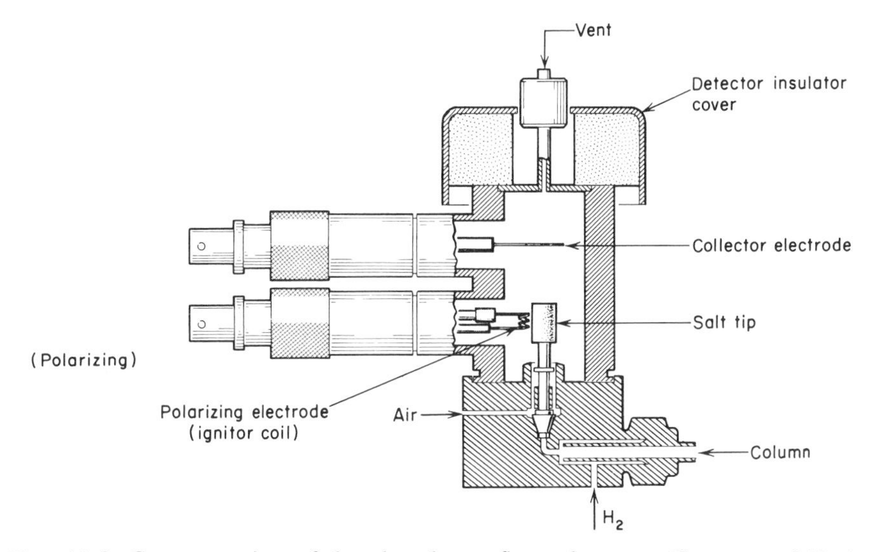

FIG. 11-8. Cut-away view of the phosphorus flame detector. (Courtesy of Varian Aerograph.)

Micro-Cross-Section Ionization Detector The basic components of this detector are an ionization chamber of small dimensions and a radioactive beta emitter. Cell geometries comprise the parallel-plate and concentric-tube design. In the latter design, Fig. 11-9, the ion chamber is formed from 1.6-mm diameter tubing which encloses the radiation source—a sheet of stainless steel foil 0.05 mm in thickness with a surface layer of 200 mCi titanium tritide. This is one electrode. The collecting electrode, made of 0.8 mm diameter stainless steel rod, is mounted axially along the chamber. The two electrodes are connected in series with a polarizing battery. Carrier gas enters the chamber through channels in a plug of ceramic Teflon

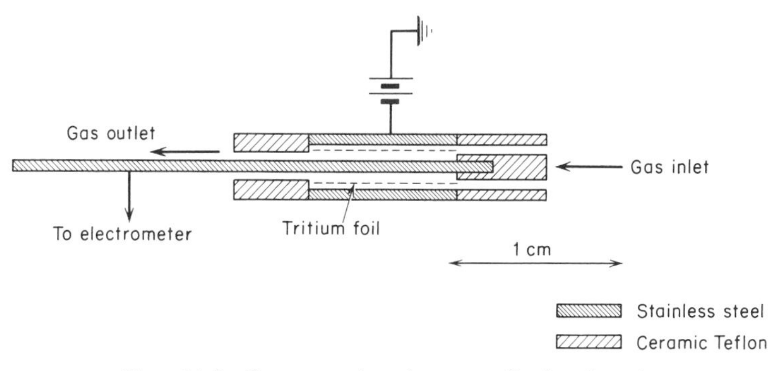

FIG. 11-9. Cross-section detector (8 μl volume).

which also serves to insulate and support the ion-collecting electrode. The gas exits through a central channel in a second Teflon plug through which the ion-collecting electrode extends for exterior electrical connection.

Radiation from the beta emitter strikes the molecules of carrier gas, causing them to ionize and thereby produces a steady concentration of ion pairs. A potential gradient of 300–1000 V between the parallel or concentric electrodes about 1 mm apart ensures collection of the ions. Hydrogen is the preferred carrier gas because of its small molecular cross section. It may be used up to 100°C; at high temperatures, helium plus 3% CH_4 is used. The detector temperature limit is 220°C. The ions formed may either be positive (the result of losing valence electrons from the energy of impact) or negative (the result of capturing free electrons). When conditions are such that essentially only positive ions are collected, the detector operates in the cross-section mode. The ionization chamber size must be related to the energy of the beta particles emitted in such a manner that only a small part of the beta-particle energy is dissipated in the gas. Electrode spacing and polarizing voltage must be chosen so that essentially all the ions formed reach the electrodes before they recombine.

Polyatomic gases and vapors are more strongly absorbing than hydrogen or methane. Thus, when molecules larger than those of the carrier gas are present in the effluent, they provide an increase in current in proportion to their concentration. Response to any substance, organic or inorganic, is linear up to 100% concentration and is absolute. It can be calculated from the values of the atomic cross-sections of the constituent atoms and ambient physical conditions.[4] Reproducibility is low, however, requiring constant recalibration. Another disadvantage is low sensitivity (10^{-4} mg per sec), less than any of the other ionization detectors.

Beta-Ray Ionization Detectors The cross-sectional view of the helium detector[5] is shown in Fig. 11-10. The detector consists of two electrodes spaced about 1 mm apart in a parallel-plate geometry with an internal detector volume of 160 μl. A 250-mCi tritium foil (1 cm square) serves as the anode. The cathode is connected to a very stable voltage supply capable of between 350 to 700 V. The exact voltage used depends on the purity of the helium carrier gas; the purer the gas the higher the applied voltage can be without increasing the background current and noise level. An operating point close to the sparking potential provides the greatest sensitivity. The anode is connected to an electrometer capable of measuring small (10^{-11} A) changes in current.

Helium passing from a chromatographic column passes underneath the tritium foil and into the small volume between the two electrodes and then

[4] J. W. Otvos and D. P. Stevenson, *J. Am. Chem. Soc.* **78,** 546 (1956).
[5] C. H. Hartmann and K. P. Dimick, *J. Gas Chromatog.* **4,** 163 (1966).

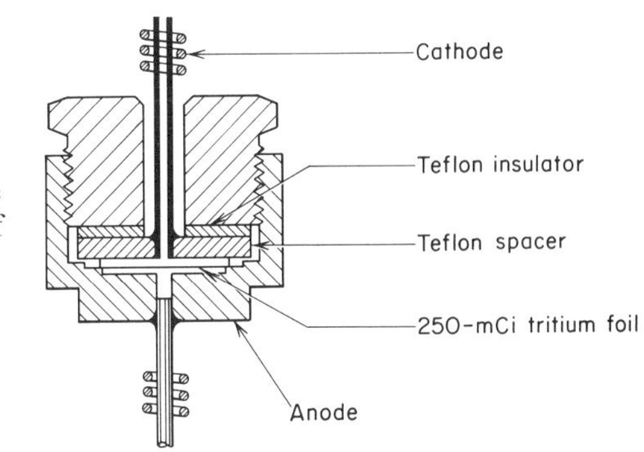

FIG. 11-10. Schematic view of the helium detector. (Courtesy of Varian Aerograph.)

exits through a hole in the cathode. Helium atoms are excited to the metastable state[6] by the beta particles emitted by tritium in a field gradient in the range 4000 to 7000 V/cm. As vapors of solute molecules emerge from the column and enter the detector volume, collisions with metastable helium atoms will result usually in the transfer of energy to the solute molecules. Since all gases except helium and neon have lower ionization potentials than metastable helium, they will produce ion particles and dissociated fragments when they collide with metastable helium. The net result is a positive increase in detector current.

The helium detector senses all compounds but it is primarily useful for the analysis of permanent gases. Non-bleeding columns, such as the molecular sieves, must be used. The detector has a different response factor for each compound; these fall in the range 0.1 to 3 parts per billion. Linear dynamic range is about 5×10^4. Upper temperature limit is 225°C.

The argon ionization detector[7] is basically the same in construction as the helium detector. However, argon is used as carrier gas. High-energy beta particles from a strontium-90 source promote ground-state argon atoms to the metastable state. A field strength of 1000 V/cm is used to provide for the production of metastable argon and the efficient collection of ionization products from column effluents. The difference between the two detectors arises primarily from the lower metastable energy level of Ar* (11.5 eV) as compared with He* (19.6 eV). As a result, the argon detector does not respond to water vapor, methane, oxygen, carbon dioxide, nitrogen, carbon monoxide, ethane, acetonitrile, or fluorocarbons. Performance is seriously impaired by the presence of air or water vapor in the carrier gas. Upper temperature limit is 400°C.

Electron-Capture Detector The principle of the electron-capture detector is based on electron absorption by compounds having an affinity for

[6] These are helium atoms with one or both electrons displaced to higher-energy metastable orbital configuration.

[7] J. E. Lovelock, *J. Chromatog.* **1**, 35 (1958).

free electrons. Unlike other ionization detectors, this detector measures the loss of signal due to recombination phenomena. The schematic diagrams of Fig. 11-11 show the cutaway of a parallel-plate cell and pin-cup design. As the nitrogen carrier gas flows through the detector, beta particles from a tritium source (or from nickel-63 for temperatures up to 400°C) ionize the nitrogen molecules and form "slow" electrons. These slow electrons migrate to the anode under a fixed potential which can be varied from 2 to 100 V. Collected, these electrons produce a steady baseline current (about 10^{-8} to 10^{-9} A) flowing across the cell from A to B. Due to the limited range of tritium beta particles, the production of secondary electrons occurs only in the area E. Nitrogen carrier gas is preferable to helium or argon for the attachment mode of detection because nitrogen forms no metastable species which would give rise to positive peaks for column bleed and organic compounds which might cover negative peaks of interest.

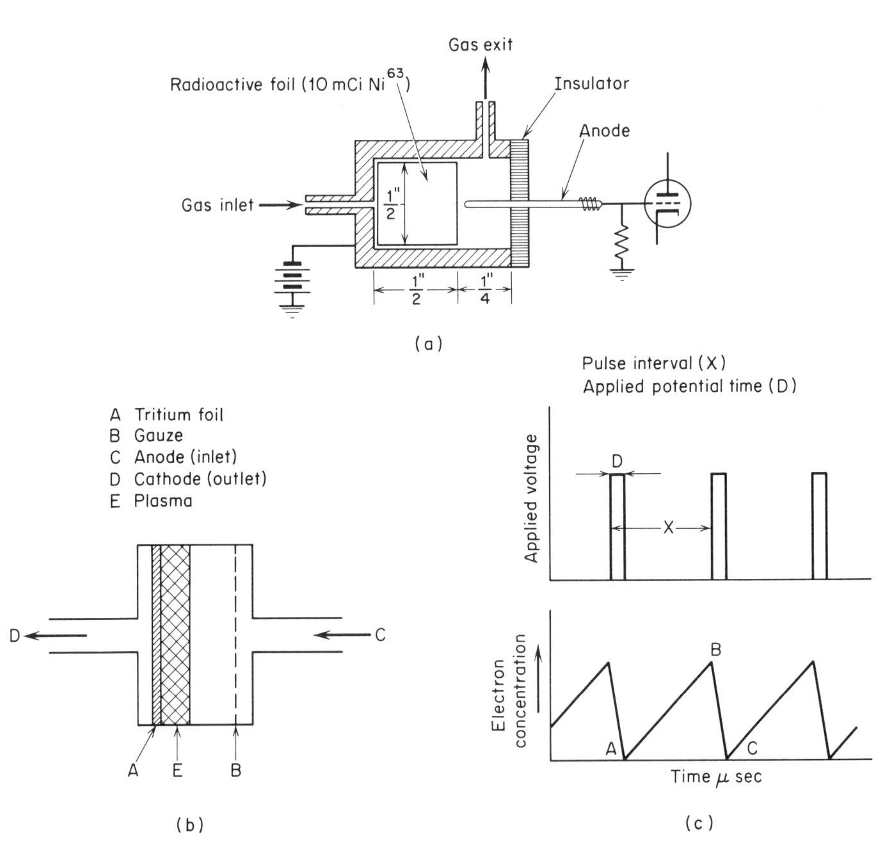

FIG. 11-11. Electron capture detector. (a) Pin cup design; (b) schematic diagram; (c) pulse sampling mode. (Courtesy of Varian Aerograph.)

When an electron-capturing component emerges from the column and is carried into the cell, it moves from *C* into area *E*. This area, having an abundance of low-energy free electrons, is an ideal electron-capture environment. The solute reacts with an electron to form either a negative molecular ion, or a neutral radical and a negative ion. Since the solute molecules have only moderate affinity for electrons at best, slow electrons attach more readily than rapidly moving electrons. The net result is the removal of an electron from the system and the substitution of a negative ion of far greater mass. The decrease in electron concentration is reflected as a negative excursion of the current trace on the recorder. The reduction in ion current when the sample enters the detector follows a law similar to Beer's law:

$$i = i_o e^{-kxC} \tag{11-1}$$

where i is the current when a concentration C of the electron-capturing vapor passes through the detector, i_o is the current when only carrier is passing through the detector, k is a constant depending on both the field strength and the absorption cross-section of the vapor, and x is a geometric factor.

The electron-capture detector can be operated under either a pulsed or a constant voltage. In the square-wave pulsed mode, sensitivity is a function of pulse interval. The electron concentration in the detector is not constant but varies in a saw-tooth fashion. With the application of a pulse, the electron concentration drops to zero—this represents total collection of the electrons in the cell. In the interval between pulses (5 to 150 μsec), the concentration of electrons builds up due to beta-particle emission from the radionuclide. Thus, the magnitude of the electron concentration is dependent on the pulse interval; as the interval decreases the detector sensitivity also decreases. Since the drift of negative ions is negligible, a negative space charge cannot occur in the pulsed mode. A positive space charge is present in the plasma (region *E*) at all times; these ions are available for recombination with negative ions. Thus, the detector measures pure electron capture to the exclusion of signals from negative ion migration.

In the d-c mode of operation, the detector is operated at a potential which is just sufficient to collect all the electrons liberated in the detector chamber. The current measured is due to the combined effect of both electron and ion components; the resulting signal peak on the chromatogram is due to both electron capture and ion migration. Since, in the d-c mode, the optimum operating voltage differs for each molecule, detector discrimination can be regulated through the potential applied to the collector electrode. The response of weakly capturing compounds can be abolished in turn by increasing the applied potential. Response for different classes ceases at well-defined applied potentials.

The electron-capture detector is extremely sensitive to certain molecules, such as organic and inorganic halogen-containing compounds, anhydrides, peroxides, conjugated carbonyls, nitriles, nitrates, ozone, oxygen, organometallic, and sulfur-containing compounds, but it is virtually insensitive to hydrocarbons, amines, and ketones. Electron affinity is difficult to predict, but it appears to be related to the ease of dissociation of a heteroelement from the compound. An elegant area of application for this detector is the analysis of agronomic materials for pesticide residues and for lead alkyls in gasolines. The detector is sensitive to temperature changes on the column; therefore temperature programming is not recommended.

Flame Emission Detector The detection of phosphorus and sulfur compounds in submicrogram quantities can be based on the photometric detection of the flame emission of phosphorus (the green HPO band system) and sulfur (the violet S_2 band system) in a fuel-rich hydrogen flame.[8] This detector consists of a standard flame ionization burner, enclosed in a light-tight housing, with a mirror and lens to focus the emitted radiation from the flame gases through a narrow-band interference filter and on to an end-on multiplier phototube. The carrier gas is nitrogen which is mixed with oxygen and hydrogen just behind the burner tip. For phosphorus compounds, an interference filter peaked at 5260 Å is used; for sulfur compounds, a filter peaked at 3940 Å is used. For both classes of compounds the detector's linear dynamic range is 10^4. The limit of detection for parathion is about 10^{-12} g/sec. Dual detectors enable both elements to be determined simultaneously.

Coupling a Beckman sprayer burner and the optical system of a Bausch & Lomb Spectronic 20 spectrophotometer with the detection sensitivity of a 1P28 multiplier phototube enables one to measure the emission of organic compounds (CH band at 4315 Å or the C_2 band at 5765 Å) and inorganic elements.[9]

Other Detectors Many relatively new non-ionization detectors and their use in gas chromatography have been recently reviewed.[10]

The electrolytic conductivity detector consists of a pyrolyzer, a gas–liquid contactor, a gas–liquid separator, and a pair of platinum electrodes in a d-c bridge circuit.[11] The pyrolyzer converts organically bound halogen, sulfur, or nitrogen to oxidized or reduced substances that form electrolytes when dissolved in water. Chlorinated hydrocarbons may be burned with moist oxygen in the pyrolysis tube to form hydrogen halides; however, the

[8] S. S. Brody and J. E. Chaney, *J. Gas Chromatog.* **4,** 42 (1966).

[9] R. S. Juvet, Jr., and R. Durbin, *Anal. Chem.* **38,** 565 (1966); F. M. Zado and R. S. Juvet, Jr., *ibid.* **38,** 569 (1966).

[10] J. D. Winefordner and T. H. Glenn, in *Advances in Chromatography,* Vol. 5 (J. C. Giddings and R. A. Keller, (Eds.), Dekker, New York, 1968.

[11] D. M. Coulson, *J. Gas Chromatog.* **3,** 134 (1965); *ibid.* **4,** 285 (1966).

most specific results for halogenated compounds are observed in the reductive mode with a platinum catalyst and hydrogen. Under these conditions there is little or no response from the elements N, S, P, C, O and H in organic molecules. Sulfur compounds give a strong response in the oxidative mode. The electrolytic conductivity methods are therefore quite selective or can be made quite selective by proper choice of combustion conditions.

The r-f discharge detector[12] involves a corona discharge which is excited at radio frequencies. The exact basis of operation has never been explained; the rectification, however, of an r-f corona discharge in helium at atmospheric pressure is affected by the presence of other gases in the helium carrier gas. Sensitivity is less than with the ionization detectors, and response is critically sensitive to changes in temperature of the detector housing. The need for ancillary equipment is also a drawback.

Fast-scanning infrared and mass spectrometric equipment has been attached to the chromatograph outlet to resolve the composition of peaks containing two or three components, and to identify components appearing as shoulders or lost in the background.

Recording the Signal

In the area of readout, the choice of recorder determines the ultimate accuracy of the chromatogram. Full-scale pen response should be 1 sec or less. Recorder sensitivity is usually 10 mV (full scale) but may range from 1–10 mV. The recorder must be equipped with a series of good-quality resistances connected across the input to attenuate large signals. Background signals are offset by a variable potential applied to the base of the attenuator resistances.

Electrometer Circuits The small electron currents obtained from the ionization class of detectors must be amplified before recording. Two different amplifier designs have been employed widely in gas chromatography. Both employ an electrometer tube to couple the high-impedance input to a high-gain operational amplifier, but they differ in two characteristics: (1) position and polarity of the polarizing voltage and (2) manner of switching of range resistors and output attenuation. Figure 11-12 shows the design (for the flame ionization detector) which features a floating jet. Two signal-level controls are used. The range switch changes the input range resistors, thus changing the gain of the electrometer amplifier by factors of 10, and the attenuator provides several binary steps of output signal-level attenuation. A virtue of the system using a floating jet is that all five ionization detectors are immediately interchangeable and may be used with the same electrometer.

[12] A. Karman and R. L. Bowman, *Ann. N. Y. Acad. Sci.* **72,** 714 (1959).

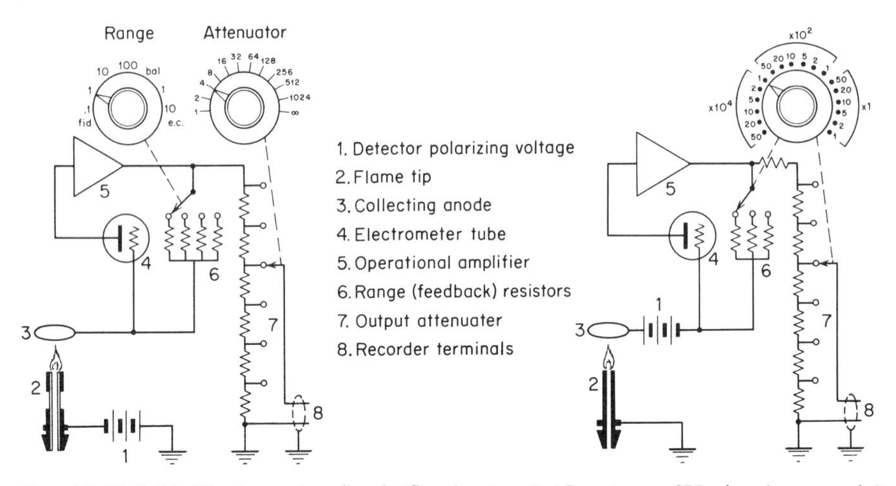

FIG. 11-12 (left). Electrometer circuit (floating type). (Courtesy of Varian Aerograph.)
FIG. 11-13 (right). Electrometer circuit (grounded type). (Courtesy of Varian Aerograph.)

In the second mode, Fig. 11-13, the flame jet is grounded; polarization is provided with batteries between the anode and the tube grid. This arrangement normally causes more noise and drift. The input range resistor switching and output attenuation is performed using a single knob.

Dual-Channel Operation For dual-channel gas chromatography two electrometers are built into a single unit and share a common power supply. Each channel can operate independently or oppose each other to give a differential output.

In dual-detector operation the effluent from a single column is split and sent to two different detectors simultaneously (Fig. 11-14a). A dual-pen recorder allows simultaneous presentation of the signals. Two detectors, each with their individual selectivity, make possible a more detailed characterization of the sample analyzed than can any single-detector run. It aids in qualitative analysis, particularly of natural products, by distinguishing components comprising overlapping peaks. In Fig. 11-15 we see a simultaneous flame ionization–thermal conductivity analysis of a trace alcohol mixture in aqueous solution: four alcohols are detected by the flame detector operated at high sensitivity while the thermal-conductivity detector reveals three inorganic constituents (NH_3, CS_2, and H_2O, respectively) that the flame detector fails to pick up. The combination of flame and electron-capture detectors often supplies useful information when sampling vapors, pesticide residues, and other micro-samples.

In differential operation mode, matched columns and identical detectors are used. This arrangement (Fig. 11-14b) eliminates baseline drift caused by bleeding of the liquid phase, particularly during temperature program-

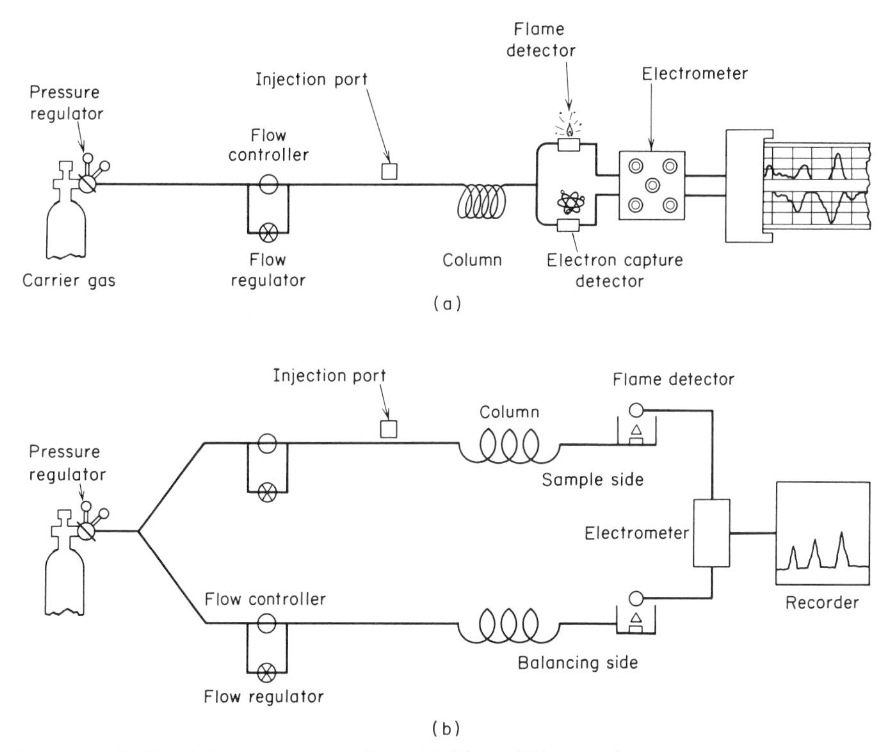

FIG. 11-14. Dual-detector operation. (a) Two different detectors operated simultaneously; (b) differential operation with identical detectors.

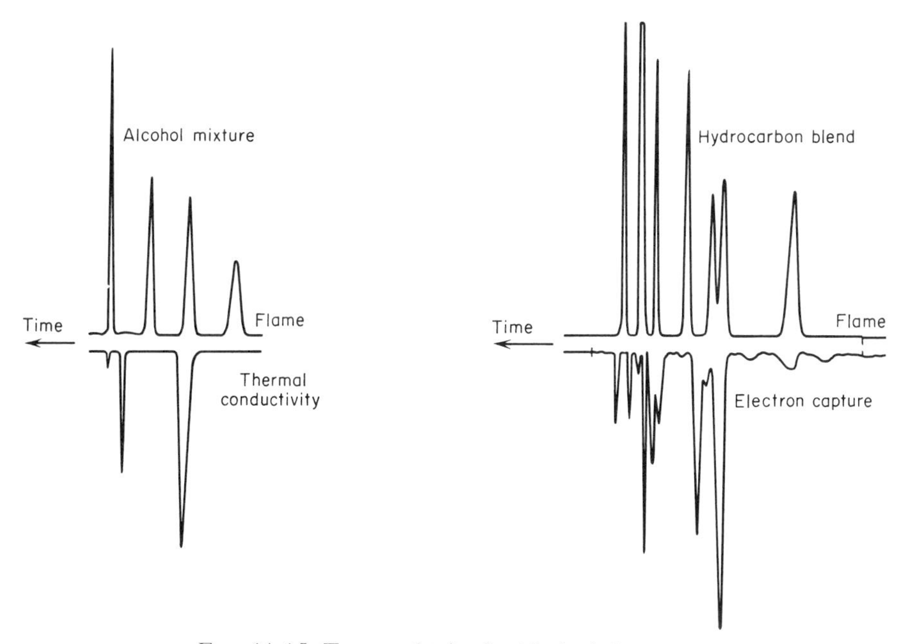

FIG. 11-15. Traces obtained with dual detectors.

ming. Sample component peaks superimposed over a drifting baseline make peak identification and quantitative evaluation of the chromatogram practically impossible. As the temperature of the columns is increased, the bleed signal from one column is canceled by the other. The output of the electrometer is a measure only of the differences in response of the twin detector units; the difference signal will be zero except when sample components enter the sensing unit.

Thermal Compartment

Precise control of the column temperature is a requisite, whether it is intended to maintain an invariant temperature or to provide a programmed temperature. It is desirable to have separate heaters and controls for the injection block, column oven, and detector unit(s). Each should be designed to operate up to 500°C from ambient temperature.

Temperature of the column oven should be controlled by a system which is sensitive to changes of 0.01°C and which maintains control to 0.1°C. Ovens should be of low mass, thin wall construction, large enough to accommodate all diameter columns of reasonable lengths, yet small enough to allow rapid heating and cooling with uniform and reproducible temperatures (Fig. 11-16). This is aided by a high-velocity, squirrel-cage fan

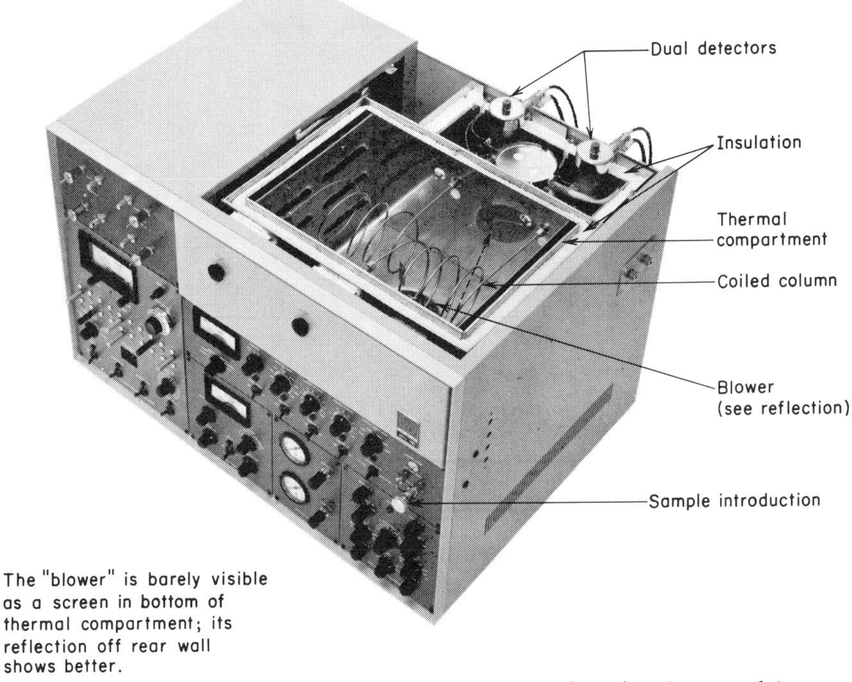

FIG. 11-16. Thermal compartment. (Courtesy of Varian Aerograph.)

to circulate the air in the oven or to vent warm air during a cooling cycle.

Programs are available which feature linear and nonlinear temperature programming of sample and reference columns. A standard cam and a program rate selector allow an operator to dial any linear program up to a maximum of 60°C/min in the lower temperature ranges and about 35°C/min at higher temperatures. By substituting cams, one may employ completely automatic nonlinear programs.

GAS CHROMATOGRAPHY THEORY

Gas chromatography theory cannot be discussed here in detail, nor can the complex interactions of all the variables be considered. For details the literature should be consulted. On the other hand, a brief treatment of basic parameters should help in understanding the technique.

Retention Behavior

The volume of carrier gas necessary to convey a solute band the full length of a column is the retention volume—the fundamental quantity measured in gas chromatography. It reflects the distribution of the solute between the eluent gas and the stationary liquid phase. For a given column operated at temperature T_c and carrier gas flow rate F_c, the length of time that each component spends in the column, called the retention time, is a constant. The retention time will be the same whether the solute is pure or in a mixture. On a chromatogram the distance on the time axis from the point of sample injection to the peak of an eluted component is called the *uncorrected retention time*, t_R. The corresponding *retention volume* is the product of retention time and flow rate (expressed as volume of gas per unit time):

$$V_R = t_R F_c \tag{11-2}$$

Gas flow rate must be corrected to column temperature and outlet pressure. With wet flow meters, allowance must be made for the vapor pressure of water, and with capillary meters for the pressure drop across the capillary.

The air spike, shown in Fig. 11-17, measures the transit time for a non-retained substance. Converted to volume (V_M), it represents the interstitial volume of gaseous phase in the column plus the effective volume contributions of the injection port and detector. Retention volume measured from the air peak provides an *adjusted retention volume*, V'_R, which is corrected for the dead space:

$$V'_R = t_R F_c - t_{\mathrm{air}} F_c = V_R - V_M \tag{11-3}$$

The adjusted retention volume changes slightly with amount of sample; consequently, the significant adjusted retention volume is one obtained by extrapolation to zero sample size.

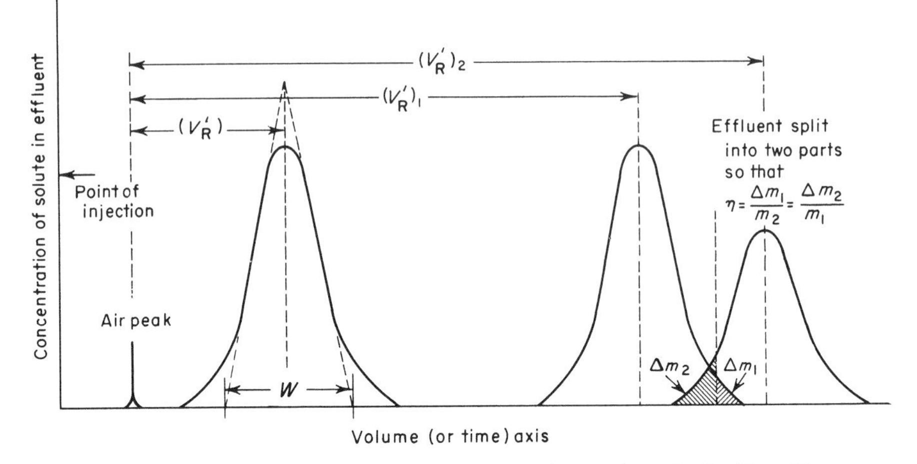

FIG. 11-17. Idealized elution peaks showing notation and method of handling overlapping bands.

In gas chromatography the mobile phase is compressible. Since gas moves more slowly near the inlet than at the exit of the column, a *pressure-gradient correction* or *compressibility factor j* must be applied to the adjusted retention volume to give the *net retention volume* V_N, namely,

$$V_N = jV_R' \tag{11-4}$$

The compressibility factor is given by the expression [13]

$$j = \frac{3}{2} \frac{[(P_i/P_o)^2 - 1]}{[(P_i/P_o)^3 - 1]} \tag{11-5}$$

where P_i is the carrier gas pressure at the inlet to the column, and P_o that at the outlet. This factor stresses the importance of a small pressure drop across the column.

When the solute enters the column it immediately equilibrates between the stationary solvent phase and the mobile gaseous phase. The concentration (or weight) in each phase is given by the *partition coefficient*

$$K_d = C_S/C_M \tag{11-6}$$

where C_S, C_M are the concentrations of solute in the stationary solvent phase and mobile gas phase, respectively. For example, when $K_d = 1$, the solute will distribute itself evenly between the two phases and thus will spend half the time in the gas phase and half the time in the liquid phase, emerging at a retention time equal to twice the retention time of the air peak. If the partition isotherm is linear, the partition coefficient will be a

[13] For a derivation of the compressibility factor, see W. E. Harris and H. W. Habgood, *Programmed Temperature Gas Chromatography*, Wiley, New York, 1966, page 49.

constant independent of the solute concentration. Generally this is true at the low concentration prevailing in gas–liquid chromatography.

For a particular column it is sometimes more convenient to consider the *partition ratio* k (also denoted the capacity ratio): $k = K_d(V_S/V_M)$.

At the appearance of a peak maximum at the column exit, one-half of the solute has eluted in the retention volume V_R, and half remains in the volume of the gaseous phase V_M, plus the volume of the stationary liquid phase V_S at the column temperature. Thus,

$$V_R C_M = V_M C_M + V_S C_S \qquad (11\text{-}7)$$

Rearranging and inserting the partition coefficient, we obtain the fundamental equation for gas chromatographic separations:

$$V_R = V_M + K_d V_S \quad \text{or} \quad V_R - V_M = K_d V_S \qquad (11\text{-}8)$$

provided the compressibility factor j is applied.[14] Retention volumes will be primarily determined by the term $K_d V_S = K_d(w_L/\rho_L)$, where w_L is the weight and ρ_L the density of the liquid phase at the temperature of the column.[15] Since the coefficient of cubical expansion of most organic liquids lies between 0.5×10^{-3} and 1.5×10^{-3}, it will usually be adequate to obtain the density of the liquid phase by measurement of the density at room temperature and estimation with a coefficient of $10^{-3}/°C$.

When the amount of liquid is increased either by using a thicker layer or a longer column, the retention volumes are increased. In order to take into account the weight of liquid phase in a column, the specific retention volume, V_g, is defined as

$$V_g = (273/T_c)(V_N/w_L) \qquad (11\text{-}9)$$

It corresponds to the volume of carrier gas required to remove half of the solute from a hypothetical column at a specified temperature (0°C unless otherwise specifically stated) which contains 1 g of a liquid phase and which has no pressure drop or apparatus dead space. Now from Eqs. (11-6) and (11-9),

$$K_d = V_g \rho_L (T_c/273) \quad \text{or} \quad V_g = (273/\rho_L)(K_d/T_c) \qquad (11\text{-}10)$$

Temperature Dependence The temperature dependence of a specific retention volume is given by the relation

$$\log V_g = \Delta H/2.3\mathscr{R}T_c + \text{const} \qquad (11\text{-}11)$$

[14] If one prefers to express Eq. (11-8) using the partition ratio, it becomes $(V_R - V_M)/V_M = k$, which expresses the number of column (or void) volumes required to elute a sample component, as is done in ion-exchange chromatography.

[15] For polar substances, the calculation of partition coefficients and other thermodynamic data must take into account the contribution of liquid surface adsorption. The modified equation is: $V_N = K_d V_S + K' A_S$ where K_d is the usual partition coefficient and K' is an adsorption coefficient for the liquid–gas interface, and V_S and A_S are the volume and surface area of the liquid per gram of packing.

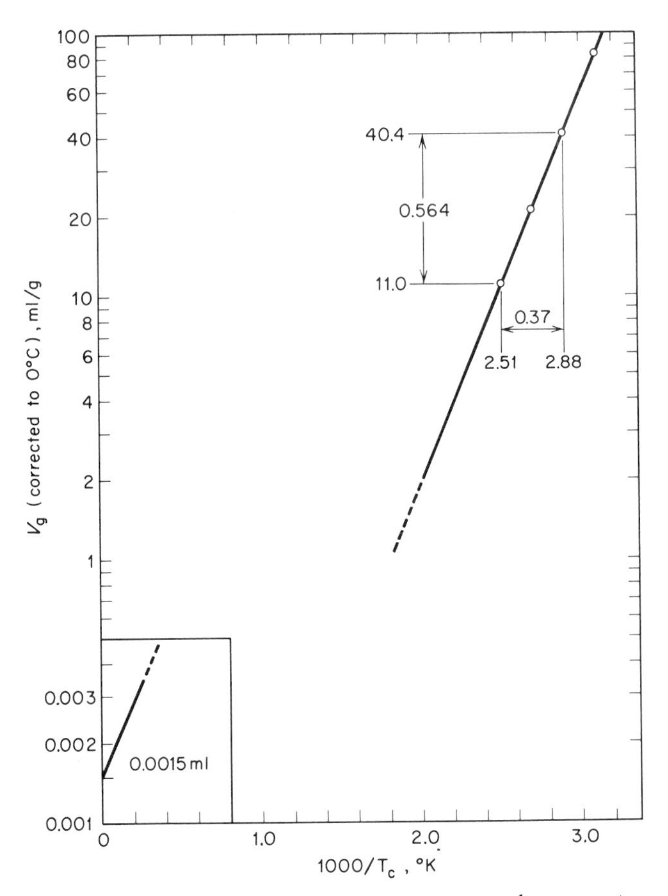

FIG. 11-18. Data for retention volumes at several temperatures.

where ΔH is the partial molar heat of solution of solute in the liquid state. By plotting log $(V_g \rho_L)$ against $1/T_c$ for several isothermal runs, $\Delta H/2.3\mathscr{R}$ may be obtained from the slope of the inverse temperature plot which is approximately a linear function, as shown in Fig. 11-18. The value of ΔH is the same for all weights of solvent whereas the value of the constant varies directly with the weight of solvent and also with the chosen standard temperature at which the volumes are expressed.

It is evident from Eq. (11-11) that the elution times of individual components can be altered by adjusting the column temperature. Lower operating temperatures lead to increased retention, although not necessarily to improved separation of components. Roughly, a decrease of 30°C will approximately double the retention volume. A knowledge of ΔH and V_g immediately makes it possible to draw the inverse temperature plot since the former quantity enables calculation of the slope of the line while the latter gives an ordinate value at one temperature. Among members of a homologous series, the solutes have the same specific retention volume at the column temperatures corresponding to their boiling points.

Example 11-1

The determination of ΔH and the constant in Eq. (11-11) from data for specific retention volumes at several column temperatures is illustrated in Fig. 11-18 for $CHCl_3$ on an Apiezon L column. The slope is 1.52×10^3 $[0.564/(0.37 \times 10^{-3})]$; therefore, ΔH is $2.3\mathscr{R}$ times this value, or 7.00 kcal/mole. The intercept is 0.0015 ml for the column containing 4.58 g of liquid; hence the constant is 0.00033 ml/g of stationary phase.

The plots of retention index (*q.v.*) versus column temperature for isomeric paraffins have the characteristic that their plots intersect. Figure 11-19 illustrates this effect. Because of this phenomenon, the order in which the peaks will emerge depends upon the temperature at which the sample is analyzed. For example, below 67°C, methylcyclopentane emerges before 2,4-dimethylpentane, while above 67°C, they emerge in reversed order. Below 33°C, methylcyclopentane would have an even shorter retention time than 2,2-dimethylpentane.

Relative Retention The influence of experimental variables is minimized by expressing retention behavior as a relative retention. Both the sample components and the reference substance are analyzed under identical conditions. In many cases the reference standard is actually part of the sample; if not, it is intentionally added. The retention times (or volumes) of sample components are expressed as a ratio of the retention time of the

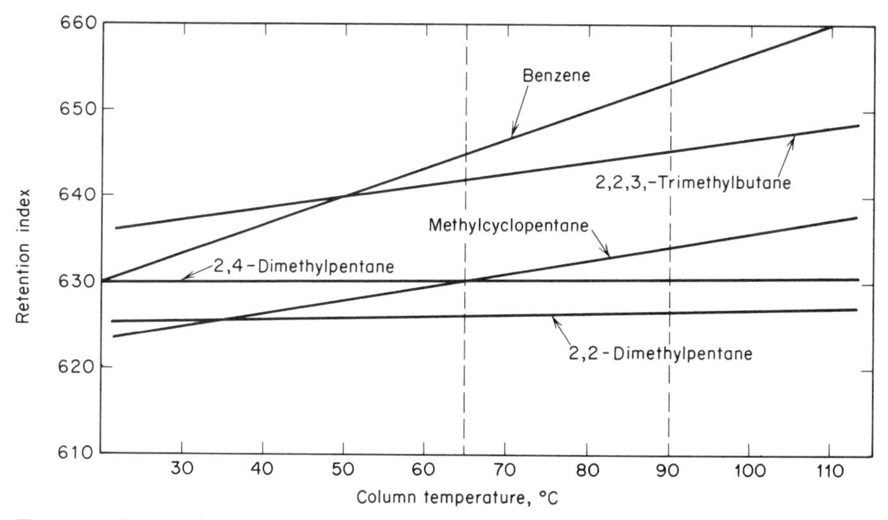

FIG. 11-19. Kováts retention index vs column temperature for several hydrocarbons, on squalane liquid substrate.

standard:

$$\alpha = t'_R/(t'_R)_{std} = (K_d)_2/(K_d)_1 \tag{11-12}$$

Relative retentions are independent of column length, carrier flow rate, the compressibility factor, and ratio of liquid substrate to solid support, but not independent of operating temperature. The basic shortcoming of relative retentions is the fact that it is almost impossible to fix one standard for all classes of compounds. Whenever possible the internal standards should be *n*-alkanes. Where it is inconvenient or impossible to use *n*-alkanes, alternative internal standards should be used and, preferably, these are stand-

TABLE 11-4. Relative Retentions (*n*-Pentane = 1) of Various Compound Types

Substrate	Convoil-20			Silicone D.C. 703			Poly(propylene-glycol)-2000		
Operating temp., °C	100			100			120		
Boiling-point level, °C	60	100	140	60	100	140	60	100	140
Hydrocarbon types									
n-Paraffins	1.80	5.6	17.0	1.67	4.3	13.5	1.60	3.8	9.6
2-Methyl paraffins	1.75	5.4	16.5	1.60	4.2	...	1.47	3.3	6.9
Type I olefins	1.90	5.6	17.5	1.90	5.2	15.4	1.80	4.4	11.4
Type III olefins	1.90	5.6	17.5	1.90	5.2	15.4	1.95	4.4	11.4
Cyclopentanes	2.33	6.7	19.8	2.37	6.0	16.5	2.3	5.2	13.0
Cyclohexanes	...	7.0	20.0	...	6.0	16.5	...	5.4	13.0
Cyclo-olefins	2.5	7.2	20.5	2.80	7.9	17.0	2.9	6.5	14.0
Diolefins	...	...	...	2.10	6.0	16.5	2.1	5.4	...
Acetylenes	1.50	4.5	15.0	2.2	6.4	19	2.8	6.6	16.3
Alkylbenzenes	...	6.7	19.8	...	7.9	23.0	...	8.6	19.5
Oxygenated types									
Alcohols									
Primary normal	0.29	1.4	10	0.76	2.6	...	2.1	6.3	24
Secondary	0.2	1.9	...	...	3.0	...	...	6.0	...
Tertiary	0.28	2.6	...	...	3.6	...	...	6.5	...
Ketones	0.72	3.1	12.0	1.85	6.0	19.7	2.3	6.5	18.0
Ethers	1.35	5.4	22	2.1	6.4	18	2.1	5.2	12.5
Esters									
Formates	1.0	3.3	13	2.0	6.2	...	2.6	6.5	...
Acetates	1.0	3.3	13	2.3	6.9	25	...	7.0	20.0
Aldehydes	1.0	3.3	13	2.05	6.2	...	2.3	6.5	...
Acetals	...	...	...	...	...	...	2.7	6.0	14.0
Retention volume for *n*-pentane (at 25°C, 1 atm), ml	77			61			20		

ards whose retentions are known relative to the n-alkanes. Wherever a relative retention is cited, it should be accompanied by information on the solute, liquid phase, operating temperature, and observed retention of the standard solute. To estimate the retention time (or volume) for any component under any conditions of flow rate, column length, or liquid phase loading, one simply measures experimentally the retention time for the reference standard under the given conditions at the column temperature given in the literature reference and multiplies this value by the relative retention ratio tabulated.

In Table 11-4 retentions (relative to n-pentane) for various solute

at Three Boiling-Point Levels for Various Liquid Substrates

Tricresyl phosphate 100			Di-2-ethylhexyl sebacate 100			Poly(diethylene glycol succinate) 100			Oxydipropionitrile 67		
60	100	140	60	100	140	60	100	140	60	100	140
1.65	4.6	13.5	2.2	5.0	13	1.5	2.9	5.6	1.4	3.2	8.8
1.60	4.3	11.5	1.8	5.0	13	1.0	1.9	...	1.4	3.0	...
2.25	5.8	16.5	2.2	5.4	16	1.7	3.4	...	2.9	6.5	14.8
2.50	5.7	...							3.5	6.9	...
2.25	5.8	16.0							3.1	6.1	13.0
...	6.7	18.3							...	6.2	14.5
3.9	9.7	23							6.4	13.7	30.0
2.6	7.1	18.3							5.4	11.5	24
3.6	9.7	23							12	25.5	60
...	16.4	43.5	...	9.3	27		18.4	50	...	57	135
3.3	11.5	...	1.1	3.7	...	33	54	140	40	100	...
2.2	11.5	...	2.0	5.0	...	30	46.8	145	24	85	...
2.2	11.5	...	2.3	6.0	...	24	45.6	160	20	77	...
4.8	13.5	37.5	1.4	6.5	...	20.5	37	80	56	115	270
...	...	...	1.9	6.2	20	9.2	56	...	7.6	15	30
4.6	12.5	43							37	77	190
4.6	12.5	43	1.7	6.4	20	16	30	61	37	77	190
5.2	16.3	...	1.8	7.0	...	14.3	30	61	40	95	...
...	...	...							23	39	65
	34			24			13			8	

classes are tabulated for seven typical stationary phases. With this type of information one can determine in advance whether a mixture can be separated with a particular stationary phase. It also shows what classes of compounds can readily be separated from other classes. Extensive tabulations of relative retentions will be found in the *Journal of Chromatographic Science* (formerly *Journal of Gas Chromatography*), *Chromatographic Reviews,* and the *Journal of Chromatography,* plus many texts.

Example 11-2

On which liquid substrate would the lower acetate esters be separated best? For the three stationary liquid phases given below, these relative retentions (n-pentane = 1) are available at a column temperature of 100°C:

	B.P., °C	Silicone oil DC 200		Di-2-ethyl-hexyl sebacate		Poly(diethylene glycol succinate)	
n-Propyl acetate	102°	4.88		6.42		30.4	
			1.95		2.10		1.50
Ethyl acetate	77°	2.50		3.05		20.2	
			1.62		1.83		1.26
Methyl acetate	57°	1.54		1.67		16.0	

On the basis of the relative retentions: propyl/ethyl and ethyl/methyl, the di-2-ethylhexyl sebacate column is slightly superior to the silicone oil substrate; however, the silicone oil column is somewhat faster. The extremely long retention times of the glycol column mitigate against its consideration.

Equation (11-11) contains the heat of solution, which is closely related to the heat of vaporization. This, in turn, determines the boiling point of a particular solute. From the regular variation of boiling point with carbon number in a homologous series, one can construct useful family plots, as shown in Fig. 11-20. Generally, a plot of log adjusted retention volume (or time) against the number of carbon atoms yields a smooth curve (except perhaps for the first member).

By utilizing the selectivity of particular liquid substrates, one can gain a great deal of information about an unknown or a mixture of unknowns. On each of two columns—one containing a polar and the other a nonpolar liquid phase—are run a series of compound classes (n-alkanes, alkylbenzenes, alcohols, ketones, etc.) to determine the relative retentions of each compound. By plotting the relative retention values for the two stationary phases against each other, lines radiating from the origin are obtained (one for each homologous series). The slopes of these lines are dependent on the relative interactions of each series with the two stationary phases. Radial areas on this diagram may then be allocated to the various types of molecular structure. A disadvantage of this plot is that points are spaced along the lines in a distribution logarithmic to molecular weight and thus tend to be

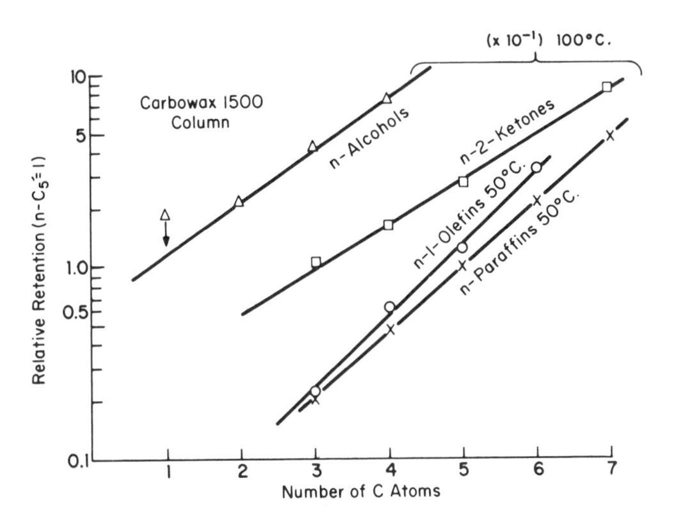

FIG. 11-20. Plot of relative retention (log scale) vs number of carbon atoms for several homologous series.

crowded into the corner near the origin. By plotting the logarithms of the relative retentions against each other, a corresponding series of approximately parallel lines is obtained, with points spaced linearly with molecular weight. This is illustrated in Fig. 11-21 for several homologous series on columns of silicone oil and tricresyl phosphate. Results from this type of plot are somewhat analogous to the readings obtained from two-dimensional chromatograms. By comparing the relative retention values for an unknown hydrocarbon on the two stationary phases, an approximate value for its molecular weight can be obtained and its structure broadly defined as paraffinic, aromatic, or the like.

Kováts Retention Index Another means of reporting isothermal data is the Kováts index system.[16] This system relates the retention volume of the compound under investigation to those of the *n*-paraffins eluting directly before and after it. Fixed reference points are obtained in this way by attaching to each *n*-paraffin the index

$$I = 100n$$

where n is the number of carbon atoms in the compound. The retention index is calculated from the relationship

$$I = 100i \left[\frac{\log R_x - \log R_n}{\log R_{(n+i)} - \log R_n} \right] + 100n \qquad (11\text{-}13)$$

[16] E. Kováts, *Helv. Chim. Acta* **41,** 1915 (1958); see also *Anal. Chem.* **36** (No. 8), 31A (1964) and *Advances in Chromatography,* Vol. 1 (J. C. Giddings and R. A. Keller, Eds.), Dekker, New York, 1965, p. 230.

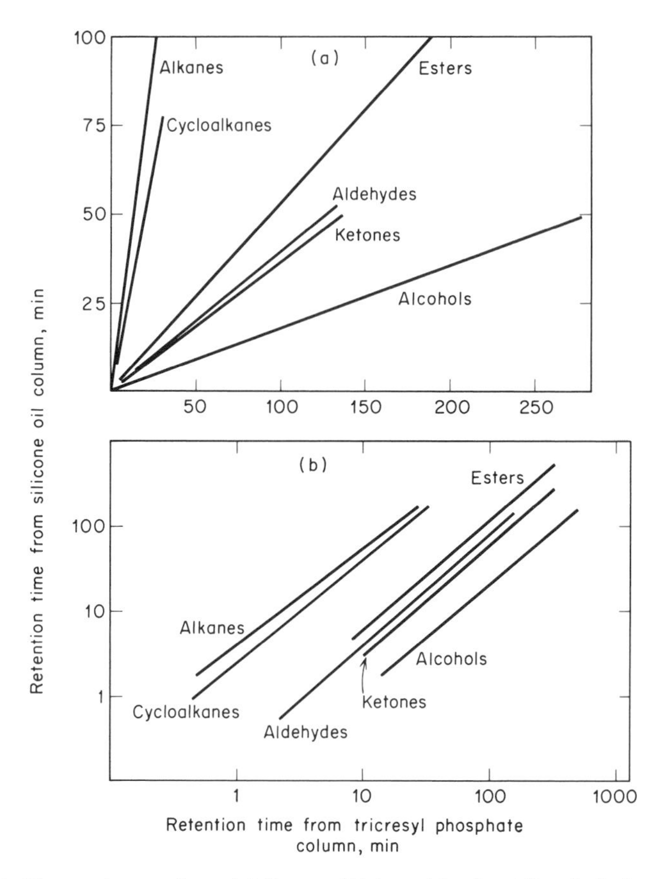

FIG. 11-21. Two-column plots: (a) linear; (b) logarithmic. After J. S. Lewis, H. W. Patton, and W. I. Kaye, *Anal. Chem.* **28,** 1370 (1956). (Courtesy of American Chemical Society.)

where R_x, R_n, $R_{(n+i)}$ are retentions of the unknown compound and normal paraffins of carbon number n and $(n + i)$. This index is based on the relationship between carbon number and logarithmic retention volume for a homologous series. It may be visualized as a carbon number corresponding to the retention volume of the compound under investigation on the carbon number versus logarithmic retention volume curve for n-paraffins. With temperature programming, the carbon number–retention volume relationship is linear instead of logarithmic. Retention temperatures may be used in place of volumes; they are substantially less affected by variations in the temperature program. A complete description of the Kováts retention index system is contained in papers by Kováts.

Column Efficiency (or Band Broadening)

Efficient columns keep elution peaks narrow. Narrow peaks offer the best chance for resolution of mixtures. Peak widths are the net result of the interaction of many variables, including all of those affecting the retention volume. When the chromatogram is obtained under linear elution conditions, ideally each solute will produce a bell-shaped or gaussian distribution, as shown in Fig. 11-17. Relative peak sharpness Q can be defined as the ratio of retention volume to the volume width W of the peak

$$Q = V'_R/W \tag{11-14}$$

where W is the volume corresponding to the intercept of the front and rear tangents to the inflection points of the gaussian with the base line of the elution curve. For a gaussian distribution, the peak width is equivalent to four standard deviations (4σ). A solute retained by the stationary phase shows greater increase in peak width than a non-absorbed solute because of the necessity for transfers of solute between the gas and liquid phases. Equilibrium is never set up instantaneously; there is always some delay during which the gas phase will have moved forward a finite distance.

The number of theoretical plates in a column is obtained from peak dimensions by

$$N = 16(V'_R/W)^2 = 16Q^2 \tag{11-15}$$

N values of 500 to 60,000 plates may be achieved with packed columns, and N values exceeding 100,000 have been obtained with capillary columns (Table 11-1). Various experimenters have suggested variants of Eq. (11-15). One of the most widely accepted is [see also Eq. (4-16)]:

$$N = 5.54(V'_R/W_{1/2})^2 \tag{11-16}$$

where $W_{1/2}$ is the width at one-half the peak height as measured from the base line. Using this method for N, the need to construct tangents to the peak slopes is eliminated and accuracy is considerably improved.

Plate Height Column efficiency is usually expressed in terms of plate height. For a gaussian distribution, plate height may be defined as

$$H = \sigma^2/L \tag{11-17}$$

where σ is the standard deviation of the chromatographic band expressed in units of distance along the column and L is the length of the column. Expressed in units of time,

$$H = L\tau^2/t_R \tag{11-18}$$

where τ^2 is the variance in time units and t_R the retention time. The standard deviation τ is the time interval from the peak center to the point of inflection of the near-gaussian profile. We may regard H as basically a measure of the extent of band spreading during the passage time of the solute through the column.

Plate height commonly is expressed at constant temperature as a function of carrier-gas velocity:

$$H = A + B/\nu + E_{liq}\nu + E_{gas}\nu \tag{11-19}$$

where A, B, E_{liq}, and E_{gas} are constants for the particular system and ν is the gas velocity at the column outlet. The velocity may be calculated by dividing the column length by the air-spike time corrected for pressure gradient, i.e., L/jt_{air}.

The A term, equal to $2\lambda d_p$ and attributed to eddy diffusion, represents the distance that a streamline persists before its velocity is drastically changed by the packing. Here d_p is the average particle diameter of the support while λ is a number indicating how well the column is packed.

The B term, $2\gamma D_{gas}$, contains the obstructive factor, γ, indicating the degree to which diffusion is hindered by the packed stationary phase, and the gaseous diffusion coefficient D_{gas} of the solute. It describes the band spreading due to diffusion in the flowing gas along the direction of flow. Diffusion in the gas phase may be diminished by using carrier gases of higher molecular weight or by operating the columns at increased pressure.

The third and fourth terms describe the resistance to mass transfer in the liquid and gas phases. The E_{liq} term is given by $(8/\pi^2)[k/(1 + k)^2] \times (2d_f^2/D_{liq})$ where k is the partition ratio of the solute between the stationary and mobile phases in the column, d_f is the effective thickness of the film of liquid phase held on the stationary support, and D_{liq} is the diffusion coefficient of the solute in the stationary liquid film. The partition ratio contains the partition coefficient in a disguised form in combination with the phase ratio, i.e., $k = K_d/\beta = K_d(V_S/V_M)$. The E_{gas} term is proportional to the square of the diffusion distance in the gas and inversely proportional to the diffusion coefficient of the solute in the gas, D_{gas}. For a porous packing, the E_{gas} term has the approximate form $0.7[1 - 0.15/(1 + k)](d_p^2/D_{gas})$. For open tubular columns, the E_{gas} term is given by $(1 + 6k + 11k^2r^2)/[24(1 + k)^2D_{gas}]$ where r is the radius of the column.

In open tubular columns retention occurs at the walls, wet with a partitioning liquid. Gas flow is characterized by parallel streamlines in the axial direction, rather than by the chaotic streamlines which control the eddy diffusion term A in packed columns. The B term becomes simply $2D_{gas}$. The very high efficiencies obtainable with open tubular columns result in important advantages. Often liquid phases which do not allow specific separations on packed columns can be used successfully for the same separation when they are used as a coating material for open tubular columns.

The E-terms show that the effective plate height depends on the partition ratio of the species. Thus, in a separation, the plate height is different for the different species. This point is often overlooked.

When the plate height is graphed against the linear velocity, as illustrated in Fig. 11-22, the $(E_{liq} + E_{gas})$ terms are the limiting slope at high velocity, and the A term is the intercept of the slope. The A term is zero, of course, for open tubular columns. At low velocities, the plate height is controlled by the diffusion in the mobile phase—the B term. At some point the curve achieves a minimum, $[A + 2(BE)^{1/2}$ at $\nu = (B/E)^{1/2}]$, which represents the optimum velocity for the smallest plate height. A curve, such as Fig. 11-22, is obtained for each solute and, consequently, when a mixture is introduced into a column, the carrier gas velocity will be ideal for only one solute. In actuality the minimum is quite broad for the usual 10–20% liquid phase loading. As a result, a single velocity will be near optimum for several closely spaced solutes. This is a fortunate situation, inasmuch as the experimenter may then set the flow rate to that which is optimum for the most difficult solute pair in his sample. In this respect, the reader is referred to the section on flow programming where column efficiency can be "traded" for speed or for the attainment of sharper peaks.

When the stationary liquid phase is supported on an adsorbent, the value of the A term should be minimized for highest efficiency. A narrow range in size of small particles (about 100 mesh) is the best. The minimum mesh size is actually limited by the increased pressure drop through the packed column. Efficiency also increases with decreasing column radius.

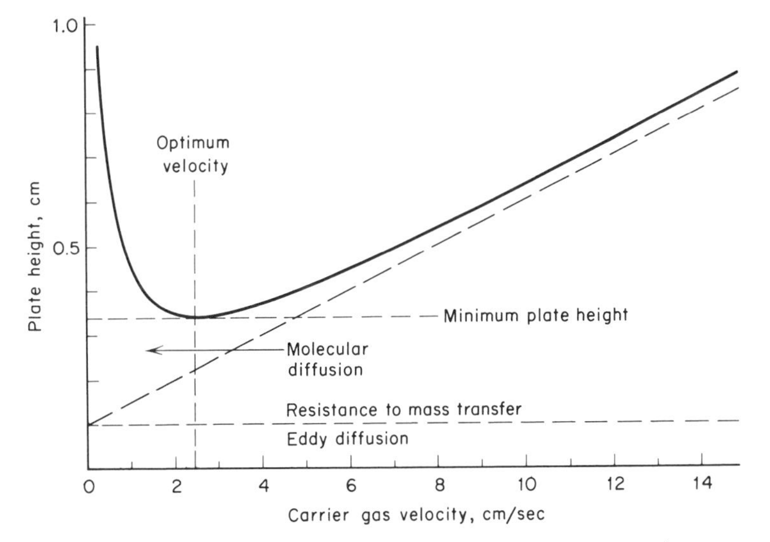

FIG. 11-22. Plot of plate height vs carrier gas velocity with schematic presentation of terms in Eq. 11-19.

Packing difficulties limit the column radius to packed columns 1/8 in. or more in diameter, except for coated, open tubular columns.

A basic column parameter is the phase ratio, or β value—the volume ratio of gas phase and stationary phase in the column. At low β values the plate height is less. Although this would seem to indicate the use of the highest possible amount of stationary phase, after the available surface area on a given support becomes coated, additional stationary phase will only result in a thicker film on the support particles. A thick film increases the E term since the liquid mass transfer component in Eq. (11-19) is proportional to the square of the average film thickness. On the other hand, the selection of the stationary phase loading concerns the sample capacity of the column. The amount of sample which can be introduced into the column is related to the amount of liquid phase present; the lower the amount of the phase, the smaller the possible amount of sample. Eventually one must compromise. The best liquid phase available having been chosen, the amount of liquid in the column must be made high enough and the temperature low enough so that the solutes are retained to a significant extent; but the temperature should not be so low nor the liquid thickness so great that peaks are excessively broadened by the mass-transfer terms.

Peak Symmetry The ideal situation in gas chromatography is to have a symmetrical peak; however, there are several factors that can cause a peak to be unsymmetrical. This dissymmetry takes two forms; a distorted forward edge, called leading, and an elongated rear edge, called tailing. Leading is caused by too low a column temperature. If the sample mixture contains compounds with a wide boiling range, those that boil above the column temperature will show a tendency to lead. This effect can be overcome by increasing the column temperature. Tailing can be caused by two factors. In one case, it is the result of interaction between the column support and the solute. If a nonpolar liquid phase is used, some compounds, if polar, will be slightly adsorbed by the support and the peak will tail. The use of a polar liquid phase will deactivate the support and minimize this cause of tailing. The second cause of tailing is insufficient heat available in the flash evaporator. The entire sample must be vaporized immediately, yet not undergo breakdown. When peaks are unsymmetrical, the adjusted retention volume may vary with the amount of solute eluted. However, the retention volume of the rear of the elution curve, $V'_{R,\mathrm{rear}}$, often remains constant. Consequently, the expression for the plate number in a column may be rewritten as

$$N = \left(\frac{4V'_{R,\mathrm{rear}}}{W} - 2\right)^2 \qquad (11\text{-}20)$$

Resolution

The efficiency for the separation of a particular pair of components is defined by resolution. It takes into account the peak-to-peak separation at the maxima of the gaussian curves and the average peak width, i.e.,

$$Rs = \frac{V_{R,2} - V_{R,1}}{0.5(W_2 - W_1)} = \frac{V_{R,2} - V_{R,1}}{4\sigma} \tag{11-21}$$

where Rs is in terms of the number of standard deviations (peak width) by which two peak maxima are separated. When back and front of adjacent peaks reach the base line at the same point, $Rs = 1.0$ which corresponds to a 2% overlap of peak areas.[17] Values less than this imply incomplete separation. Separation will be improved if the difference in peak-to-peak separation can be made significantly larger than the average peak width. Thus, a chromatogram with narrow peaks offers the best resolution.

With the help of Eqs. (11-8), (11-15), and (11-21), and if two adjacent peaks are not greatly different in concentration so that $W_2 \simeq W_1$,

$$Rs = \frac{1}{4}\left(\frac{\alpha - 1}{\alpha}\right)\left(\frac{k_2}{1 + k_2}\right)\sqrt{N_2} \tag{11-22}$$

where α is the relative retention ratio, and k_2 and N_2 are the partition ratio and the number of theoretical plates, respectively, for the second component. Rearranged, Eq. (11-22) becomes

$$N_{\text{req}} = 16\left(\frac{\alpha}{\alpha - 1}\right)^2 \left(\frac{k_2 + 1}{k_2}\right)^2 \tag{11-23}$$

where N_{req} is the number of theoretical plates required to make a separation of resolution $Rs = 1.0$. [For the completely general case, the factor 16 may be replaced by $4(1 + n)^2$, where n is the peak width ratio.] Thus, as two adjacent solutes move through the column, their peaks separate to a degree that is determined by their relative retention ratio, the capacity factor for the final component, and the number of theoretical plates for the final component.

The relationship between the relative retention ratio and resolution is clear. The more different two components are in their specific retention volumes (which depends in turn on the solute vapor pressure, i.e., solubility in the stationary phase), the better should be the separation. The relative retention ratio pertains only to the relative rates of migration and thus to the peak-to-peak separation. The extent of temperature dependence of α

[17] In the literature, resolution is sometimes defined as $Rs = 4(V_{R,2} - V_{R,1})/(W_2 - W_1)$. By this definition, $Rs = 4.0$ when back and front of adjacent peaks reach the base line at the same point.

with change in temperature can be deduced from Eq. (11-11):

$$\alpha = V_{g,2}/V_{g,1} = a \exp\left[-(\Delta H_2 - \Delta H_1)/\mathscr{R}T_c\right] \qquad (11\text{-}24)$$

Since the transfer of solute molecules from the moving to the stationary phase is an exothermic process in many cases, ΔH_1 and ΔH_2 are often negative. Moreover, the exponent will tend to be positive because the final component of the pair will have the higher ΔH. Thus in general the lower the temperature, the larger the value of α and the better the resolution. Another condition which greatly affects α and thus resolution is the type of stationary liquid phase used. For high resolution, phases of high selectivity for the given components are a necessity. A final point concerns the $[(\alpha - 1)/\alpha]$ term in Eq. (11-22). Changes in α affect resolution of a pair of solutes to a great extent only when α is close to one, a fortunate result, since a low α reflects components difficult to resolve.

Examination of Eq. (11-23) reveals that N_{req} depends quite markedly on the relative volatility. At an α value of 1.02 (and k_2 large), roughly 42,000 theoretical plates are necessary to produce resolution of 1.0. A 2% increase in α to 1.04 decreases the number of required plates to approximately 10,000, a factor of 4. The number of theoretical plates is directly related to the plate height and length L of the column:

$$N = L/H \qquad (11\text{-}25)$$

Clearly, to double resolution the column length must be increased by a factor of 4 or plate height reduced by a factor of 4. In fact, when the peak-to-peak separation is large, but resolution is poor, chromatographic conditions should be changed to narrow the peak widths. A longer column may improve the separation but at the expense of increased analysis time.

The influence of the partition ratio on the resolution is evident from Eq. (11-22). If plate number and relative retention remain constant, resolution will be best when the capacity ratio k is large and will diminish as k declines. Because of this, large values of k are desirable. This, however, implies large loading factors and long analysis time. In columns of high β, solutes of small K_d (short retention time) would be difficult to separate since the quantity in brackets $[k_2/(1 + k_2)] = [K_d/(\beta + K_d)]$ would be near zero, whereas with solutes that are considerably retained (either high K_d values or low temperature), β values can be increased without substantial resolution penalty. For high-temperature separations, where the partition coefficients are made small, columns with small β are most effective. Ordinary packed columns usually operate with relatively large values of k, while those for capillary columns range from 0.1 to 5. That is, the phase ratio, or β value (V_S/V_M), in open tubular columns may vary from 1:100 to even 1:500, whereas in packed columns it ranges from 1:3 to 1:20. For any pair of K_d values, more plates are required with the open tubular column to

TABLE 11-5. Relative Effect of Varying β and K_d on Resolution

$(K_d)_2$	$(K_d)_1$	β	*Percent liquid phase*	*N required for* $Rs = 1.0$
200	100	5	25 [a]	64
200	100	100	3 [b]	144
2	1	5	25 [a]	784
2	1	100	3 [b]	160,000

[a] Packed column on typical diatomaceous support.
[b] Open tubular column.

obtain equivalent separation. Fortunately, lengthening the column to achieve the necessary increase in plate number is quite feasible with open tubular columns. Table 11-5 shows the relative effect of different values of β and K_d on resolution. The whole advantage of the high plate numbers given by open tubular columns may be lost if they are operated under conditions in which k is fractional. For this reason, open tubular columns are operated at appreciably lower temperatures than are packed columns to obtain suitable values of the partition coefficient, and thus k. Open tubular columns are particularly adaptable to the analysis of samples having components whose volatilities vary widely and whose late peaks become very flat when using packed columns at isothermal temperatures.

Degree of Cross Contamination In practical chromatography only enough theoretical plates are required to reduce the degree of cross contamination in the adjacent peaks to a desired experimental level η. Glueckauf[18] has published curves (Fig. 11-23) relating the cross contamination of two adjacent peaks as a function of the number of theoretical plates for various values of the relative retention ratio when using packed columns, which usually operate with a relatively large value of the partition ratio ($k = 5$ to 100).

The extent of cross contamination can be estimated from the experimental elution curves. The effluent from the column (peak areas) is divided into two portions to give products of equal purity, i.e., $\eta_1 = \Delta m_2/m_1$, $\eta_2 = \Delta m_1/m_2$, and $\eta_1 = \eta_2$. When the solute concentrations are unequal, the fractional impurity η must be multiplied by the factor: $(A_1^2 + A_2^2)/2A_1A_2$, where A_1, A_2 are the areas of elution peaks. In general, the division of effluent should be made not half-way between the peaks, but nearer the peak involving the least solute (see Fig. 11-17).

[18] E. Glueckauf, *Trans. Faraday Soc.* **51,** 34 (1955).

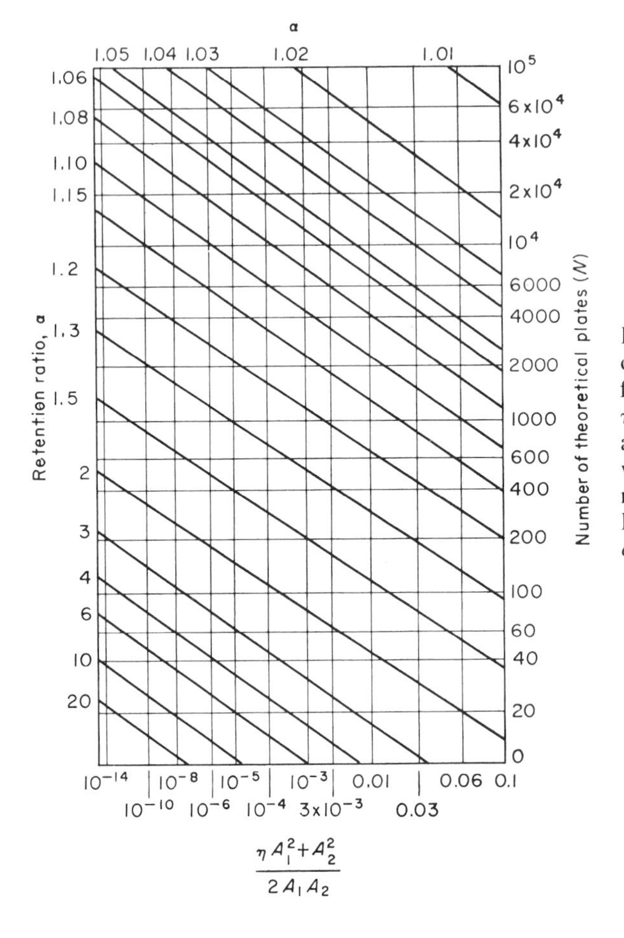

FIG. 11-23. Glueckauf plot of plate number required for any fractional impurity, η, and relative amounts of adjacent solutes, A_1, A_2, with the relative retention ratio, α, as parameter. After E. Glueckauf, *Trans. Faraday Soc.* **51,** 34 (1955).

Example 11-3

From Fig. 11-17, assume that the normalized areas are $A_1 = 0.6$ and $A_2 = 0.4$, and that the relative retention ratio is 1.8. To resolve these peaks with fractional impurity 0.01 (1.0%) in each band, first calculate the product:

$$\eta(A_1^2 + A_2^2)/2A_1A_2 = 0.01(0.36 + 0.16)/0.48 = 0.0108$$

The required number of theoretical plates is obtained from the Glueckauf plot (Fig. 11-23) by extending a vertical line from the value 0.0108 on the abscissa to the corresponding diagonal line for $\alpha = 1.8$. The ordinate corresponding to this intersection is approximately 60 plates.

From the peak dimensions for the second component, the actual number of theoretical plates in the column is obtained by means of Eq. (11-15) or (11-16). If less than the required number of plates, a longer column is indicated or some other adjustment in the parameters affecting the resolution must be made.

APPLICATION OF GAS CHROMATOGRAPHY

No single stationary phase will serve for all separation problems or be usable at all column temperatures. Indeed it is rarely possible to separate all components of any complex mixture by use of a single column. One approach is to make separate chromatograms on columns of distinctly different types of solvents. Reaction gas chromatography is another approach. Programming the temperature of the column or the carrier gas flow rate enables one to better handle mixtures containing solutes covering a wide range of boiling points.

Another difficult problem is the separation of close boiling compounds of the same general molecular class. It is solved only by employing a large number of theoretical plates and exploiting some small structural differences through the use of more selective stationary liquid phases. Knowledge of the selectivity of liquid substrates is essential to the intelligent solution of specific analytical problems. Unfortunately, except for a general characterization of liquid phases, column selection is often a matter of trial and error unless one is following published directions. Here chemical intuition is often the best guide.

Reaction Gas Chromatography

In reaction gas chromatography the injected substances pass through a reaction zone somewhere within the closed system between the introduction point and detector.[19] The reaction may occur at any point in the chromatographic pathway—ahead of the injection port, within the injection port, in the chromatographic column, or on a post column. Independent temperature control of the reaction zone is frequently required.

In an ideal reaction gas chromatographic process the reaction takes place instantaneously and the products have normal retention times, i.e., the same as if the products themselves were injected. In actual practice reaction times of 6 to 10 sec are usually satisfactory. Because the sample is processed in a closed system, transfer or handling losses or difficulties are not usually experienced. Minute amounts of substances are readily analyzed.

Subtractive Processes One of the simplest gas chromatographic reactions is the subtraction of a specific group of compounds by including in the chromatographic train a reagent that selectively retains these compounds. By comparing chromatograms obtained with and without these agents one can often simplify considerably the evaluation of a complex mixture. A rapid means of acquiring such data is with a dual-column, dual-

[19] M. Beroza and R. A. Coad, *J. Gas Chromatog.* **4**, 199 (1966).

recording instrument having a subtraction agent in one column but not in the other.

In hydrocarbon analysis the removal of straight-chain alkanes from branched and cyclic hydrocarbons with molecular sieves, especially 5Å, has been applied extensively. This retention appears to be due to physical adsorption and does not occur above the critical temperature. Thus, only hexane is not adsorbed from a mixture of hexane, heptane, and octane passed through molecular sieve 5Å at 243–268°C.

Olefins may be removed with packings containing 20% $HgSO_4$ in 20% sulfuric acid, while 4% Ag_2SO_4 in 95% sulfuric acid retains both olefins and aromatics. Water may be removed, most often in a precolumn, with a variety of dehydrating chemicals in mesh sizes between 30 and 100; these include molecular sieve 4Å and $Mg(ClO_4)_2$, and packings containing P_2O_5 or $CaSO_4$.

In addition to direct subtraction from a carrier gas stream, a compound may be converted to another with a markedly different retention time. For example, water is readily converted by passing it over calcium carbide to form acetylene or over hot iron filings to form hydrogen, and CO is readily changed to CO_2 by including a zone of iodine pentoxide.

Pyrolysis Pyrolysis is especially useful in the analysis of nonvolatile materials. The substance is cracked or degraded in the carrier gas stream, and the volatile degradation fragments are carried directly into the gas chromatograph for analysis.[20] Under controlled conditions the method gives repeatable results which are useful for identification or quantitative analysis, particularly in the analysis of polymers and copolymers and in the study of polymer degradation mechanisms and polymer structure. For example, polystyrene breaks down to yield a simple trace and easily identified fission products (Fig. 11-24). Many materials, however, produce complex traces with many diverse peaks representing literally dozens of fission products. These traces, although not amenable to chemical interpretation, are nevertheless characteristic and are useful for comparative purposes, much as a fingerprint.

Pyrolysis has been accomplished in a variety of units. In flash pyrolysis the substance is coated on a filament and then the coated filament is heated electrically for a 15–20 sec period in the carrier gas stream or in a closed loop for sudden release. In the reaction-chamber technique, the substance is placed in a tiny ceramic boat which is inserted into a tubular reaction chamber maintained at the desired pyrolytic temperature. This technique enables the pyrolysis temperature to be measured accurately and permits the weight of sample and residue to be determined. In the third method, a high-

[20] G. M. Brauer, *J. Polymer Sci., Pt. C* **(8)**, 3 (1965); G. C. Hewitt and B. T. Whitham, *Analyst* **86**, 643 (1961).

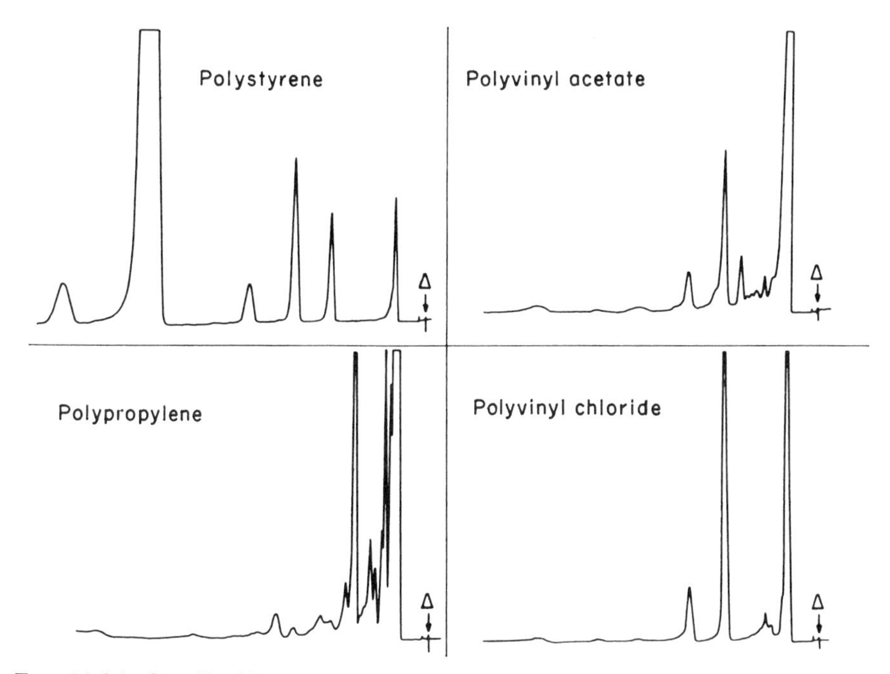

FIG. 11-24. Gas–liquid chromatograms of pyrolysis products of various polymers.

voltage corona discharge fragments the sample within a ceramic cylinder. In this manner progressive heating difficulties are avoided.

Elemental Analysis When pyrolysis is carried to the extreme, e.g., by contact with a catalyst and source of oxygen at very high temperatures, organic compounds break down to CO_2, water, and small molecules with other elements. The exposure of organics to hydrogen under similar drastic conditions produces methane, water, and small molecules containing other elements. Radioactive carbon and tritium are determined by combusting a chromatographic effluent containing these elements to labeled carbon dioxide and water. The carbon-14 is counted directly as $C^{14}O_2$, or by reducing it to methane. Tritium is determined by converting the labeled water of combustion back to tritium for counting.

Class Reactions Where a class of compounds is not readily chromatographed because of its thermal instability or lack of volatile members, conversion to derivatives that are more readily chromatographed is often possible. Fatty acids can be converted to the corresponding methyl esters by BF_3-methanol. Hexamethyldisilazane and trimethylchlorosilane are widely used for preparation of trimethylsilyl ethers of sterols, sugars, and a variety of compounds containing hydroxyl groups. Acetylation of fatty alcohols, sterols, and other hydroxy compounds can also be accomplished with acetic anhydride and pyridine. Amino acid samples are converted to methyl ester hydrochlorides; these are transesterified to butyl ester hydrochlorides, and the latter are converted to *N*-trifluoroacetyl *n*-butyl esters. Kits of appropriate reagents in vials are available commercially for the examples cited. Strictly speaking, of course, the foregoing reactions are completed before the products are injected into the gas chromatograph.

Programmed-Temperature Gas Chromatography

The separation of constituents in samples composed of compounds with a wide range of boiling points can be improved and accelerated by raising the temperature of the entire column at a uniform rate during the analysis. The particular advantage in raising the temperature is that temperature has a greater effect on the chromatographic process than any other single variable. The sample is injected into a relatively cool column in the normal manner. Earlier peaks, representing low-boiling constituents, emerge essentially as they would from an isothermal column operated at a relatively low temperature. However, the solubilities of the higher-boiling materials in the stationary phase are so large that these substances are almost completely immobilized at the inlet of the column. As the temperature is raised, solubilities will decrease, and the remaining compounds will successively reach temperatures at which they have significant vapor pressures and will begin to migrate and ultimately emerge from the column. On a low-temperature isothermal column, these higher-boiling materials would emerge as flat peaks, often undetected, whereas by temperature programming they will be kept bunched by the rapidly increasing temperature. As a consequence, an extremely wide boiling range of compounds may be separated in less time, and the peaks on the chromatogram are sharper and more uniform in shape (Fig. 11-25). In fact peak heights may be used as the basis for accurate quantitative analysis with programmed-temperature gas chromatography.

Programming of column temperature is achieved by electrically or mechanically raising the set point of a temperature controller. Using a linear temperature program, the temperature rises at a rate of r°/min. Heating from ambient to 400°C within 8 min is possible, with cooling from 400° to 100°C in 3 min. A low thermal mass of the oven is essential and is achieved by constructing the inner compartment from thin-gauge stainless steel. It is separated from the outer insulated compartment by an air space through which air is drawn rapidly on the cooling cycle. In one arrangement the heater is suspended around the blower. Temperature gradients may be kept within 1°C with commercial units. Subambient programming extends the technique to extremely volatile components.

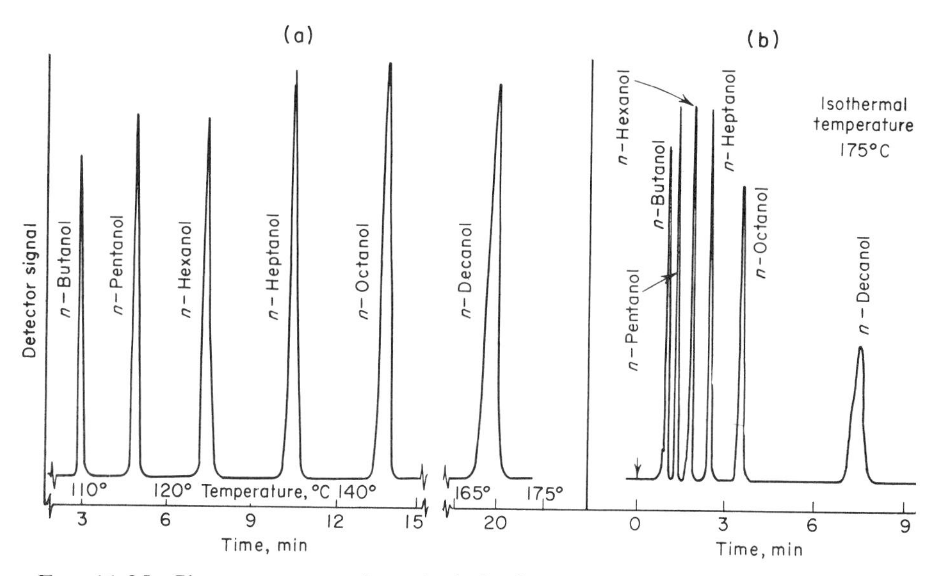

FIG. 11-25. Chromatograms of an alcohol mixture. (a) Programmed temperature from 100° to 175°C. (b) Isothermal operation at 175°C.

For each component eluted the significant temperatures are (1) the temperature of the column when the sample was injected T_0 and (2) the retention temperature T_R, i.e., the temperature reached by the column as the solute peak emerges. The initial temperature is chosen to give optimum resolution of the low-boiling components. For any given column the retention temperature depends on the ratio of heating rate to carrier gas flow (rather than on either one alone), and is given by the expression

$$\int_{T_0}^{T_R} \left(\frac{1}{V_T}\right) \partial T = \frac{r w_L}{F_c} \tag{11-26}$$

where V_T is the retention volume at temperature T, T_0 is the initial temperature, and w_L is the weight of liquid stationary phase in the column.

An analytic evaluation of the integral is difficult and a graphical method is preferable. The isothermal retention volumes for the solutes of interest are summarized as plots of log $(V_R - V_M)_T$ against $1/T$. From such a family of lines and from the values of V_M, determined for the particular apparatus, plots of $(1/V_T)$ against temperature are constructed as in Fig. 11-26. The required integrals are the areas under these curves from an arbitrary initial temperature to higher temperatures, as plotted in Fig. 11-27 against the upper temperature limit. The necessary isothermal retention volumes may be obtained by direct measurement or from the literature. Published values are usually the volumes of carrier gas at column temperature and mean

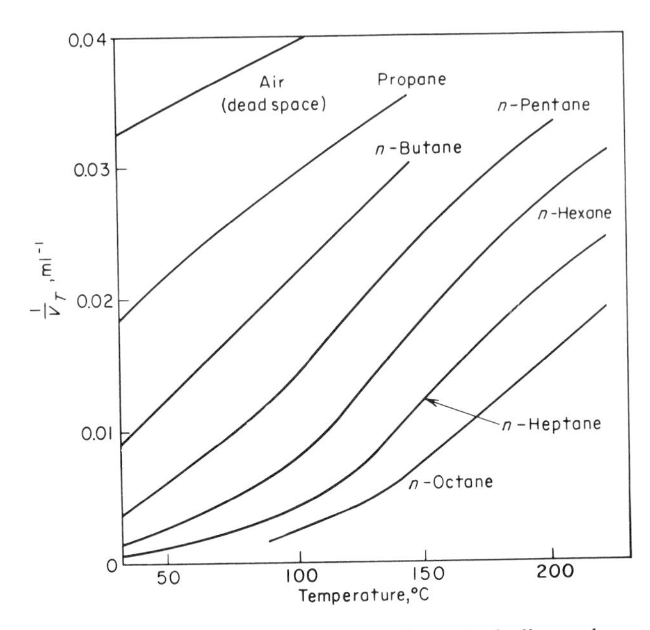

FIG. 11-26. Reciprocal of corrected retention volume including column dead space as a function of column temperature. After H. W. Habgood and W. E. Harris, *Anal. Chem.* **32,** 451 (1960). (Courtesy of American Chemical Society.)

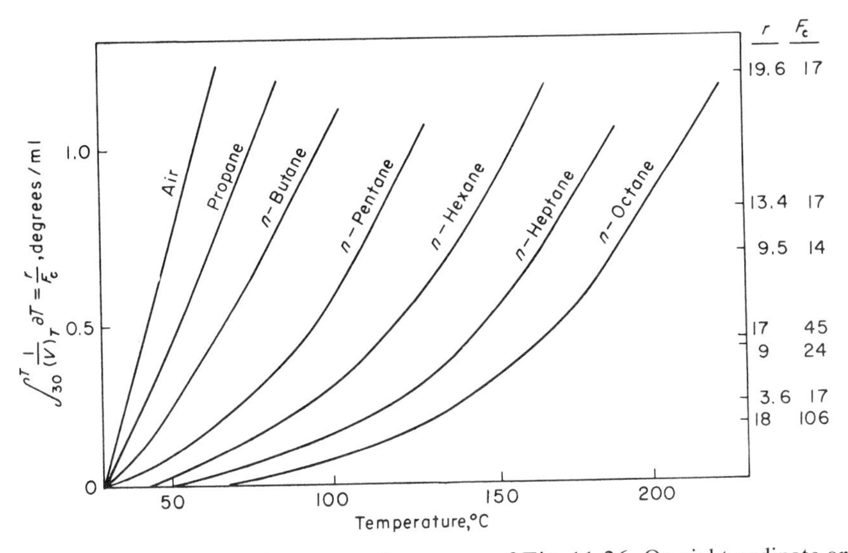

FIG. 11-27. Plot of integral—areas under curves of Fig. 11-26. On right ordinate are various programs whose r and F_c values are given. After H. W. Habgood and W. E. Harris, *Anal. Chem.* **32,** 451 (1960). (Courtesy of American Chemical Society.)

pressure. They must therefore be corrected to some chosen standard temperature, since the flow rate is constant only when expressed at constant temperature.

The number of theoretical plates for programmed-temperature gas chromatography is given by the expression

$$N = 16[(V_T)_R/W]^2 \tag{11-27}$$

which resembles the isothermal temperature relationship. Indeed, a column used for programmed-temperature gas chromatography has essentially the same number of theoretical plates as it does in isothermal gas chromatography.

Choice of Practical Parameters Simplified calculations of programmed-temperature gas chromatography are handled as follows. First, we will determine the average increase in temperature needed to increase the R value any desired amount. Recall that $R = 1/(1 + k) \simeq V_M/K_dV_S$. From the relationship of the partition coefficient and temperature [Eq. (4-13)], an expression for the ratio of R values can be derived, and which is

$$2.3 \log (R_2/R_1) = (\Delta H/\mathscr{R}T)(\Delta T/T) \tag{11-28}$$

On rearrangement, this gives

$$\Delta T = [2.3 \log (R_2/R_1)](\mathscr{R}T^2/\Delta H) \tag{11-29}$$

In these expressions T is the geometric mean of the two temperatures in the step, and $\Delta T = T_2 - T_1$. Although the chromatographic process involves vaporization from a solution rather than from pure solute, Trouton's rule is still approximately valid, and $\Delta H/T_b \simeq 23$. Since the process is operated near the solute boiling point, $T \simeq T_b$, and the ratio $\Delta H/T$ may be approximated by 23. Substituting $\mathscr{R} = 1.99$ cal deg^{-1} mole^{-1},

$$\Delta T = [2.3 \log (R_2/R_1)](0.087T) \tag{11-30}$$

where T is now the operating temperature (under isothermal conditions). For example, at a typical operating temperature of 227°C (500°K), the temperature increase that typically doubles the R value is 30°. This value will shift somewhat depending on the operating temperature and its departure from the boiling point of the solute.

A step-function approximation can be used to follow the solute migration as a function of temperature. Let us assume that the temperature increases in steps to halve the partition ratio k each unit time which, as has been shown, corresponds closely to a rise of 30°. Further, let us assume the dead space (air peak) time to be 0.5 unit and $k = 40$. The isothermal retention time is given by

$$t_R = (k + 1)t_{\text{air}}$$
$$= (40 + 1)0.5 = 20.5 \text{ units}$$

In unit time the fraction traveled by the band is $L/20.5 = 0.049L$ of the column. During the next step, when the effective k is 20,

$$t_R = (20 + 1)0.5 = 10.5$$

and the fraction of column traveled by the band is $1/10.5 = 0.095$; the total distance being $0.049 + 0.095 = 0.144$ column length. During the succeeding steps,

$$t_R = (10 + 1)0.5 = 5.5$$

and the fractional distance is 0.18 for a total distance of 0.32 column length;

$$t_R = (5 + 1)0.5 = 3.0$$

and the fractional distance is 0.33 for a total distance of 0.65 column length; and finally

$$t_R = (2.25 + 1)0.5 = 1.75$$

and the fractional distance traveled is 0.57. However, only 0.35 column length remains to be traversed, or $0.35/0.57 = 0.61$ of a time unit. Thus, total retention time is 4.61 units.

The distance migrated by the peak in successive 30° temperature increments is shown in Fig. 11-28. In the last 30° interval the distance migrated is one half the column length, $L/2$. The time required to migrate through this interval is the interval length, $L/2$, divided by the peak velocity, $R\nu$. Thus, the interval migration time is $L/2R\nu = t_{\text{air}}/2R$, since $t_{\text{air}} = L/\nu$. The time for the air peak to emerge is equal to the 30° divided by the programmed rate of increase of temperature with respect to time, i.e., the interval migration time equals $30°/r$ in our example. Equating the two times we find that the mean R_i value in the last segment of the column is

$$R_i = rt_{\text{air}}/60 \tag{11-31}$$

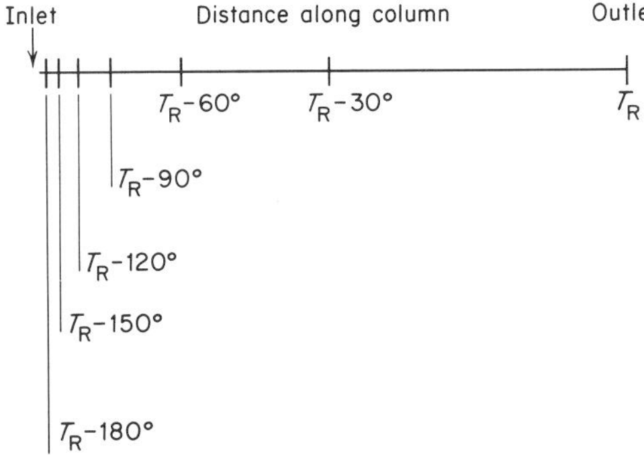

FIG. 11-28. Distance migrated by a peak in successive 30° increments in programmed temperature chromatography from 80° to 260°C (*schematic*).

However, the elution occurs right at the end of the final segment rather than in the middle where mean values are applicable. If R_e denotes the R value at the moment of elution, then $R_e/2$ would characterize the beginning of the segment, and the "mean" value would be $R_i = (R_e + R_e/2)/2 = 0.75R_e$. The terminal R_e value is then

$$R_e = rt_{\text{air}}/45 \tag{11-32}$$

Following through with our example, k can be assumed to be 3.5 at the moment of emergence, and therefore $R_e \simeq 0.22$. The heating rate can then be estimated to be 20° per unit time. Generally, $rt_{\text{air}} \leq 12°$.

The actual temperature, T_R, at which elution occurs is difficult to calculate unless several pieces of data are available; it is given by the expression

$$T_R = (\Delta H/\mathscr{R})/[2.3 \log (45a/rt_{\text{air}})] \tag{11-33}$$

where a is related to the entropy of vaporization. Reference to Eq. (11-33) is useful when it is necessary to manipulate the parameters of programmed-temperature gas chromatography in order to improve some aspects of the analysis. However, detailed theory shows that a temperature does exist which can be used to characterize the programmed operation. This temperature is called the significant temperature, T'. Its value is about $1.5°\Delta T$ below the elution temperature; or, in our example, $T' = T_R - 45°$. Thus, in programmed-temperature operation, one would choose a heating rate which would lead to an elution temperature $1.5°\Delta T$ above the best isothermal temperature—the significant temperature in our example. One would then obtain essentially the same degree of resolution as found in the isothermal run. The significant temperature provides the common link between isothermal and programmed-temperature methods.

Flow Programming

In flow programming, an alternative to temperature programming, carrier-gas flow rate is progressively increased during an analysis, thus sweeping components through the column more rapidly. Peak height is directly related to retention time, thus to flow rate. The height of late-emerging peaks would therefore be raised by increasing gas flow as an analysis proceeds.

Since a flow-programmed analysis involves no temperature rise, this approach has advantages over temperature programming when thermally unstable samples or volatile liquid phases are analyzed.

The flow programmer is a pneumatically controlled system which permits the pressure along the column to rise logarithmically, between preset limits, during a predetermined time interval. The controlling component is a pneumatic, differential-flow valve which is installed in the carrier-gas supply line upstream of the column. Any flow or pressure controllers built into the existing chromatograph should be bypassed, because inlet column pressure must be controlled by the flow programmer itself.

Carrier-gas pressure is set at the regulator on the carrier-gas cylinder supply. This pressure is the maximum which the programmer will supply to the column inlet. By means of a bypass toggle valve, the initial inlet-column pressure is next set. The program is initiated by connecting the column inlet to the carrier-gas cylinder supply via the time-delay element, consisting of a capillary tube or other restrictor. Interchangeable capillaries of various lengths can be fitted into the time-delay element; these determine the rate at which column inlet pressure is returned to cylinder regulator pressure.

A drawback to the flow-program approach is that higher gas-flow rates mean lower column efficiency – plate height for a given separation increases, and resolution is somewhat poorer than it is at optimal gas velocities. This effect is more pronounced with packed columns than with open tubular columns. On the other hand, flow programming reduces some of the base-line drift problems that may occur during temperature-programmed analyses. Vapor pressure of the liquid phase, resulting in column bleeding, rises exponentially with column temperature, but increases only linearly with flow rate. The extreme sensitivity to flow of some detectors may limit their use in flow-programmed analysis; the flame ionization detector has been used with flow-programmed capillary columns, as have the hot-wire and thermistor detectors with packed columns.

Gas–Solid Chromatography

The gas chromatographic separation of the permanent gases and certain light hydrocarbon gases is achieved with gas–solid chromatography and temperature programming. Using a silica gel column, carbon dioxide is separated (also acetylene, and from each other) but the other gases emerge as a single composite peak. The separation of hydrogen, oxygen, nitrogen, and carbon monoxide can be achieved on a 4-ft column packed with molecular sieve 5Å (Fig. 11-29). On molecular sieve 5Å the straight–chain alkanes alone are adsorbed; all the isomers pass through as a composite peak superimposed on the air peak.

Quantitative Evaluation[21]

Ordinarily the chromatogram is made by a strip-chart recorder connected to the output of the detector. Two conditions must prevail: (1) The output of the detector–recorder system must be linear with concentration. The linear dynamic range expresses this range and, coupled with sensitivity,

[21] See H. W. Johnson, Jr., "The Quantitative Interpretation of Gas Chromatographic Data" in *Advances in Chromatography*, Vol. 5 (J. C. Giddings and R. A. Keller, Eds.), Dekker, New York, 1968.

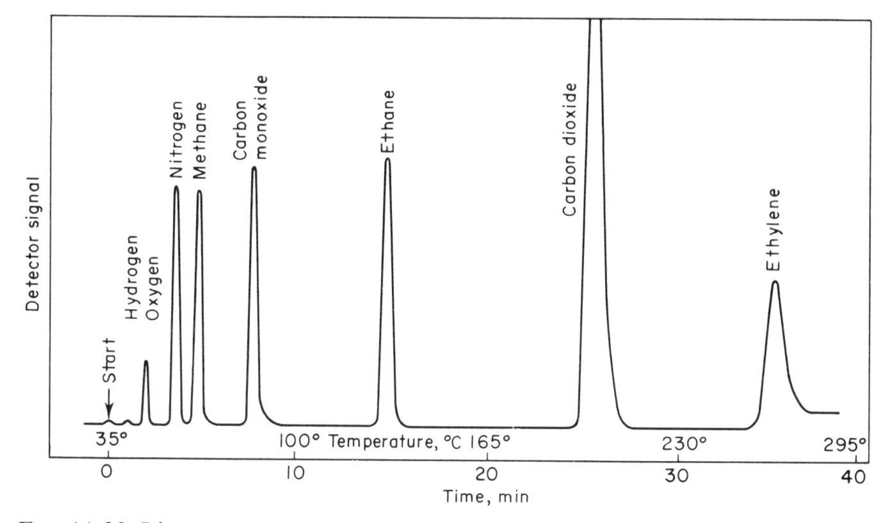

FIG. 11-29. Linear programmed temperature chromatogram on 4-ft column of molecular sieve 5A. Temperature programmed at 6.4°C/min from 35° to 291°C. After J. M. Slater, L. Mikkelsen, and M. G. Beck, in *Lectures on Gas Chromatography 1962*, H. A. Szymanski, Ed., p. 121, Plenum Press, New York, 1963.

the concentration limits. (2) The carrier-gas flow rate must be constant so that the time abscissa may be converted to volume of carrier gas. Under these conditions, peak area can be used as a quantitative measure of component present. Peak area can be measured by triangulation methods, by automatic integrator devices, or by planimetry. Precision is compared in Table 11-6.

TABLE 11-6. Comparison of Quantitation Methods

Integration technique	*Standard deviation (%)*
Planimetry	4.0
Triangulation	4.0
Height × width (at half height)	2.6
Cut and weigh peaks	1.7
Ball and disk	1.3
Electronic digital	0.44

Triangulation Methods Condal-Bosch[22] has discussed triangulation methods which are very simple and precise if certain precautions are taken. These methods consist of extending the approximately straight segments about the inflection points of the elution peak. These two lines, together

[22] L. Condal-Bosch, *J. Chem. Educ.* **41,** A235 (1964).

with the base line, constitute a triangle the area of which is obtained as one-half the product of the base times either the actual peak height or the vertical height of the apex of the triangle formed by the tangents. In this way one obtains an area that is equal to about 97% of the actual area of the chromatographic peak. The area can also be obtained as the product of the peak height times the width at half the peak height. The only geometric construction necessary consists of drawing the base line, measuring the peak height, noting its midpoint, drawing a horizontal through this point, and measuring the peak width on this horizontal. The area calculated in this way corresponds to 91% of the actual area for a gaussian curve. All these methods involve careful construction of necessary tangents in addition to the base-line placement. Serious errors arise in the measurement of narrow peaks and in peaks whose shape is not gaussian in form.

For asymmetric peaks, a method which gives the area without requiring any correction factor at all, and applicable to curves of every degree of symmetry, is accomplished by resorting to a trapezoid construction. One width is taken at 0.15 and the other at 0.85 of the peak height. The formula for the area is half of the sum of the widths multiplied by the peak height.

Planimetry The peak is traced manually with a planimeter, a mechanical device which measures area by tracing the perimeter of the peak. The area is presented digitally on a dial. The planimeter requires considerable experience in its manipulation. All measurements should be made in duplicate, but this makes the procedure very time-consuming. The sensitivity (smallest division of the nonius) of a normal planimeter is insufficient in many cases for analytical purposes.

Cut and Weigh The peak area is determined by cutting out the chromatographic peak and weighing the paper on an analytical balance. The accuracy of the method depends on the care used in cutting and on the constancy of the thickness and moisture content of the chart paper (or preferably a Xerox copy of the chromatogram to preserve the original chromatogram).

Automatic Integrators Two types of integrators will be discussed. The digital integrator is an all-electronic integrator. It consists of an electronic decimal counter preceded by an impulse generator, the output rate of which is proportional to the input voltage signal for the detector–amplifier combination. The incoming signal is received from beyond the zero adjustment in the case of a thermal-conductivity detector, or from the amplifier in the case of an ionization-type detector. The signal passes through an input-signal attenuator, then to the preamplifier and finally is supplied to a peak-and-valley sensor and voltage–frequency converter. The latter is the key component which develops an output pulse having a frequency that is directly proportional to the magnitude of the incoming d-c signal. Thus, the total number of counts over a period of time is proportional to the peak area. Elapsed time for each peak from the start of the upward curvature to peak crest is measured, displayed, and printed out on paper tape alongside the associated peak-area value. A finite time is required to recognize peak start and peak end; therefore it is necessary to keep the signal to the integrator large. To secure high accuracy on small peaks the count rate must be high, at least 1000 counts/sec/mV. At the same time the counter capacity needs to be high to accommodate the wide linear range.

A digital differentiator-type slope detector must sense valleys and shoulders of unresolved peaks, also peaks varying in width from less than 1 sec to several minutes. Peak-and-valley sensors, sensitive to signal rates of change as small as 0.1 μV/sec, accomplish this feat. Adjustment of slope sensitivity enables the operator to discriminate between noise and actual gas chromatographic peaks.

The ball-and-disk integrator (illustrated in Fig. 11-30) is an automatic mechanical type of integrator. A ball positioned on a rotating flat disk will rotate at a speed proportional to its distance from the center of rotation. The ball is positioned on the disk at a distance from the center in the same relationship as the position of the recorder pen to the base line. If the disk is rotated at a constant speed (time), then the ball will rotate at a speed proportional to the position of the recorder pen from zero – provided the recorder zero and the center of the rotating disk have been exactly aligned. This speed is then transmitted to a roller through a second ball which, by means of "spiral in" and "spiral out" cam, actuates the integrator pen at a speed directly proportional to the position of the recorder pen. The drive between the disk and the ball is by traction through an oil film. Although this hydrostatic phenomenon is not clearly understood, the oil film acts similar to an induction motor where slip is proportional to the driven load.

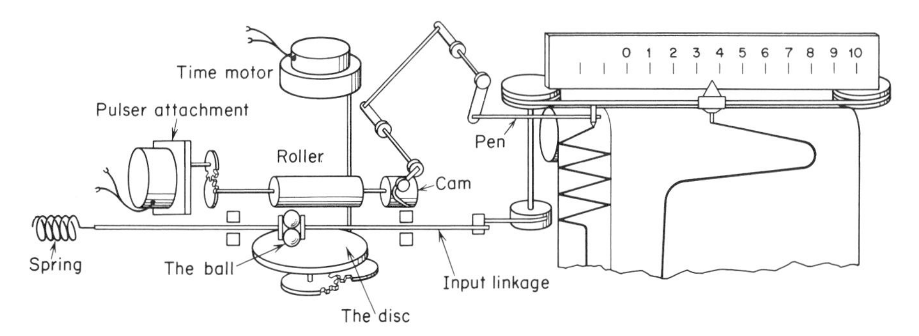

FIG. 11-30. Ball-and-disk integrator (*schematic*).

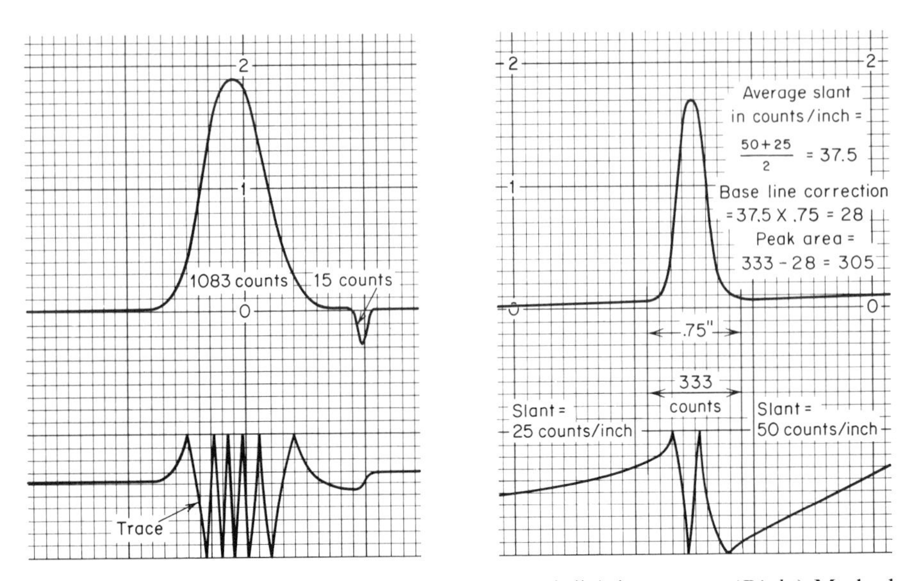

FIG. 11-31. Estimation of peak areas with ball-and-disk integrator. (*Right*) Method for handling baseline correction.

To read the integrator pen trace (refer to Fig. 11-31), first establish the desired chart time interval from the recorder pen trace and then project directly down to the integrator pen trace (*arrows*). The value of an interval is obtained by counting the chart graduations (not time lines) crossed by the integrator pen trace. A full stroke of the saw-tooth pattern in either direction represents 100 counts. Every division has a value of 10. Values less than 10 are estimated. On some models the space between "blips" is equivalent to 600 counts, making it possible to record up to 24,000 counts per inch of chart. For example, the interval for the main peak (left) on Fig. 11-30 is 1083 counts. With care the pattern can be read within two counts. The pattern (on Fig. 11-31, right) illustrates the means for estimating the baseline correction when the peak base line does not coincide with the recorder base line.

Correlation of Area and Quantity Areas under a gas chromatographic peak can be correlated with similar areas obtained with known concentrations of the particular solute run under conditions that are as similar as possible. This is usually done by the construction of a graph of quantity versus peak area. Errors in the injection technique normally limit the accuracy to 0.5% at best.

To express the result of the analysis in percentages, one can assume that all of the components of the sample are present in the chromatogram. The sample size can then be obtained by summing up the individual components. Individual quantities are referred to this sum. This procedure is termed "normalization." It fails for minor components under conditions that the major peaks do not register adequately, and in the use of detectors that are not sensitive to certain components. All substances do not exhibit equal molar response with detectors. Response factors, obtained by analyzing standard samples, serve to calibrate the normalization method to some extent and serve to correct the detector response.

In the internal standard method, a reference substance in a known amount is added to the mixture to be analyzed before taking the sample to be injected. The requirements for the internal standard method are stringent but, as a result, this method permits the best precision of any of the methods. Adjacent peaks must be completely resolved. The internal standard must elute near the peaks of interest, must be similar in concentration to the peaks of interest, and must be chemically similar but not present in the original sample. This technique is most often used where some sample may be lost during preparation or may not be completely eluted.

Preparative-Scale Gas Chromatography

Gas chromatography can be applied to the preparative-scale production of substances. Now one is interested in trapping the separated components of the sample in amounts sufficient for the particular requirements. Two classes of application may be distinguished: (1) The preparative separation for subsequent additional identification (infrared, NMR) and investigation, and (2) preparative separation of substances for purification from intermediate or end products. For the first class the apparatus retains essentially the dimensions of normal, analytical gas chromatography with columns 1.0 to 1.5 cm in diameter. Samples ranging from 1 to 1000 mg are obtained by repeated separations of small quantities. As instrumental accessories, a special injection device and a repetitive collection arrangement are necessary. In the second case the requirement is for the separation of quantities of from 1 to 200 g. Isolation of these larger quantities of material introduces problems.

The amount of mixture that can be separated in one run is limited by the danger of overloading the column. The large sample size may cause the partial vapor pressures of the components and the corresponding concentrations in the liquid substrate to become so high that they fall outside the linear range of the sorption isotherms for the entire column or a large part of it. With laminar flow, overloading occurs if the vapor-phase volume of the sample is larger than about $0.5(V_R/\sqrt{N})$. Also the vaporization chamber must be designed to flash-vaporize efficiently the larger quantity of injected sample.[23]

[23] E. Bayer, *J. Chem. Educ.* **41,** A755 (1964); G. W. A. Rijnders in *Advances in Chromatography,* Vol. 3 (J. C. Giddings and R. A. Keller, Eds.), Dekker, New York, 1966, pp. 215–258.

Lengthening the column with packings containing the same percentage of liquid substrate permits the use of larger samples. However, column pressure drop increases. Consequently, in a large part of the column the linear gas velocity deviates from the optimum, which results in a decrease in efficiency. Moreover, the time of separation is increased.

The usual way to raise column capacity is to increase the column diameter. Column capacity increases in proportion to the square of the diameter so that a column of 3.2 cm diameter can handle 10 times more sample than a 1 cm column, and there is no increase in time of separation. There are drawbacks, however. Flow inequality arises due to segregation of support particles which tend toward the walls of the column, thus increasing the linear velocity at the walls. The column material must be able to dissipate the amount of heat involved in the transfer of a passing solute between the liquid and gas phases. This is not insignificant. The temperature in a 2.5-cm column for a 5-ml sample can vary as much as 25°C during the passage of a solute band. Due to poor thermal conductivity of the support material, these temperature swings are greater, and thus the larger the column diameter, the greater is the peak broadening. Annular columns have been used to overcome both the wall effect and the temperature effect; columns provided with fins extending from the wall into the packing have also been used.

Another approach to preparative gas chromatography involves the repetitive (and automatic) injections of small samples on narrow-diameter columns. Columns that are 3/8 to 3/4 in. in diameter are used. Sample size ranges from 5 to 30 ml. A high percentage of stationary liquid phase is necessary to accommodate the sample without overloading the column.

The relative merits of the two approaches to preparative columns must be assessed in terms of capacity, resolution, initial cost, and operating cost. Short fat columns represent the most obvious approach to high-volume preparative work. However, the sample capacity, to a first approximation, is proportional to the amount of solid support, regardless of the mesh size or column dimensions. Since long thin columns can contain the same weight as short fat columns, they can provide the same capacity at much higher resolution. With repetitive sample injection, quite impressive throughputs can be achieved, even with difficult separations. On the other hand, since the object of preparative-scale work is to obtain pure materials, not a quantitative analysis, one can tolerate a certain amount of overlapping of the peaks. By taking the "heart" out of a peak and reprocessing the portions of the peak that overlap others, one can obtain a reasonable degree of component purity at lower resolution.

The economics of carrier-gas usage must also be considered. Narrow columns use flow rates of approximately 200 ml per min versus 8–10 liters per min for a 4-in. column. The cost per day for nitrogen is negligible, but for helium it would amount to $18 (at present prices) for the large-diameter column versus $0.35 for a 100 ft × 3/8-in. column. In larger-bore columns nitrogen, because of its lower thermal conductivity and higher viscosity, causes serious peak broadening and loss of resolution. Trapping efficiency may be a problem with high flow rates of carrier gas.

The sample injection system must permit a large sample to be injected in a short period of time. In one unit the sample is placed under pressure in a cylinder. The rate of injection, about 3–4 ml per sec, is controlled by the pressure above the sample. The carrier gas sweeps the vaporized sample into the column. Another instrument uses a quartz preheater to bring the temperature of the injected material near the vaporization temperature of the high-boiling component. Whatever the rate of injection the capacity of the vaporizer must be adequate to prevent band broadening due to slow volatilization of the sample.

Temperature programming is generally used to vaporize the material off the column. This technique is used in large- and narrow-diameter columns to give a better separation of component peaks and in a shorter time.

LABORATORY WORK

Numerous commercial gas chromatographs are available. A simple but very useful gas chromatograph in kit form is available from Gow–Mac Instrument Co., Madison, N. J.

General Directions for Use of Gas Chromatograph

Close the needle valve on the carrier gas cylinder. Turn on the gas by opening the main cylinder valve. Adjust the diaphragm valve so that the pressure on the second gauge reads about 14 psig. Slowly open the needle valve. Check the flow with a flow meter. Adjust the flow to the desired value with the needle valve; 80–100 ml/min is good for many columns. Increase the cylinder pressure if necessary. Purge at least 5 min.

Rotate the "current control" to its OFF position for thermal-conductivity units. Turn the filaments ON. Adjust the current to the recommended value:

Cell temperature, °C	Bridge current, mA
25°	250
100°	230
200°	200
300°	170

with the "current control." Recommended values are for helium carrier gas; for nitrogen carrier gas, use bridge currents that are one-half the above values. Turn on the recorder. Use the "zero control" to adjust the pen to 5% of full-scale deflection. NEVER OPERATE THE FILAMENTS OF A THERMAL-CONDUCTIVITY UNIT UNLESS CARRIER GAS IS FLOWING.

Turn on the heater for the injection block. Adjust the oven temperature to the desired value. Wait until the base line (recorder trace) is stabilized, then inject the sample and record the chromatogram. AVOID SAMPLES CONTAINING WATER.

Use of Syringes in Injecting Samples

In filling a microliter syringe with liquid, it is usually desirable to exclude all air initially. This can be accomplished by repeatedly drawing liquid into the syringe and rapidly expelling it. Viscous liquids must be drawn into the syringe slowly; very fast expulsion of a viscous liquid could split the syringe. Draw up about twice as much liquid into the syringe as you plan to inject.

Hold the syringe vertically with the needle pointing up. Push the plunger until it (or the included air pocket) reads the desired value. Wipe off the needle with tissue. Draw some air into the syringe now that the exact volume of liquid has been measured. This will serve two purposes: first, the air will give a signal from some detectors which can be used to calculate "adjusted" retention volumes; second, the air prevents any liquid from being expelled if the plunger is accidentally pushed during sample injection.

Injection Procedure

Hold the syringe in two hands. Use one (normally the left) to guide the needle into the septum, and the other to provide force to pierce the septum and to prevent the plunger from being forced out by the pressure in the gas column.

Insert the needle in the septum, depress the plunger, and withdraw the needle (keeping the plunger depressed) as rapidly as possible—the three steps should be done in a continuous, smooth motion. Insert the needle well into the injection block before releasing the sample to assure instantaneous vaporization.

More detailed directions usually accompany each microliter syringe and should be consulted for handling special liquids and for cleaning the syringe after use.

Preparation of Packed Columns

Diatomaceous earth supports are available under various trade names. Generally, a suffix to the trade name indicates different grades corresponding to certain chemical treatments which include, among others, acid washing and silanization. Acid washing reduces surface activity. Silanization reduces peak tailing and surface activity, but also decreases the surface area of the support; therefore, silanized supports should not be used with more than about 10% stationary phase loading.

The following rules will help in the selection of the proper diatomaceous-earth type support:

Nonpolar Stationary Phases Untreated supports can be used with nonpolar samples. For polar samples an acid-washed and silanized support is recommended. In all cases, a silanized support should be used if the stationary phase loading is less than 5%.

Moderately Polar Stationary Phases Acid-washed supports are usually adequate. For low phase loadings (less than 5%), a silanized support is preferable.

Polar Stationary Phases An acid washed-support is adequate except for lower phase loadings when a silanized support is preferable.

Mesh Size A 60/80 mesh support is frequently employed. Finer mesh size will require a larger pressure drop—about 34 psig when using 80/100 mesh in a 3-meter column as compared with about 3 psig when using 30/35 mesh. Pressure drop is proportional to column length, and inversely proportional to the square of the particle diameter. Efficiency of the column with 80/100 mesh support will be about three times the efficiency of the 30/35 mesh column. However, a 9-meter column packed with 30/35 mesh will have about the same efficiency as a 3-meter column packed with 80/100 mesh; the pressure drop of the former is only 9 psig.

Coating the Support In the slurry procedure, the support is weighed and covered with a volatile solvent containing the exact amount of liquid phase desired to be coated on the support. The support and solution are then slurried and the solvent is evaporated on a steam bath.

In the filtration procedure the weighed amount of support is slurried with an excess of solution consisting of a volatile solvent and a known amount of liquid phase. The excess solution is filtered off the support in a Büchner funnel (under vacuum) and the wet support air dried.

The moist packing, prepared by either of the above procedures, is (a) spread out in a tray and dried in a low-temperature oven or under infrared lamps in a hood or (b) placed on a porous plate in a tube and an inert gas such as nitrogen blown up through the plate to fluidize the packing.

Packing the Column First the bottom of the straight metal tubing is plugged with glass wool. The dry coated support is slowly poured through a funnel into the tubing. Uniform packing is facilitated by tapping or with the aid of a mechanical vibrator. Plug the upper end with glass wool, and coil the copper tubing on a mandril to the desired configuration. Attach Swagelok fittings to each end.

For packing cylindrically coiled glass columns, a conventional aspirator on the water faucet is used to create a partial vacuum and pull the first packing material down to the glass wool plug. The rest of the column is filled up with the aid of gentle tapping.

EXPERIMENT 11-1 *Efficiency of Chromatographic Column*

Inject 1 μl of some pure compound (and include some air) into the column. Operate the column at some convenient temperature. Adjust the flow of carrier gas to about 140 ml/min for the run (assuming 0.25-in. i.d. column diameter).

When the peak has eluted, lower the flow rate to 120 ml/min, inject a fresh sample, and record the elution curve. Repeat the experiment at flow rates of 100, 80, 60, and 40 ml/min.

Calculate the plate number for your compound at each of the flow rates. From the column length, calculate the plate height at each flow rate. Plot the values of plate height against the linear gas velocity (which is calculated by dividing the length of the column by the retention time for the air peak). Estimate the coefficients of the terms in the plate height–velocity equation. Determine the optimum linear gas velocity and the minimum plate height.

EXPERIMENT 11-2 *Effect of Temperature on Retention Behavior*

Stabilize the column at 150°C (if compatible with the liquid-phase packing). Adjust the carrier gas flow to the optimum value, and inject a sample containing several adjacent members of a homologous family, or members separated by two carbon atoms for higher molecular weights.

Repeat the experiment at different column temperatures: 120°, 100°, 80°C.

Tabulate the retention time for the components at each column temperature. Graph the logarithm of the retention time (corrected for the air peak) vs the carbon number.

If sufficient information is available, calculate the partition coefficient of each component. Graph the logarithm of the partition coefficient vs the reciprocal of the absolute temperature. From the slope of the graph, calculate the heat of vaporization for each component.

EXPERIMENT 11-3 *Qualitative Analysis*

Stabilize the column temperature and adjust the carrier gas flow. Inject a 2-μl sample of some commercial fluid (such as lighter fluid) or a synthetic mixture of organic liquids. Also run chromatograms of appropriate known compounds. For lighter fluid, these compounds should be run: benzene, toluene, and at least two normal paraffins.

Establish the identity of each peak in the chromatogram of the unknown sample by reference to runs of known materials and by comparison with tabulated relative retentions. To use tabulated retention values, it may be advisable to adjust the column temperature and carrier flow rate to conform with the tabulated conditions. Use of two columns packed with a polar and a nonpolar substrate will materially aid in the identification.

EXPERIMENT 11-4 *Resolution of Mixtures*

Stabilize the column temperature and adjust the carrier gas flow to the optimum value. Inject a mixture known to give overlapping peaks under the particular operating conditions.

Repeat the experiment using unequal amounts of the overlapping components in the mixture.

Determine the plate number for each component and calculate the resolution between the two peaks. Estimate the fractional impurity for each peak. Revise operating conditions to improve the resolution.

Problems

1. The following data were obtained on a 25% dinonyl phthalate, 80/100 mesh Chromosorb P column, 91.5 cm in length, operated at 53°C, which contained 2.50 g of dinonyl phthalate (density = 0.9712 g/cm^3 at 20°C) and 7.50 g of oven-dried Chromosorb. Recorder speed was 2.54 cm/min.

Compound	F_c (ml/min)	t'_R (cm)	P_o (cm)	P_i (cm)
Benzene	48.0	52.0	75.8	122.2
Cyclohexene	51.7	42.1	75.2	123.9
Cyclohexane	52.6	31.0	75.2	124.4

For each compound, calculate (a) the adjusted retention volume, (b) the net retention volume, (c) the specific retention volume (at 0°C), and (d) the partition coefficient.

2. In Problem 1, the base width of the peaks (in cm) was: benzene, 9.1; cyclohexene, 7.4; and cyclohexane, 5.4. Calculate (a) the average number of plates and (b) the plate height for the column.

3. From the information in Problems 1 and 2, what plate number would be needed to resolve adjacent compounds within (a) 1.0% and (b) 0.1%? What length column, packed as in Problem 1, would provide the separation?
4. The following data were obtained from a 360-cm column packed with 30/50 mesh Chromosorb P and various amounts of hexadecane as the liquid substrate. Carrier gas was hydrogen. Column operating temperature was 30°C. Solute was propane. Chart speed was 61.0 cm/hr.

Column description	Flow rate, liters/hr	P_i, mm Hg	P_o, mm Hg	Adjusted retention volume, mm	Peak width (base), mm
31% (*w*/*v*) substrate	1	799	766	204.2	25.2
$V_S = 22.7$ ml	2	785	722	108.1	12.7
$V_M = 59.5$ ml	4	788	667	57.3	7.7
Column volume, 100 ml	6	787	613	40.9	6.1
Liquid cross section, 0.0629 cm^2	10	799	532	26.7	4.7
Free gas cross section, 0.165 cm^2					
23% (*w*/*v*) substrate	1	794	768	157.4	19.2
$V_S = 15.3$ ml	2	783	731	78.0	8.1
$V_M = 68.6$ ml	3	785	711	54.1	5.3
Column volume, 102 ml	4	785	686	40.6	4.0
Liquid cross section,	5	779	655	32.8	3.3
0.0426 cm^2	10	785	563	18.1	2.1
Free gas cross section, 0.191 cm^2					
13% (*w*/*v*) substrate	2	769	715	60.1	6.3
$V_S = 9.37$ ml	4	769	661	31.2	2.7
$V_M = 77.4$ ml	6	788	626	21.4	1.8
Column volume, 105.5 ml	10	790	542	14.5	1.25
Liquid cross section, 0.0233 cm^2					
Free gas cross section, 0.215 cm^2					

(a) For each column, calculate the plate number for each flow rate. (b) From column length and plate number, compute the plate height for each value of linear carrier-gas velocity at column temperature and average column pressure. (c) Graph the values of plate height against the linear velocity. (d) For each column, estimate the terms in the equation of plate height vs linear gas velocity. (e) Estimate the optimum linear gas velocity for each column. Comment on the change in shape of the plate-height plot as the amount of liquid substrate is decreased.

5. From the data in Problem 4, calculate the values of the partition coefficient and the partition ratio for propane on each of the columns.
6. The following data illustrate the influence of support mesh size on plate height. A column, 360 cm in length, and packed with 23% (*w*/*v*) hexadecane on Chromosorb P, was operated at 30°C with nitrogen as carrier gas.

Column description	Linear velocity, cm/sec	Plate height, cm	
		Propane	*n*-Butane
30/50 mesh support	0.71	0.198	0.189
$V_S = 15.3$ ml	1.38	0.145	0.125
$V_M = 68.6$ ml	2.72	0.152	0.132
Column volume, 102 ml	6.12	0.214	0.180
20/30 mesh support	2.74	0.276	0.354
$V_S = 13.7$ ml	5.28	0.299	0.395
$V_M = 70.3$ ml	7.60	0.337	0.436
Column volume, 100 ml	11.70	0.405	0.506

(a) Graph plate height against linear gas velocity. Note the effect of mesh size on the A term. (b) Compare the optimum plate height and v_{opt} obtained on the 30/50 mesh support with nitrogen as carrier gas and with hydrogen (Problem 4) as carrier gas.

7. Under a particular set of operating conditions, with helium as carrier gas, the terms in the plate-height equation of a packed column were: $A = 0.1$ cm, $B = 0.30$ cm²/sec, and $E = 0.05$ sec. (a) Graph the equation and sketch with dashed lines the contribution of each term in the equation. (b) Calculate the minimum plate height and the optimum carrier-gas velocity.
8. The performance of methylnaphthalenes on columns of 80/100 mesh Chromosorb P and on 200/230 mesh microbeads is given. All columns were 152 cm in length and 0.6 cm in diameter. Liquid substrate was silicone oil 710 [density = $1.0 - (0.0008T_c)$].

Column number	1	2	3	4
Solid support	Glass beads	Chromosorb P		
Liquid loading, %(*w*/*w*)	0.16	3	10	30
Weight of liquid, g	0.060	0.40	1.44	5.55
Temperature, °C	90°	100°	142°	182°
Pressure ratio, P_i/P_o	5.33	4.33	3.26	3.67
Flow rate, ml/min	155	415	192	208

(*continued*)

Column number	1	2	3	4
Solid support	Glass beads	Chromosorb P		
Linear velocity, cm/sec	41.7	49.3	27.2	36.1
V_N (in ml) 1-MeN	365	1350	870	1020
V_N (in ml) 2-MeN	325	1200	780	910
Base width (in ml) 1-MeN	32	127	72	92
Base width (in ml) 2-MeN	34	115	64	77
V_M, ml	10	22	18	15

For each methylnaphthalene on each column, calculate (a) the partition coefficient at the column temperature, (b) the partition ratio, (c) the plate number, (d) the phase ratio, and (e) the resolution obtained.

9. If two peaks have relative areas of $A_1 = 0.7$ and $A_2 = 0.3$, and their relative retention ratio is 1.04, how many theoretical plates would be needed to obtain a fractional impurity content of 0.01 (1.0%)? What type of column is indicated?
10. What number of plates would be required to effect a separation with 0.1% impurity in adjacent bands ($\alpha = 1.1$) in these 3 cases: Case 1, using ordinary packed columns with relatively large values of partition ratio ($k = 5$–100). Case 2, using open tubular columns with $k = 0.5$. Case 3, using open tubular columns with $k = 3$.
11. Tabulated are elution data obtained on a liquid paraffin column:

Solute	ΔH_S (in kcal/mole)	V_g (at 100°C)
n-Hexane	6.80	17.8
n-Heptane	7.32	40.0
1-Hexene	7.45	18.2
1-Heptene	8.78	39.5
2-Hexene	7.60	26.0
2-Heptene	8.52	55.0

(a) Graph the inverse temperature plot for each solute. (b) Select an isothermal operating temperature that provides an optimum separation of all solutes from each other. To achieve a fractional impurity of 0.01 (1%), how many plates would be needed? (c) On the *n*-alkane curves, note the specific retention value corresponding to their boiling points. (d) Given that V_g (at 100°C) is 8.2 ml/g for *n*-pentane (whose boiling point is 36°C), calculate the ΔH_S value.

12. Listed are the retention times for a number of compounds on a column (360 cm × 3 mm) of porous styrene–divinylbenzene copolymer that contains approximately 800 plates.

Compound	t'_R (in min)		Compound	t'_R (in min)	
	150°C	200°C		150°C	200°C
Isopropyl alcohol	1.11	0.47	Hexane	2.55	0.8
n-Propyl alcohol	1.50	0.60	Benzene	3.85	1.1
t-Butyl alcohol	1.73	0.60	Heptane	5.60	1.4
Isobutyl alcohol	2.90	0.90	Toluene	7.6	1.9
n-Butyl alcohol	3.50	1.0	Octane	12.5	2.4
			Ethylbenzene	15.8	2.3

(a) What column temperature would permit the separation of *n*-propyl alcohol from *t*-butyl alcohol within a fractional impurity of 0.01, assuming about equal amounts of each alcohol? (b) How high may the column temperature rise before the separation of *t*-butyl alcohol and isopropyl alcohol becomes unsatisfactory? (c) To separate isobutyl alcohol from *n*-butyl alcohol within a fractional impurity of 0.01, what column temperature is required? (d) At what isothermal column temperature would it be possible to separate the *n*-alkanes and the alkylbenzenes from each other?

13. The relative retentions (*n*-pentane = 1.00) of *n*-alkanes and alkylbenzenes at three boiling-point levels are given for convoil-20 (a saturated hydrocarbon oil) and tricresyl phosphate substrates on columns operated at 100°C:

Boiling-point level, °C	Convoil-20			Tricresyl phosphate		
	60	100	140	60	100	140
n-Alkanes	1.80	5.6	17.0	1.65	4.6	13.5
Alkylbenzenes	...	6.7	19.8	...	16.4	43.5

(a) Construct a plot of boiling point (°C) versus log relative retention for each hydrocarbon type and for each substrate (4 graphs in all). (b) Construct a graph of log relative retention on convoil-20 vs log relative retention on tricresyl phosphate for each class of compounds. (c) Utilize these graphs to identify the specific *n*-alkane or alkylbenzene from these relative retentions:

Solute	Relative retentions	
	Convoil-20	Tricresyl phosphate
A	2.3	2.10
B	63.0	130
C	8.7	21
D	11.2	9.0
E	1.0	0.95
F	17.5	39
G	5.2	4.4
H	34	70

14. On a column whose liquid substrate is convoil-20, the retention volume for *n*-pentane is 77 ml under the conditions employed in Problem 13. The retention volume for *n*-pentane on tricresyl phosphate is 34 ml. Estimate the adjusted retention volumes that would be observed for each *n*-alkane from pentane through octane on the two substrates; do the same for the alkylbenzenes.
15. On a tricresyl phosphate column, what *n*-alkanes will have emerged before benzene is eluted? Before *n*-propylbenzene is eluted?
16. Adjusted retention volumes are given for a series of *n*-alcohols and acetates. Column temperature was 77°C; helium flow was 89 ml/min.

	V'_R (in ml)	
	15% Carbowax 400	15% Nujol
Acetates		
Methyl	43.2	24.0
Ethyl	60.5	48.4
Propyl	106	120
Alcohols		
Methyl	72.3	8.0
Ethyl	110	14.1
n-Propyl	207	32.7
n-Butyl	408	85.5

(a) Prepare a graph of log retention volume against carbon number for each homologous series on each column substrate. (b) Plot the log retention volume on one column against the log retention volume on the other for each family. (c) Estimate the adjusted retention volume on each column for *n*-butyl acetate and for *n*-amyl alcohol.

17. Specific retention volumes (in ml) for some chlorinated hydrocarbons and benzene at several column temperatures on three column substrates are listed:

Solute	CH_2Cl_2	$CHCl_2$—CH_3	$CHCl_3$	CCl_3CH_3	CCl_4	CCl_2=$CHCl$	CCl_2=CCl_2	C_6H_5Cl	C_6H_6
Temp., °C				*Paraffin*					
74	29.5	51.3	74.3	108	141	189	568	677	136
97	17.0	29.5	41.5	57.8	76.5	98.2	273	323	74.3
125	8.08	14.2	19.6	28.0	34.5	42.6	105	124	...
				Tricresyl phosphate					
74	56.1	71.8	137	96.3	99.8	181	359	826	133
97	30.3	38.1	68.5	51.6	52.7	92	170	372	69.4
125	13.9	17.9	29.6	24.2	24.7	38.8	69.5	143	...
				Carbowax 4000					
74	69.6	54.8	138	57.1	55.8	122	165	548	89
97	31.9	26.3	59.3	28.5	27.4	57.1	78.7	235	43.3
125	13.5	12.1	24.2	13.5	13.5	23.1	31.1	83.6	...

All columns were made of coated Chromosorb packed into 200 cm × 60 mm copper tubing. The paraffin column contained 4.58 g on 13.93 g of support; the tricresyl phosphate column, 4.46 g on 13.71 g; and the carbowax 4000 column, 4.38 g on 13.1 g. The densities of the liquid phases at 74°/4° in g/cm^3 are: paraffin, 0.768; tricresyl phosphate, 1.128; and carbowax 4000, 1.081. (a) Graph the data and compute the heat of vaporization for each solute on each of the stationary liquid phases. (b) Calculate the partition coefficient at each temperature for each solute on each column substrate.

18. These questions pertain to the chlorinated hydrocarbons of Problem 17. (a) Comment on the resolution and separation of the mixture on columns of the separate substrates at the different operating temperatures. (b) Contrast the order of elution observed with the nonpolar paraffin column and the polar tricresyl phosphate and carbowax 4000 columns.
19. Adjusted retention times at various column temperatures were obtained for *N*-methylaniline on a 20% by weight of Ucon 50-HB-5100 as stationary liquid phase. These were: 17.2 min at 117°C, 7.1 min at 138°C, and 2.3 min at 175°C. (a) Calculate the heat of vaporization of *N*-methylaniline. (b) Estimate the retention times at 150°, 200°, and 225°C.
20. Given, at various column temperatures, are retention times relative to *N*-methylaniline. Other pertinent data are contained in Problem 19.

	Retention times (in min)		
Compound	117°C	138°C	175°C
Ethylenediamine	0.057	0.078	
1,3-Propanediamine	0.114	0.141	
1,4-Butanediamine	0.229	0.251	
1,5-Pentanediamine	0.406	0.433	
1,6-Hexanediamine	0.74	0.74	
1,7-Heptanediamine		1.26	1.12
1,8-Octanediamine		2.11	1.76
1,9-Nonanediamine		3.51	2.72
Triethylenediamine	0.328	0.388	
Diethylenetriamine		0.81	0.84

(a) Graph the log retention times against carbon number for terminal diamines. (b) Graph log t'_R vs $1/T$ for each compound. (c) How many plates are required for separation of each adjacent peak at 138° and 175°C? The difficulty centers around the 1,5-, 1,6-, and 1,7-terminal diamines and diethylenetriamine and triethylenediamine. (d) If the 1-meter column had an efficiency of 1000 plates, would one expect the peaks for triethylenediamine and diethylenetriamine to be adequately resolved from the adjacent terminal diamines?

21. After esterification of a series of dicarboxylic acids, the ethyl esters were chromatographed on a 3-meter column packed with polyester succinate at a temperature of 195°C with a helium flow of 65 ml/min (see Fig. 11-32). (a) From a log plot of these esters, and knowing that peak 3 is ethyl succinate and peak 6 is ethyl sebacate, identify the remaining peaks. (b) Estimate the plate number and plate height for each solute.
22. From the information tabulated, graph the relation between log relative reten-

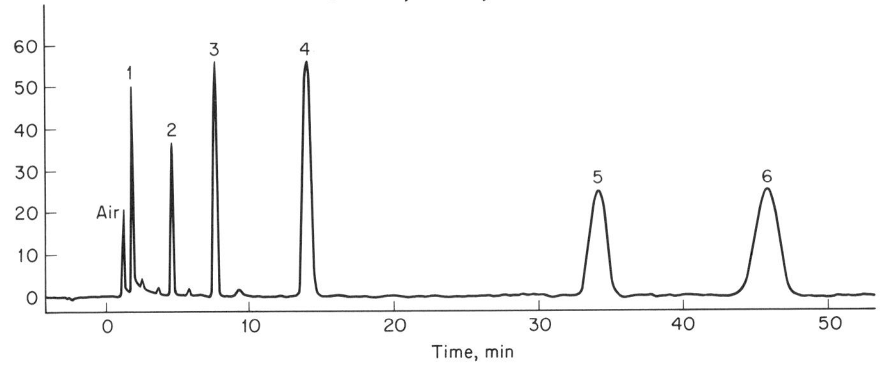

FIG. 11-32. Illustration for Problem 21.

tion volume on poly(ethylene glycol adipate) and log relative retention volume on Apiezon M for the methyl esters of saturated and unsaturated fatty acids.

RELATIVE RETENTION VALUES (16:0 = 1.00) OF FATTY ACID METHYL ESTERS

Common name of parent acid	Code [a]	Poly(ethylene glycol adipate) at 197°C	Apiezon M at 197°C
Lauric	12:0	0.30	0.18
Palmitic	16:0	1.00	1.00
Stearic	18:0	1.82	2.39
Oleic	18:1	2.04	2.0
Linoleic	18:2	2.44	1.9
Linolenic	18:3	3.13	1.9
	20:0	3.31	5.6
	20:1	3.67	4.7
	20:2	4.46	4.5
	20:3	5.02	3.9

[a] Number of carbon atoms: number of double bonds.

23. Using the double logarithmic graph prepared in Problem 22, identify the specific compound from the pair of relative retention values obtained on poly(ethylene glycol adipate) and on Apiezon M at 197°C. Compound *A* (0.549, 0.42); compound *B* (1.15, 0.89); compound *C* (0.300, 0.18); compound *D* (0.62, 0.38); compound *E* (1.43, 0.85); compound *F* (5.95, 13.0); compound *G* (1.64, 0.77); compound *H* (1.33, 1.55); compound *I* (1.97, 0.74); compound *J* (11.2, 31.0); compound *K* (0.155, 0.069).
24. Predict relative retention values for these methyl esters of fatty acids from the graph prepared in Problem 22, plus a graph of log relative retention volume on poly(ethylene glycol adipate) vs carbon number for the particular homologous series: Compound *A* (15:0); compound *B* (18:4); compound *C* (20:4); compound *D* (22:0); compound *E* (22:1).

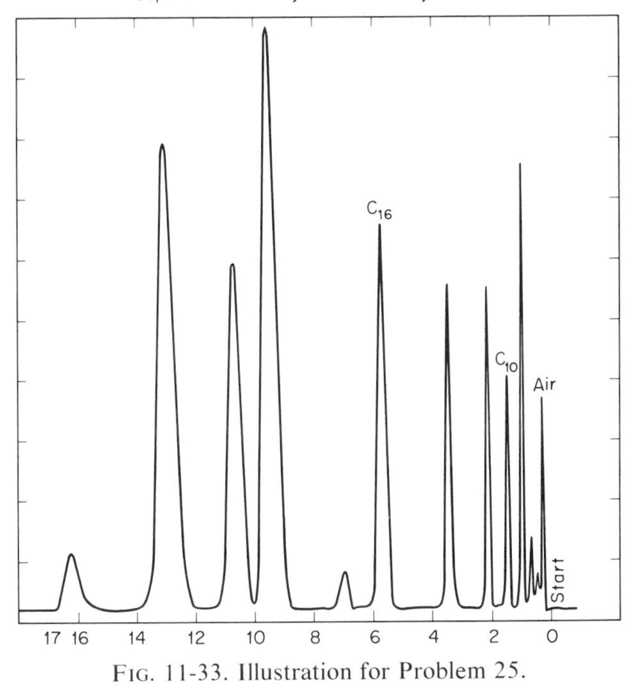

FIG. 11-33. Illustration for Problem 25.

25. The accompanying chromatogram of the methyl esters of fatty acids was run on a column of poly(diethylene glycol succinate) at 210°C (see Fig. 11-33). Although the temperature was slightly higher and the stationary liquid substrate differed slightly from the column used in Problem 22, identify the peaks from the position of the C_{10}, C_{16} esters, the graph of log relative retention vs carbon number, and the retention differences for members of different homologous ester families in terms of number of double bonds in the ester.
26. The accompanying chromatogram shows the routine C_5-cut from a petroleum sample (see Fig. 11-34). (a) For each component peak, estimate the number of plates. (b) To separate the 2-methyl-butene-2 peak from the pentene-2 peaks (*cis* plus *trans*) within 1.0%, what plate number is needed? (c) Would any problem arise in the separation of the *n*-pentane from the isopentane if the amount of *n*-pentane should increase?
27. The following data were obtained on an *o*-nitrophenetole column, 4.25 meters in length of 60 mm diameter coiled copper tubing containing 4.0 g of substrate on 10 g of Celite 545. Carrier gas was helium; flow rate, 50 ml/min.

Component	Temp., °C	t_R, min [a]	Δt_b, min [b]
Air	25	1.32	...
Butene-1	25	5.95	0.46
trans-Butene-2	25	7.56	0.57
cis-Butene-2	25	8.55	0.65
1,3-Butadiene	25	10.42	0.76

(*continued*)

Component	Temp., °C	t_R, min [a]	Δt_b, min [b]
Air	16	1.34	...
Butene-1	16	7.14	0.56
trans-Butene-2	16	9.24	0.70
cis-Butene-2	16	10.55	0.80
1,3-Butadiene	16	12.99	0.95

[a] Retention time from injection point to peak maximum.
[b] Δt_b is baseline intercept cut by tangents to peak inflections.

For each compound, calculate (a) the adjusted retention volume, (b) the plate number and plate height, and (c) the resolution of adjacent pairs of peaks.

28. The accompanying chromatogram shows the separation of an aldehyde mixture on a column containing 10% silicone oil 200 plus 1% polyethylene glycol on 60/80 mesh Chromosorb W operated at 60°C with hydrogen as carrier gas (see Fig. 11-35). (a) Graph log adjusted retention time vs. carbon number for the *n*- and iso-aldehydes. (b) For each aldehyde, estimate the plate number.
29. The separation of glycols on Porapak Q at 227°C is shown in the chromatogram (see Fig. 11-36). Peak 1 is water and peak 7 is glycerol. Identify the remaining diols of the linear C_2-C_4 glycols.
30. The separation of C_1-C_4 formates and acetates on Porapak Q is made according to total carbon number. Except for peak 1 (methanol), identify the numbered peaks in the chromatogram (see Fig. 11-37); among the formate esters, only *n*-alkyl members are present.
31. On the accompanying chromatogram (Fig. 11-38), the C_1-C_6 alcohols were run on a column of Porapak Q with a temperature program from 135°–200°C. Identify the components from among the *n*-alcohols, the *sec*-alcohols, and the 2-methyl-l-alcohols. [Hint: Compare elution times with boiling points. The secondary and the 2-methyl derivatives emerge ahead of the *n*-alcohols.]

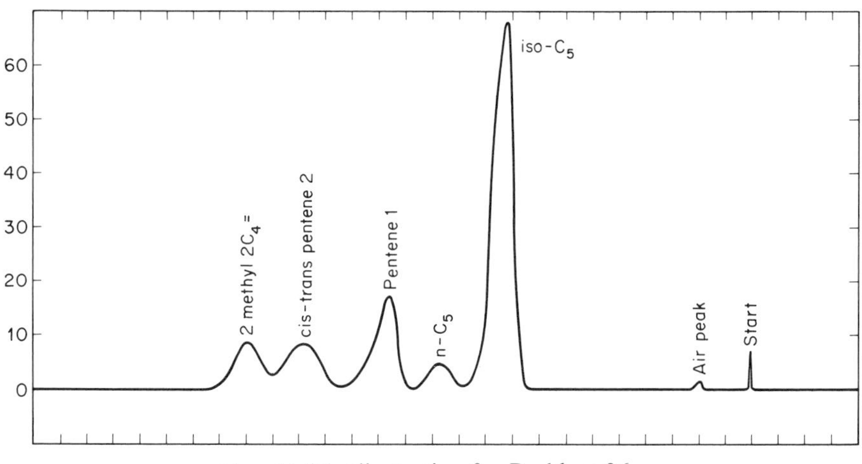

FIG. 11-34. Illustration for Problem 26.

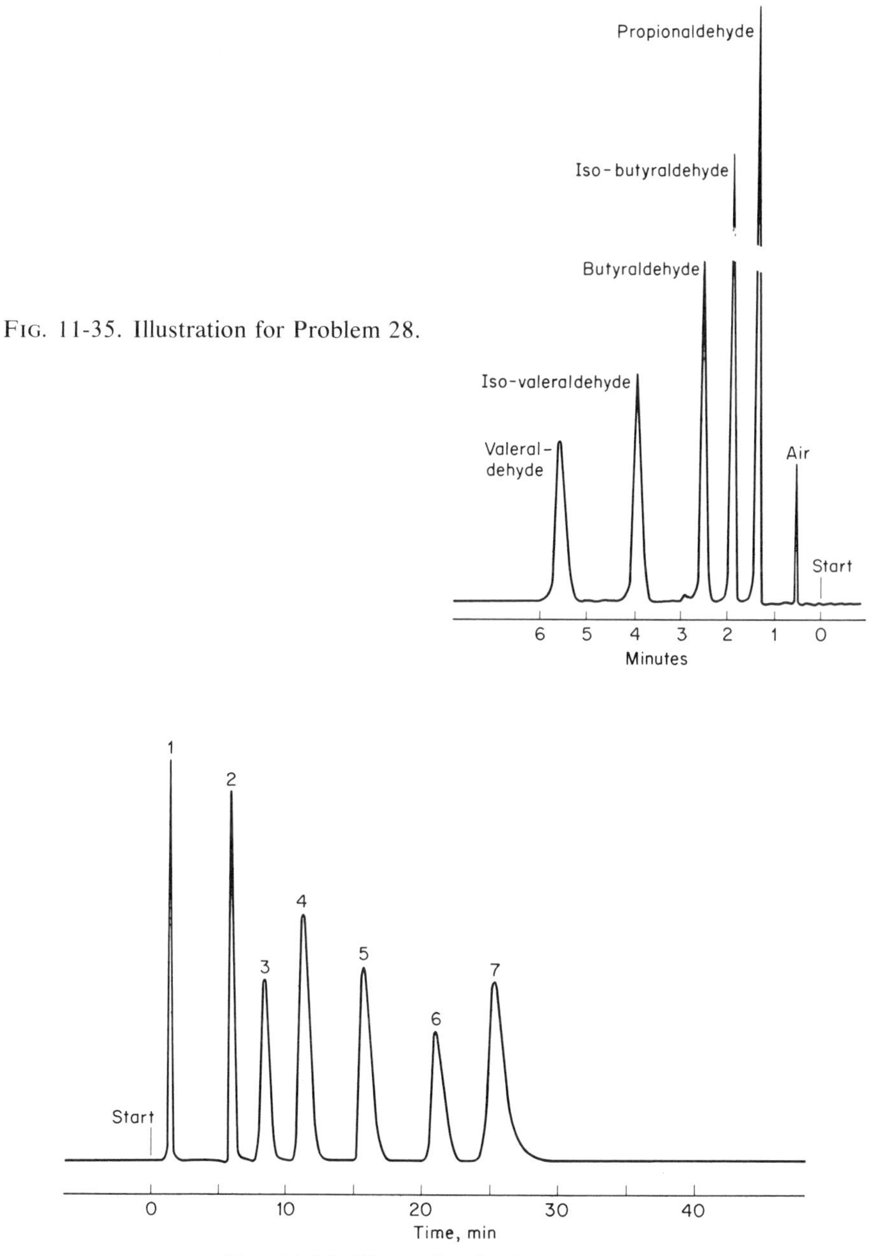

FIG. 11-35. Illustration for Problem 28.

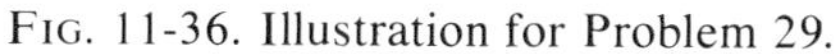

FIG. 11-36. Illustration for Problem 29.

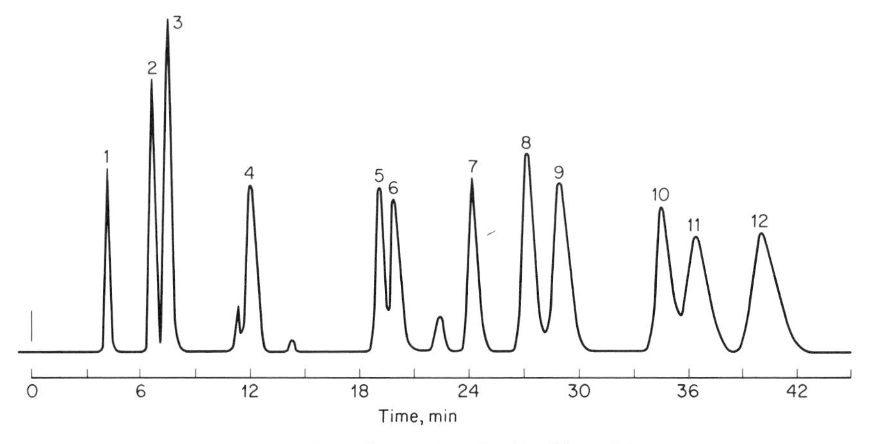

FIG. 11-37. Illustration for Problem 30.

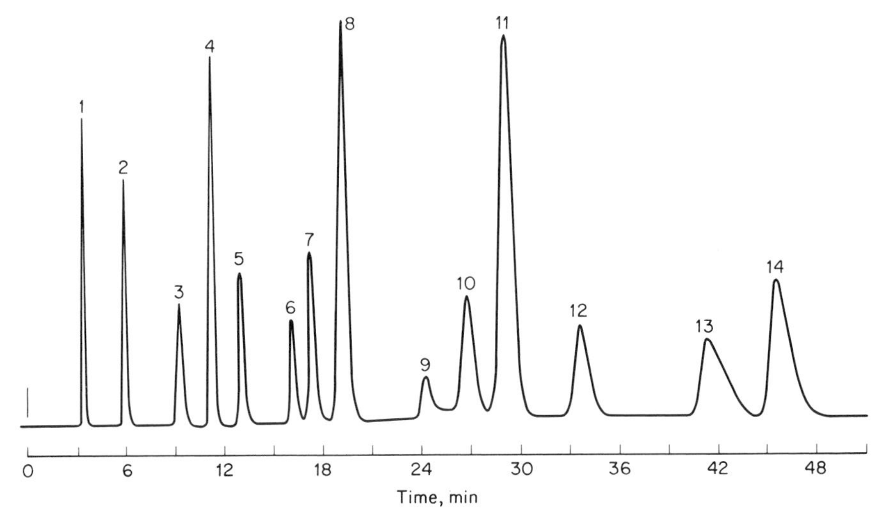

FIG. 11-38. Illustration for Problem 31.

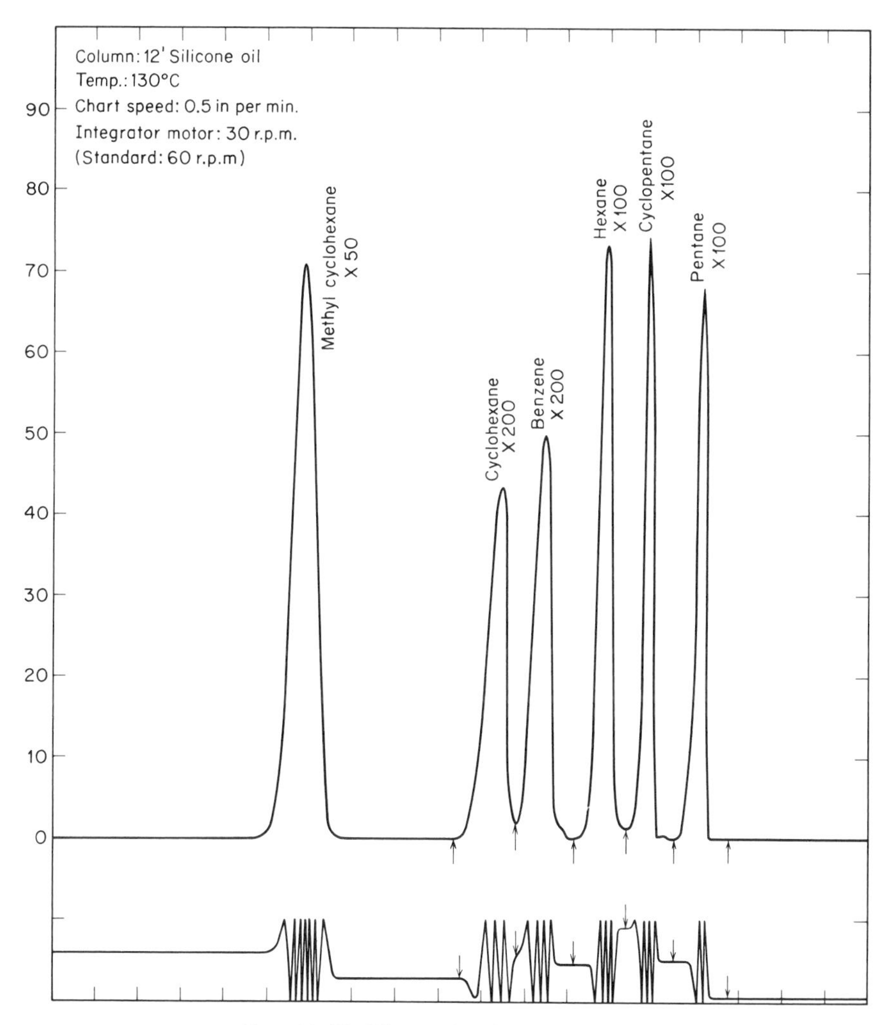

FIG. 11-39. Illustration for Problem 32.

32. Estimate the readings of the chart integrator for each component peak, as indicated by the arrows (Fig. 11-39).
33. On the accompanying chromatogram and chart integrator (Fig. 11-40), estimate the readings of the chart integrator for each component peak, including the negative peak.
34. Castor oil contains principally glycerides of ricinoleic acid. After conversion to the methyl esters by transmethylation, the accompanying chromatogram was obtained (Fig. 11-41). From the lower trace made by the disk integrator, determine the percentage composition for this mixture.
35. With the aid of a millimeter rule, estimate the readings of the integrator for each component peak on the accompanying chromatogram (Fig. 11-42).

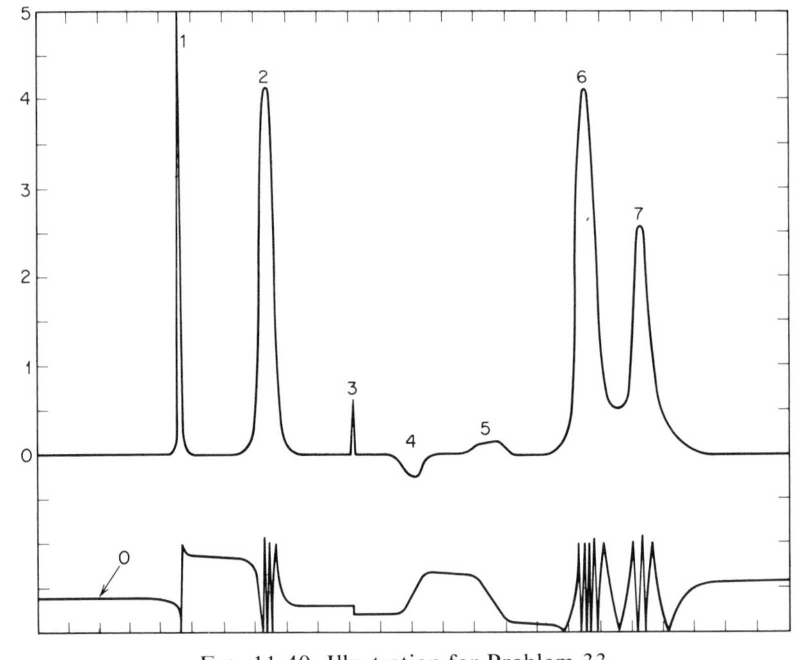

FIG. 11-40. Illustration for Problem 33.

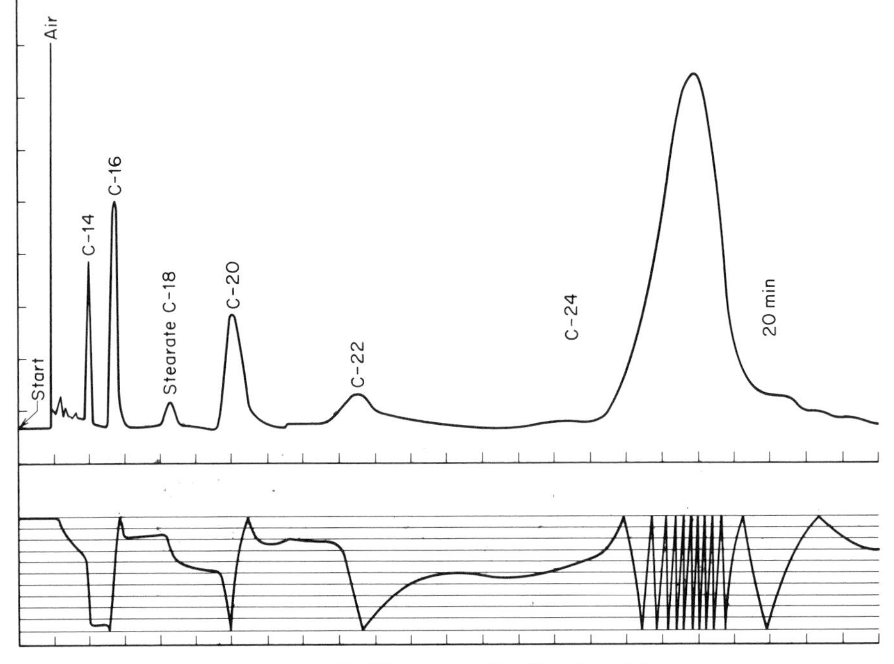

FIG. 11-41. Illustration for Problem 34.

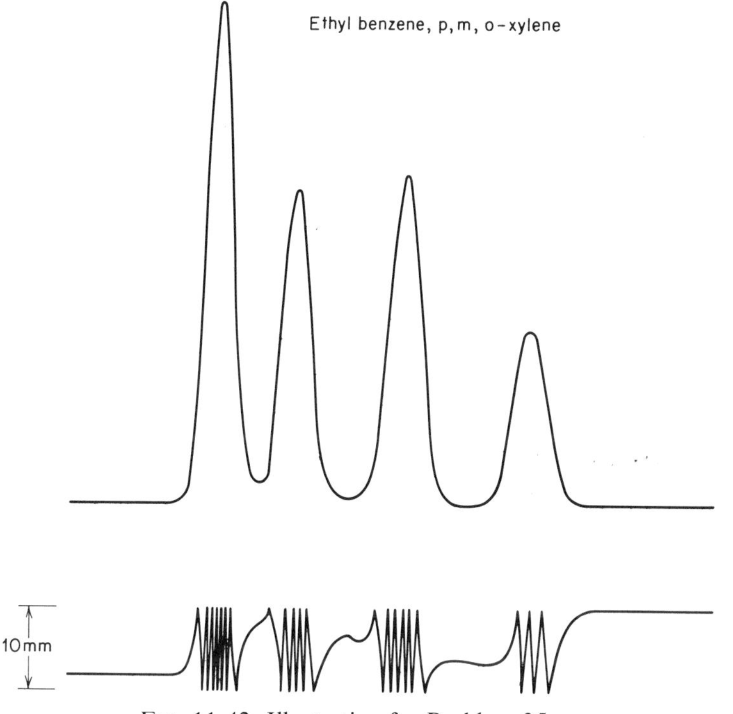

FIG. 11-42. Illustration for Problem 35.

36. Ascertain the upper limit on R values that exists because of the resolution loss suffered when solute is mostly in the form of vapor. Hint: $N = N_0/(1 - R)^2$ where N_0 is the plate number necessary to effect the separation at low R values.
37. For a particular column, assume that the effective dead space equals 2 units. By means of bar graphs, illustrate the isothermal distribution within the column phases when the partition ratio is 0.25; also when $k = 4.0$.
38. For a programmed temperature elution, assume $k = 0.25$ and $t_{air} = 2$ units. Calculate the retention time if the partition ratio is halved in each unit time. Do the same when $k = 1.0$ and 4.0. Compare the times with those obtained in Problem 37 for isothermal column operation.
39. If a tank of helium gas, used as carrier, contained 1.0 ppm oxygen, what type of response would be observed with a sample gas of higher oxygen content when using a helium ionization detector? Of lower oxygen content? Of any other impurity present in one cencentration range in the carrier gas, and a different concentration range in the column effluent?

BIBLIOGRAPHY

Dal Nogare, S., and R. S. Juvet, Jr., *Gas–Liquid Chromatography,* Interscience, New York, 1962.

Harris, W. E., and H. W. Habgood, *Programmed Temperature Gas Chromatography,* Wiley, New York, 1966.

Keulemans, A. I. M., *Gas Chromatography,* 2nd ed., Reinhold, New York, 1959.

Purnell, H., *Gas Chromatography,* Wiley, New York, 1962.

12 *Exclusion Chromatography*

Chemical selectivity based upon the sterical interrelationship between adsorbate and adsorbent is possessed by molecular sieves. These comprise two categories: the inorganic zeolites and the organic xerogels. The dissimilarity of the two categories is reflected in the fields of application. The zeolites have proved useful in handling small molecules while the xerogels have been used to fractionate substances of quite high molecular weight, particularly solutes of biological origin and polymeric materials.

GEL-PERMEATION (FILTRATION) CHROMATOGRAPHY

Gel-permeation (filtration) chromatography is a technique that fractionates substances largely according to their molecular size. It is based upon inclusion and subsequent elution of the solutes through a stationary phase consisting of a heteroporous, cross-linked polymeric gel. Different porosity gels present a gradation of size barriers from 10 up to 440 Å to the molecules.

In a sense, gel-permeation chromatography is a special case of liquid–solid elution chromatography. The entire separation is envisioned to take place between the liquid phase within the gel particle and the liquid surrounding the particle. The primary mechanism of retention is a different penetration (or permeation) by each solute molecule into the interior of the gel particles. As the liquid phase containing the sample passes through the gel phase, molecules diffuse into all parts of the gel not mechanically barred to them. Molecules whose size is too great will be mechanically barred from certain openings into the gel network. These larger molecules will also penetrate less into the comparatively open regions of the gel. Consequently, they will pass through the column chiefly by way of the interstitial liquid volume. Smaller molecules are better able to penetrate into the interior regions of the gel particles depending, of course, upon their size and upon the distribution of pore sizes available to them in the gel. As a result, the larger molecules emerge in the effluent first, followed by smaller molecules which must follow a more circuitous and tortuous path as they travel through the column packing.

Ideally, the entire separation is due to the size barriers that exist within the gel particles. With larger molecules, the molecular size seems to be the prime factor with polarity becoming more important at lower molecular weights. The degree of cross-linking in the gel essentially determines the discreteness of separation. With polymers, a close approach to the true molecular weight distribution is obtained by using long columns having appropriate permeability. Valid standards of known molecular weight must be available.

Properties of Xerogels

Xerogels fall into two categories: the hydrophilic gels, such as agar, agarose, polyacrylamide, and cross-linked dextrans, and the hydrophobic gels based on polystyrene. Tables 12-1 and 12-2 list a number of these gels.

Polyacrylamide gels (trade name Bio-Gel) are obtained from the copolymerization of acrylamide with *N,N'*-methylene-bisacrylamide. The latter is the cross-linking agent. Dextran gels (trade name Sephadex) are obtained by cross-linking, via ether linkages, the linear polyglucose chains of dextran with epichlorohydrin. Both series of gels are water insoluble, but because of their abundant hydroxyl or amide groups, they retain their hydrophilic character and will swell in aqueous media and in solvents like glycol, dimethylsulfoxide, formamide, and mixtures of water and the lower alcohols. Acylation or alkylation of the free hydroxyl groups of the dextran gel results in the formation of derivatives which display excellent swelling properties in many organic solvents and possess the corresponding resolving power. Sephadex and Bio-Gel series of gels are supplied in the form of minute beads whose size can be specified in the ranges of particle diameter from 10 to 80 μ, 50 to 150 μ, or 100 to 300 μ.

Modified polystyrene gels (trade name Styragel) are prepared by suspension polymerization of styrene with a difunctional monomer (the cross-linking agent), such as divinylbenzene, in the presence of an inert diluent, such as toluene alone or diluted with *n*-dodecane. A sponge-like gel structure is produced in the diluent which must be a good solvent for the monomer but somewhat of a mismatch in solvent power for the polymer. The gels are very rigid despite their high porosity. Consequently, the beads prepared from these gels are barely swollen. Polar solvents will cause some shrinkage and may damage a prepacked column. Among eluents recommended tetrahydrofuran is the preferred solvent because of its low viscosity.

TABLE 12-1. Hydrophilic Type Gel Materials

Gel type		*Fractionation range in molecular-weight units*	*Hydrated bed volume, V_b, in ml per g dry gel*	*Water regain, ml per g dry gel*
Bio-Gel®	P-2	200 to 2,000 [a]	3.8	1.6
	P-4	500 to 4,000	6.1	2.6
	P-6	1,000 to 5,000	7.4	3.2
	P-10	5,000 to 17,000	12	5.1
	P-30	20,000 to 50,000	14	6.2
	P-60	30,000 to 70,000	18	6.8
	P-100	40,000 to 100,000	22	7.5
	P-150	50,000 to 150,000	27	9
	P-200	80,000 to 300,000	47	13.5
	P-300	100,000 to 400,000	70	22
Sephadex®	G-10	up to 700 [b]	2.5	1.0
	G-15	up to 1,500 [b]	3.0	1.5
	G-25	100 to 5,000 [a]	5	2.5
	G-50	500 to 10,000 [a]	10	5.0
	G-75	3,000 to 70,000 [c]	13	7.5
	G-100	4,000 to 150,000 [c]	17	10
	G-150	5,000 to 400,000 [c]	24	15
	G-200	5,000 to 800,000 [c]	30	20

[a] Determined with polysaccharides.
[b] Determined with polyethylene glycols.
[c] Determined with globular proteins; values considerably lower with polysaccharides.

However, a number of other solvents have been used, such as benzene, toluene, *o*-dichlorobenzene, chloroform, perchloroethylene, carbon tetrachloride, and cyclohexanone.

Gels based on agarose, the non-ionic constituent of agar, extend the molecular exclusion limit to over 150 million molecular-weight units. Due to their porosity, these gels are capable of separating virus particles by size. Prepared in a beaded form,[1] they are sold under trade names such as Sepharose and Bio-Gel A. By altering the concentration of agarose during preparation, gels with different fractionation ranges are produced. These are indicated in the Sepharose series by a number that indicates the percent agarose (e.g., Sepharose 2B implies 2% agarose) and in the Bio-Gel series by the upper exclusion limit (e.g., Bio-Gel A–0.5*m* implies 0.5 million is

[1] S. Hjerten, *Biochim. Biophys. Acta* **79,** 393 (1964); *Arch. Biochim. Biophys.* **99,** 466 (1962).

the upper exclusion limit). The gel structure is due to hydrogen bonding and is not a result of cross-linking. These gels should be used at temperatures above freezing but below 30°C. Above 30°C the gel softens and gradually dissolves. The optimum pH range is pH 5 to 8. Almost any buffer or salt solution can be used for elution *except* borate buffers which form complexes with the hydroxyl groups of the agarose.

Chemical stability of the xerogels is largely determined by the linkages in the framework. The hydrophilic gels of the acrylamide and dextran series are hydrolyzed by strong acids at elevated temperatures; also strong alkaline media cause degradation.

Although the name xerogel implies an uncharged gel, the hydrophilic gels do contain 10 to 20 microequivalents of charged groups (per gram of

TABLE 12-2. Properties of Agarose and Styragel Materials

	Type	*Fractionation range in molecular-weight units*	*Approximate exclusion limit in molecular-weight units (average porosity in Å)*
60 [a]	Styragel	800 [b]	1,600
100		2,000	4,000
400		8,000	16,000
1×10^3		20,000	40,000
5×10^3		100,000	200,000
10×10^3		200,000	400,000
30×10^3		600,000	1,200,000
1×10^5		2,000,000	4,000,000
3×10^5		6,000,000	12,000,000
5×10^5		10,000,000	20,000,000
10×10^5		20,000,000	40,000,000
0.5 m	Agarose (10%) [c]	10,000 to 250,000	400,000 (75 Å)
1.0 m	(8%)	25,000 to 700,000	930,000 (100 Å)
1.5 m		10,000 to 1,000,000	
2.0 m	(6%)	50,000 to 2,000,000	2,600,000 (150 Å)
5.0 m		10,000 to 5,000,000	
15.0 m	(4%)	200,000 to 15,000,000	15,600,000 (200 Å)
50.0 m		100,000 to 50,000,000	
150 m	(2%)	500,000 to 150,000,000	181,000,000 (440 Å)

[a] The gel types are identified by their respective average porosity in Å. Data supplied by the manufacturer: Water Associates, Inc., Framingham, Mass.
[b] For vinyl polymers.
[c] In the Sepharose gels (Pharmacia Fine Chemicals, Uppsala, Sweden) 10B implies 10% agarose; Bio-Rad gels are identified by numbers in left column (Bio-Rad Laboratories, Richmond, Calif.).

gel) capable of acting in the capacity of an ion exchanger. In equilibrium with an electrolyte solution the charge effect is cancelled out; an eluent with an ionic strength as low as 0.01 is sufficient.

Apparatus

Differences exist in equipment needed to handle open columns of the non-rigid gels used in gel-filtration studies and the closed columns of polystyrene gels used in gel-permeation chromatography. For relatively coarse fractionation according to molecular size, the eluent is permitted to percolate under the normal force of gravity through a column of gel particles suspended in a suitable solvent. Flow rate increases with increasing particle size and decreasing gel porosity; it is almost inversely proportional to the height of a gel bed. Pressure cannot be used on beds of non-rigid gels since compression of the bed would occur which would mean a reduction in interstitial volume. Optimum results seem to be obtained using columns with a length/diameter ratio of 10:1. As the pore size of the gel increases, the stability of the bed decreases. Columns containing soft gels can be stablized by applying a layer of a hard gel at the top of the column bed. A solvent pump can be connected to either the inlet or the outlet end of the column.

To achieve the precision needed in molecular-weight studies on columns of hydrophobic gels, the flow rate and temperature must be closely controlled. This requires a closed system through which the eluent is pumped from a reservoir through a heater to degas the solvent and then a filter to the chromatographic column. The sample, as a 0.1 to 0.5% solution in the same solvent to be used as eluent, is injected with a hypodermic syringe through a septum into the solvent stream just ahead of the column as is done in gas chromatography. The schematic diagram of a gel-permeation chromatograph is shown in Fig. 12-1.

The usual sample size in analytical-scale gel-permeation chromatography is only a few milligrams; the sample emerges from the column over a 25 to 100 ml elution volume. This means that each milliliter of eluent contains less than 1 mg of sample. Hence it becomes necessary to use a very sensitive detection method having a high signal-to-noise ratio. Advantage can usually be taken of a physical property of the solutes, such as refractive index, ultraviolet absorption, or fluorescence. A differential refractometer[2] provides a very sensitive detector. It is used in the Waters Associates chromatograph (Fig. 12-1). Differential refractometry is applicable to a wide variety of organic compounds because the method does not depend upon the presence of any particular functional group or structural feature in the solute.

[2] See H. H. Willard, L. L. Merritt, and J. A. Dean, *Instrumental Methods of Analysis*, 4th ed., Van Nostrand, Princeton, N. J., 1965, Chapter 14.

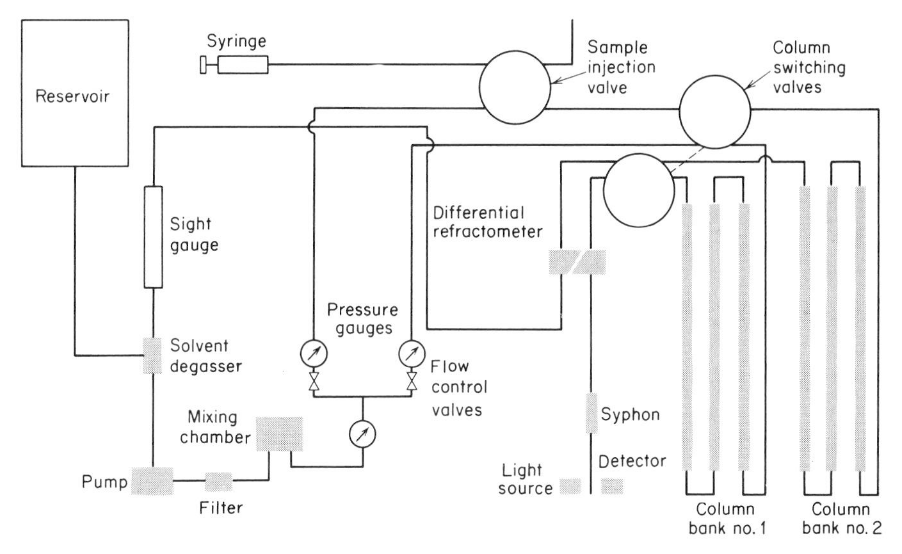

FIG. 12-1. Flow diagram of the Waters Model 200 gel permeation chromatograph. (Courtesy of Waters Associates.)

Retention Behavior

The portion of the column which is filled by the gel bed (V_b) consists of two phases: the gel phase with an inner volume (V_i) of solvent within the gel network and an outer free solvent (V_o) between the gel particles—the void volume. Three components contribute to the total volume of the bed:

$$V_b = V_o + (V_i + V_r) = V_o + V_S \tag{12-1}$$

where V_r is the volume of the gel matrix and V_S is the total volume of the stationary gel phase.

The void or outer volume may be established by chromatographing on the gel bed a high-molecular-weight substance which cannot enter the gel particles. The solvent volume which leaves the column between the time of injection of this substance onto the top of the bed and the appearance of this substance in the effluent corresponds to the void volume. Blue dextran (mol. wt. 2 million) is often used to characterize hydrophilic gels.

The inner volume, V_i, is related to the porosity of the gel. For those gels which imbibe solvent and swell, the inner volume may be calculated from the dry weight of the gel (m) and the solvent uptake or regain on swelling, S_r (column 4 of Table 12-1), and the density of the solvent, ρ_s,

$$V_i = mS_r/\rho_s \tag{12-2}$$

The looser the gel network, the larger the solvent regain. When the dry weight of the gel is unknown, the inner volume may be calculated from the wet density (ρ_r) of the swollen particles and Eq. (12-3):

$$V_i = \frac{S_r \rho_r}{\rho_s(1 + S_r/\rho_s)} \tag{12-3}$$

To characterize the behavior of a solute, its elution volume (V_e) is measured. Figure 12-2 illustrates two different criteria that have been used for the determination of elution volume. When very small samples are applied, the position of the maximum of the peak in the elution chromatogram should be taken as the elution volume. On the other hand, when large samples are used (giving a plateau region on the chromatogram), the volume eluted from the start of sample application to the inflection point (or the half height) of the rising part of the elution peak should be taken as the elution volume. On a given bed the same substance will always be eluted at the same elution volume; the latter is related to the partition coefficient (K_d) as follows:

$$V_e = V_o + K_d V_i \tag{12-4}$$

The partition coefficient is a constant for a given gel; its value is determined by the molecular dimensions of the substance under study. It is largely independent of the flow rate and of the concentration of the sample. The elution volume is independent of temperature. Upon rearranging Eq. (12-4), one obtains the expression

$$V_e/V_o = 1 + K_d V_i/V_o \tag{12-5}$$

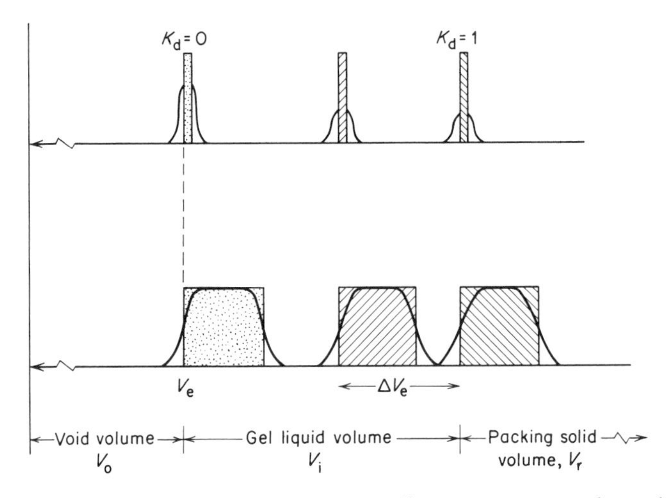

FIG. 12-2. Schematic elution curves. The top diagram corresponds to the application of a small sample. The bottom diagram corresponds to the maximum sample volume to be applied to obtain complete separation. The shaded areas correspond to the elution profiles that would be obtained if no zone broadening were to take place. The volume parameters are illustrated below the diagrams.

for the relative elution volume. This expression is similar to the "free column volume" which was used in ion-exchange chromatography.

Uncertainties in the determination of V_i have led to the definition of the constant K_{av} (av = available):

$$K_{av} = (V_e - V_o)/(V_b - V_o) \tag{12-6}$$

This constant differs from the K_d value in that the total volume of the gel phase ($V_i + V_r = V_s$) instead of the inner volume is employed in the calculation. The two constants are related as follows:

$$K_{av}V_s = K_d V_i \tag{12-7}$$

In Eq. (12-6) all variables are easily measured. The difference between K_{av} and K_d is smaller the more strongly swollen is the gel under consideration.

Solutes that are completely excluded from the gel phase, i.e., solutes for which $K_d = 0$ or $K_{av} = 0$, move outside the gel particles and appear in the effluent after passage of a volume of solvent equal to the void volume, i.e., $V_e = V_o$ or $V_e/V_o = 1.0$. A wide fractionation range is to be expected for the hydrophilic gels; the random distribution of cross-linking points leads to a wide distribution of pore sizes. Exclusion limits will differ among different classes of solutes and among gels of different types.

A value of unity for the partition coefficient implies that all of the solvent in the column is available to the solute. The solute emerges from the column after a volume ($V_o + V_i$) has passed through the gel bed. Affinity of the solutes to the gel phase can lead to K_d values exceeding unity. For example, aromatic substances show greater affinity to dextran gels than to aqueous solutions, and are thus usually retarded to varying extents compared with non-aromatic substances of like size. Also, solutes containing hydroxyl and carboxyl groups are retarded more than would be expected from their molecular size on dextran and polyacrylamide gels.

Theories on the behavior of molecules of different size on a gel bed have been advanced in papers by Akers,[3] Flodin,[4] Giddings and Mallik,[5] Laurent and Killander,[6] Porath,[7] and Squires.[8] This is not an exhaustive listing but it does provide a cross-section of various models and principles. The independence of the elution volume from variations in the flow rate or temperature offers a strong argument in favor of an exclusion mechanism as opposed to a concept of restricted diffusion or a partition mechanism.

[3] G. K. Akers, *Biochemistry* **3,** 723 (1964).
[4] P. Flodin, Dissertation, Uppsala, Sweden, 1962.
[5] J. C. Giddings and K. L. Mallik, *Anal. Chem.* **38,** 997 (1966).
[6] T. C. Laurent and J. Killander, *J. Chromatog.* **14,** 317 (1964).
[7] J. Porath, *J. Appl. Chem.* **6,** 233 (1963).
[8] P. G. Squires, *Arch. Biochem. Biophys.* **107,** 471 (1964).

Resolution

The volume which separates the front of the elution curve of one substance from that of another substance, i.e., the difference between their elution volumes, is called the separation volume:

$$\Delta V_e = V_{e,2} - V_{e,1} \tag{12-8}$$

Theoretically, the sample size can be as large as the separation volume were no zone broadening to take place. However, due to microturbulence, nonequilibrium between the stationary phase and the mobile phase, and longitudinal diffusion in the bed, the zones will always be broadened. The sample size must therefore be smaller than the separation volume. This is illustrated in Fig. 12-2. Giddings has discussed resolution and optimization.[8a]

Diffusion between the gel phase and the interstitial volume is the rate-determining step. The rate of attainment of diffusion equilibrium is governed by the size of the gel beads and the diffusion rate of the solute within the gel network. High flow rates and coarse-grade bed material invariably give large zone broadening. Improved resolution can be obtained by operating at elevated temperatures and using longer columns or columns packed with superfine or fine-grade bed material (Fig. 12-3). The flow rate should be kept as low (7 to 9 ml/cm^2/hr) as convenience permits. However, for preparative purposes the use of a higher flow rate, 45 to 50 ml/cm^2/hr (and consequently faster separation), often outweighs the loss of resolution in the chromatographic run. By contrast with liquid–solid chromatography, improved resolution manifests itself as a decrease in peak width rather than as an increase in the separation volume. The elution volume of individual solutes remains almost constant.

A critical variable is the viscosity of the sample relative to the eluent. A high viscosity of the sample causes instability of the zone and an irregular flow pattern. A lower flow rate will not improve the situation. The zones corresponding to low values of the partition coefficient will be less broadened than zones corresponding to high K values. Eluents should be solvents with low viscosity, preferably less than 4 centipoises.

Selection of Gel Type

The choice of the appropriate gel depends on the molecular size and the chemical properties of the substances to be separated. Each type of gel fractionates mainly within a particular molecular-weight range, as determined by the range of pore sizes within the gel network. Matrices now available commercially can be used to fractionate mixtures of substances whose molecular weight may vary from a few hundred to over 150 million

[8a] J. C. Giddings, *Anal. Chem.* **40,** 2143 (1968).

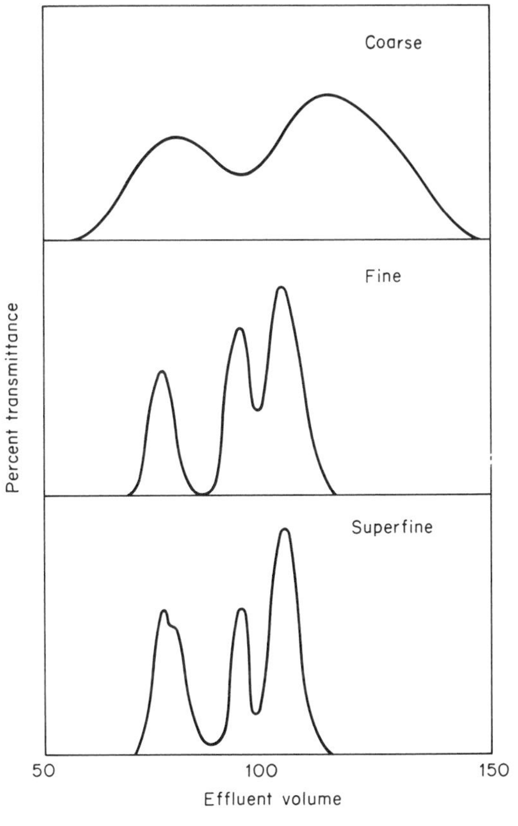

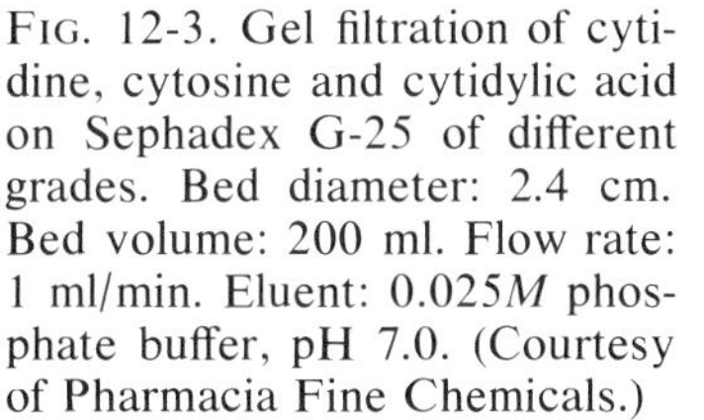
FIG. 12-3. Gel filtration of cytidine, cytosine and cytidylic acid on Sephadex G-25 of different grades. Bed diameter: 2.4 cm. Bed volume: 200 ml. Flow rate: 1 ml/min. Eluent: 0.025*M* phosphate buffer, pH 7.0. (Courtesy of Pharmacia Fine Chemicals.)

(Table 12-2). For example, Bio-Gel P-10 possesses an effective operating range between 5000 and 17,000 molecular-weight units. Molecules of a molecular weight above the upper limit of this range—the *exclusion limit*—are essentially excluded from the gel. Molecules of a molecular weight below the fractionation range are usually eluted at an elution volume approximately equal to the total bed volume. To separate two substances, when the molecular weights are known in advance, it is best to select a gel that has an exclusion limit between the two. If this is not possible or is impractical, then select the gel with the lowest applicable operating range.

The relationship between elution behavior and molecular weight (on a logarithmic scale) for various substances is graphically shown in Fig. 12-4 for several Sephadex gels, in Fig. 12-5 for the Bio-Gel P-series of gels, and in Fig. 12-6 for agarose columns. The linear rising portion of each curve defines the molecular weights between which each gel gives optimum separation. The complete curve at which the slope is other than zero indicates the extreme limits of the useful working range of each gel. For Sephadex G-75,

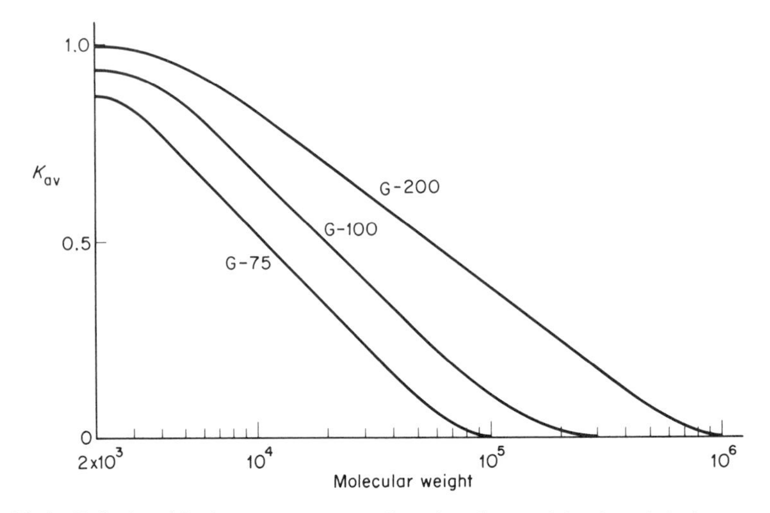

FIG. 12-4. Relationship between K_{av} and molecular weight for globular proteins. After P. Andrews, *Biochem. J.* **91,** 222 (1964); *ibid.* **96,** 595 (1965).

the optimum fractionation range extends from 3000 to about 50,000; the extreme limits are 2000 and 100,000. The slope of each rising segment of the gel curves is directly related to resolving power.

Example 12-1

To use the plots in Fig. 12-5, one needs to know only the total volume of the gel bed, the void volume, and the molecular weights of the substances to be separated.

For example, if you are using a bed volume of 100 ml and are looking for the elution volume of a substance with a molecular weight of 10,000, the ratio V_e/V_o is read from the graph for gels P-10 through P-100. For P-10 the calculations are as follows:

Total bed volume: $V_b = 100$ ml.

The estimated void volume is determined from the hydrated bed volume (Table 12-1) of 12 ml/g dry gel and the bed volume as follows: Weight of gel apparently used was 100 ml ÷ 12 ml/g = 8.3 g. From the solvent regain, $V_i = (8.3\text{ g})(5.1\text{ ml/g}) = 43$ ml. Estimating that $V_g = 0.6(8.3) =$ 5 ml, then $V_s = 43 + 5 = 48$ ml, and $V_o = 100 - 48 = 52$ ml.

From Fig. 12-5, $V_e/V_o = 1.46$ (at 10,000 molecular weight), and therefore, $V_e = (1.46)(52) = 76$ ml.

The void volume could be determined directly by chromatographing blue dextran (mol. wt. 2 million) which is completely excluded from the gel.

Glass packings from Corning with a controlled pore size offer several unique advantages: the material is rigid and withstands pressure changes without pore-size deformation. It is easily sterilized by heat alone or with hot nitric acid. It is not subject to bio-degradation. Packing density is 0.4

gram per cm³. Five types are available with these exclusion limits in millions of molecular weight units: 0.120, 0.400, 1.200, 4.000, and 12.000.

For work in nonaqueous systems in a broad sense, one turns to lipophilic gels. Sephadex LH-20, made by alkylating most of the hydroxy groups of Sephadex G-25, swells in polar organic solvents, water, and mixtures of the two. Solvent regain values, and therefore the fractionation range, vary with the particular solvent (Table 12-3). The gel can fractionate molecules having molecular weights up to 4000. Similarly, Bio-Beads S, a series of polystyrene-based gels, is used for the separation of hydrophobic materials within the exclusion limits from 1000 to 2700 molecular weight.

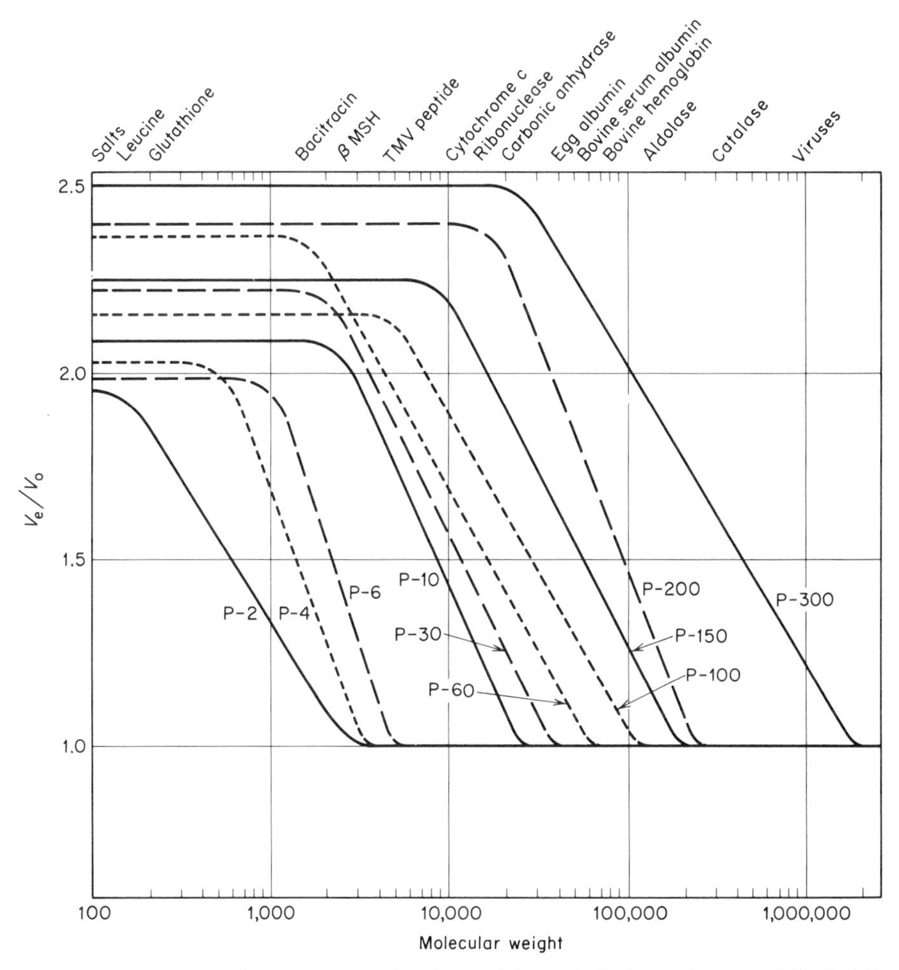

FIG. 12-5. Relationship between molecular weight and elution volume on Bio-Gel P. (Courtesy of Bio-Rad Laboratories.)

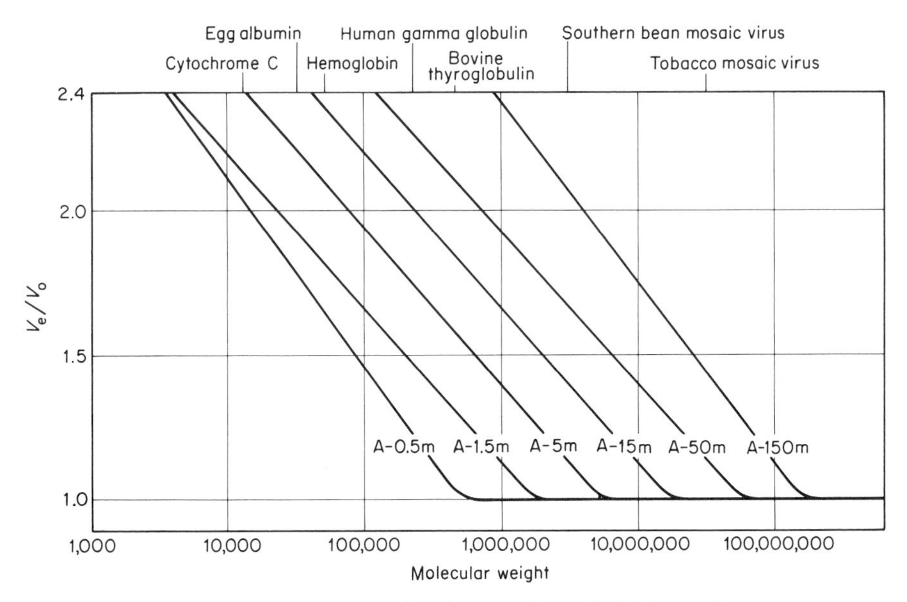

FIG. 12-6. Relationship between molecular weight and elution volume on agarose columns. (Courtesy of Bio-Rad Laboratories.)

In use, the hydrophobic gels are first swollen in the selected solvent, packed into a chromatographic column and washed with the same solvent. The sample should be applied from the same solvent. The rigid polystyrene structure of the styragels does not swell; they can fractionate molecules up to 20 million in molecular weight.

TABLE 12-3. Characteristics of Sephadex LH-20 Gel

Solvent	*Apparent solvent regain, ml solvent per gram dry gel*	*Solvated bed volume, V_b, ml/g dry gel*
Toluene	0.2	0.5
Ethyl acetate	0.4	0.5 – 1.0
Acetone	0.8	1.5
Tetrahydrofuran	1.4	2.5 – 3.0
Dioxane	1.4	2.5 – 3.0
n-Butanol	1.6	3
Chloroform (1% ethanol)	1.8	3.0 – 3.5
Ethanol	1.8	3.0 – 3.5
Methanol	1.9	3.5 – 4.0
Water	2.1	4
Dimethylformamide	2.2	4

Applications

A very common application of gel-exclusion chromatography is the analysis of mixtures of molecules of different molecular weight. An example of this application is given in Fig. 12-7, showing the separation of a tri-, di-, and monosaccharide from each other and from KCl. The elution volume for KCl can be taken as equal to $V_o + V_i$. Very often molecules of similar molecular weight can be separated by proper selection of gel type and bed length.

Desalting was one of the first applications and is still one of the most important. The term "desalting" in its broadest meaning refers not only to the removal of salts, but also to the removal of other low molecular-weight compounds from solutions of macromolecules. Gels with a small pore size will rapidly and efficiently desalt a wide variety of substances. The high molecular-weight substances to be desalted usually move with the void volume, while the components of low molecular weight are distributed between the mobile and stationary phases. As the separation between compounds of high and low molecular weight is very good, large sample volumes (up to approximately 25% of the total bed volume) may be used. Desalting is a faster and more efficient technique than dialysis. It is useful for desalting

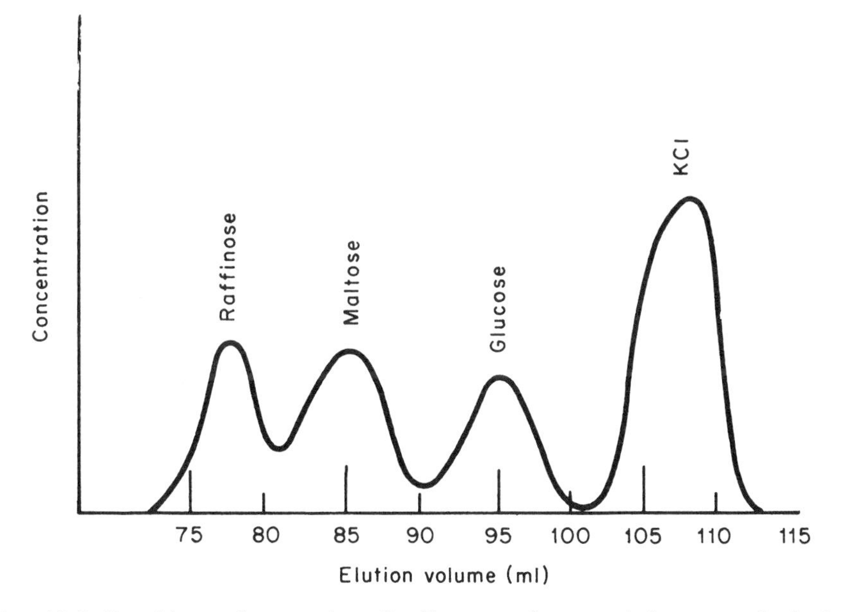

FIG. 12-7. Desalting and separation of raffinose, maltose, and glucose on Sephadex G-15. Sample volume: 0.5 ml; bed dimensions: 1.4 cm × 100 cm; flow rate: 6 ml/hr. (Courtesy of Pharmacia Fine Chemicals.)

labile substances of biological origin. The desalting of saccharides is shown in Fig. 12-7.

In a method related to desalting, one electrolyte can be substituted for another by washing the column with new electrolyte solution. The macromolecules migrate faster than the low molecular-weight electrolytes and are therefore recovered in the effluent in the new electrolyte system. It is equivalent to exhaustive dialysis, yet the internal volume need only be 1.5 times the sample volume, thus keeping dilution to a minimum. It is an excellent method for substituting one buffer system for another.

Molecular weights can be estimated (within 10%) by molecular exclusion chromatography. The fractionation of oligomers of ethylene glycol on Sephadex G-10 is shown in Fig. 12-8, lower graph. On the upper graph of the same figure is shown the relationship between elution volume and molecular weight (on a logarithmic scale). Similar behavior is shown by globular proteins, dextrans, and other macromolecules. Causes of error in molecular weights estimated by exclusion chromatography are the inaccurate estimation of elution volumes, differences in shape between one solute and another in the same generic family, and density differences between solvated molecules, resulting in different ratios between size and molecular weight. For proteins, the molecular-weight markers listed in Table 12-4 are available. For the determination of molecular-weight averages of other polymers, it is common practice to use a polystyrene calibration curve because polystyrene is the only polymer for which there are several commercially available standard samples. The molecular weights of these "standards" are plotted against their peak elution volumes obtained from the molecular exclusion chromatograms. The steps involved in the determination of the molecular-weight distributions and molecular-

TABLE 12-4. Protein Molecular Weight Markers

Protein	*Molecular weight*
Cytochrome C (horse heart)	12,400
Ribonuclease A (bovine pancreas) Cryst	13,680
Myoglobin (horse heart) 2× Cryst	17,800
Chymotrypsin (beef pancreas)	22,500
Trypsin (beef pancreas) 2× Cryst	23,800
Pepsin (hog stomach) 3× Cryst	35,500
Hemoglobin (human) 2× Cryst	68,000
Gamma globulin (human)	160,000
Thyroglobulin (pig)	670,000

P. Andrews, *Nature* **196,** 36–39 (1962).

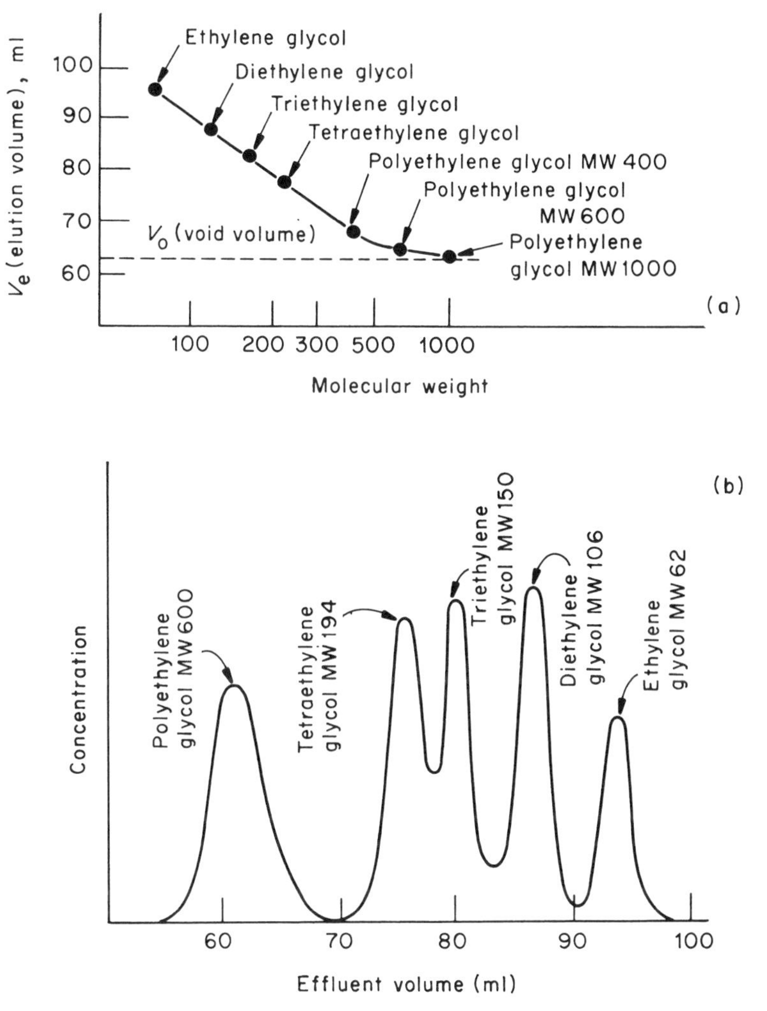

FIG. 12-8. (a) The elution volume vs molecular weight for oligomers of ethylene glycol. (b) Fractionation of oligomers of ethylene glycol on Sephadex G-10. Bed volume: 1.27 × 102 cm; sample volume: 1 ml; flow rate: 6 ml/hr. (Courtesy of Pharmacia Fine Chemicals.)

weight averages of polymers are outlined by Cazes.[9] For this type of work, the Waters Associates gel-permeation chromatograph (Fig. 12-1) is widely used. A unique advantage of this method is that it shows not only whether the distribution is broader than expected, but whether more than one peak appears in the distribution. This method may lead not only to early detection of poor, out-of-specification production batches of polymers, but also

[9] J. Cazes, *J. Chem. Educ.* **43**, A625 (1966).

may give some indication of wherein the trouble might lie. In polymer blending operations it provides a rapid method for ascertaining whether the product contains the proper proportions of the constituents and whether extensive degradation has occurred during the blending. In copolymerization studies it may answer the question of whether one has produced a pure copolymer or a mixture of homopolymers.

Substances with a molecular weight above the exclusion limit for a particular gel can be rapidly concentrated by adding dry gel beads until a thick suspension is obtained. The components remaining in the external solution can then be recovered by ordinary filtration or centrifugation. In this manner nucleic acids have been separated from nucleotides, polysaccharides from sugar, and proteins from low molecular-weight substances.

Molecular exclusion chromatography can conveniently be applied to isolation and purification processes on a preparative or an industrial scale. Column dimensions are increased proportionally; often recycling techniques are employed. An industrial example is the production of low-lactose milk: a column 45×60 cm and holding 96 liters can treat 1500 liters of milk per day.

INORGANIC MOLECULAR SIEVES

The natural[10] and synthetic zeolites (metal–alumino silicates) form a class of molecular sieves suitable for the fractionation of the permanent gases and the smaller organic molecules. Internally, the zeolite volume consists, in large part, of cavities connected by channels. Screening and sieve action of these channels combined with surface adsorption activity of the crystal matrix makes it possible to separate molecules smaller than the size of the channels from those which are larger. Separation thus results from surface activity plus molecular geometry.

Structure of Zeolite Crystals

In silicates each silicon atom is linked to four oxygen atoms which are situated at the four corners of a tetrahedron. A characteristic of silicates is the incorporation of these anions in more complex systems, the oxygen of one anion is linked to other anions, forming oxygen bridges between silicon atoms. Long chains, or cylindrical groupings, are thus formed.

In silicates aluminum is also present as a cation or as an anion. In alumino-silicates, ions of aluminum with three charges replace silicon ions with four; the extra charge holds a metal ion. The latter might be sodium or calcium ions.

[10] R. M. Barrer, *Ann. Rep. Chem. Soc.* **41,** 31 (1944); *Proc. Chem. Soc.* (1958) 99.

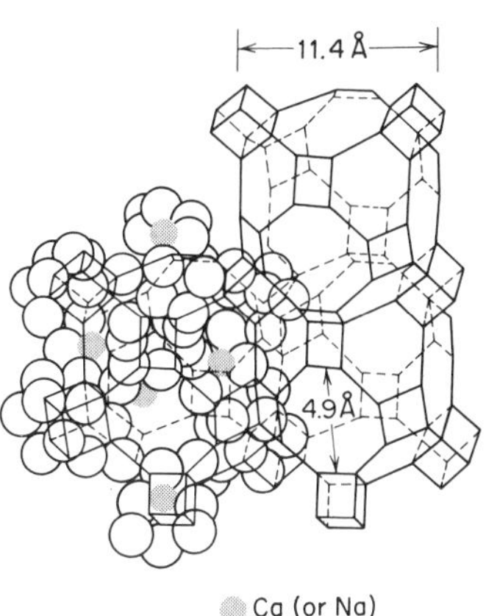

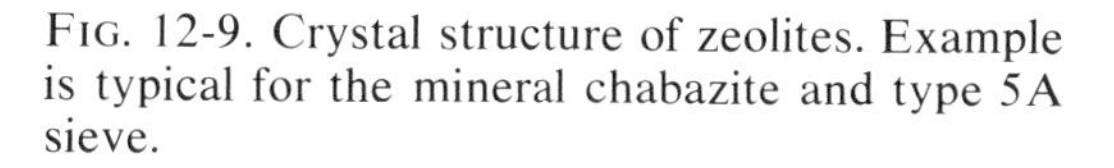
Fig. 12-9. Crystal structure of zeolites. Example is typical for the mineral chabazite and type 5A sieve.

The basic structure of zeolites resembles a skeleton of tetrahedra forming a honeycomb with comparatively large cavities connected with one another by small apertures. The honeycomb structure is three-dimensional and each cavity has a communicating aperture to six neighboring cavities so that all the cavities are interconnected. The shape and dimensions of the cavities depend on the nature and the characteristics of the zeolite. Their structure is shown in detail in Fig. 12-9. The maximum cross-section of an aperture determines the size of molecules which can penetrate into the larger cavities. One gram of chabazite, for example, has 3×10^{20} cavities. The maximum cross section of the cavity measures 11.4 Å, of an aperture 4.9 Å. The volume of each cavity is 925 cubic Å (roughly half the volume of the entire crystal). It can hold 24 molecules of water. The connecting channels are unvarying in diameter and therefore are capable of preventing molecules greater than a critical diameter from entering the sieve and being adsorbed within the internal structure. It must be remembered that for organic molecules, the effective cross-sectional area perpendicular to the hydrocarbon chain axis is of greater importance than the chain length.

Synthetic metal–alumino silicates possess a similar three-dimensional framework of silica and alumino tetrahedra which are linked together to form rings comprising 4, 6, and 8 members, leading to three sizes of cages linked together by three sizes of channels. The crystals obtained are too small for practical use (a few microns) and require granulation. This is done by adding a small quantity of clay as a binding agent. Granules or pellets of 2 to 6 mm diameter are obtained by extrusion.

The size and position of the metal ions (sodium or calcium) in the crystal and the particular type of alumino-silicate, control the effective diameter of the interconnecting channels. A range of molecular sieves can thus be prepared.

Types of Sieves and Applications

Type 4A molecular sieve is a sodium aluminum silicate [whose empirical composition is $Na_{12}(AlO_2)_{12}(SiO_2)_{12}$] with a pore opening of 4.2 ± 0.2 Å (α-cage, 8-membered rings)[11] which leads into an 11.4 Å-diameter cavity in the center of the unit cell where adsorption occurs. This type of sieve is capable of adsorbing only those molecules smaller than about 4 Å in diameter; these include materials such as H_2O, CO_2, H_2S, SO_2, and all hydrocarbons containing one or two carbon atoms. Propane (with a diameter of about 5 Å) and higher hydrocarbons are physically excluded with the exception of propylene which is adsorbed much more strongly and can enter the pore system. Adsorption of a molecule with dimensions close to the diameter of the entry aperture of the zeolite requires additional energy. The deformation of a molecule containing two or more atoms can, within narrow limits, allow the adsorption of a molecule somewhat larger in critical diameter than the aperture. A molecule of ethane, with a critical diameter of 4.4 Å, passes through type 4A (the passage is aided by thermal pulsation of the crystal lattice).

Type 5A sieve is produced from Type 4A through ion exchange of 75% of the sodium ions with calcium and potassium ions. The cation substitution alters the diameter of the entrance pores to about 5 Å. Type 5A sieve will admit straight-chain paraffins, olefins, and alcohols up to at least C_{14}. With the exception of cyclopropane, it will exclude all branched paraffins, naphthenic and aromatic molecules. For example, it adsorbs *n*-butane (whose diameter is 4.9 Å) but excludes isobutane (5.6 Å).

Types 10X and 13X are based on a different alumino-silicate ratio [empirical composition is given by the formula $Na_{86}(AlO_2)_{86}(SiO_2)_{106}$]. These sieves will adsorb molecules with diameters up to about 9 Å and 10 Å, respectively. For example, di-*n*-propylamine with a critical diameter of 9.1 Å is excluded from Type 10X.

Molecular sieves are available in granules or pellet form. Granules are used in instances where it is possible to agitate the system as in batch operations. Pellets are used where it is desired to have a liquid or gas phase flow freely through the adsorbing material. The pellets are powder bonded with about 20% inert binder and formed into extruded pellets or beads. Molecular sieves have a pH of about 10 and are stable in the 5 to 12 pH range.

In addition to separating molecules based on their size and configuration, molecular sieves show a strong affinity for polar molecules and polarizable compounds in preference to nonpolar molecules even though both

[11] Plus β-cage, 6-membered rings, channel diameter 2.5 ± 0.2 Å, cage diameter 6.6 Å; and γ-cage, 4-membered rings, too small to permit diffusion of ions.

may be of equal size. The more polar or the more unsaturated the molecule, the more tightly it is held within the crystal. This preferential adsorption is due primarily to the cations which are exposed in the crystal cavities. These cations act as sites of strong, localized positive charge which electrostatically attract the negative end of polar molecules, or induce dipoles in polarizable molecules. This property is of value in the removal of unsaturated materials from refinery and synthesis gases or in the drying of industrial gases. Should two chemically different molecules have the same size, the crystalline sieves show a preference for absorbing the more asymmetric molecules.

Crystalline sieves are used as carriers for a wide variety of chemical compounds. These are held within the pores of the sieve by the unusually strong adsorptive forces there, until they are released by heat or vacuum or by displacement with a more adsorbable material, such as water. Molecular sieves are usually reactivated at about 200° to 350°C and can be used repeatedly. While purging, the bed temperature is raised gradually, then held at temperature until desorption is complete.

Molecular sieves are utilized as the stationary phase in gas–solid chromatography (*q.v.*).

ION EXCLUSION

Ion exclusion is a process for separating ionized materials from non-ionized or slightly ionized materials. It relies on the preferential diffusion of small non-ionic molecules into the exchanger phase.

Principle

Conventional ion-exchange resins are used with the counter ion of the resin similar to the ion of one substance but dissimilar to the other. A cation-exchange resin in its hydrogen form or an anion-exchange resin in its hydroxide form can be used.

Because of the Donnan membrane effect, non-ionic solutes tend to exist at the same concentration in both the resin phase and the interstitial liquid whereas ionic materials exist at a considerably lower concentration in the resin phase than in the interstitial liquid due to repellent action caused by the common-ion effect. The total concentration of A^+ in a cation exchanger (and X^- in an anion exchanger) includes the mobile counter ions of the fixed ionized matrix plus the concentrations of electrolyte which has diffused into the exchanger phase. The "exclusion" is better the lower the external solution concentration. No such exclusion should occur for non-ionic solutes. Thus, if a solution containing both ionic and non-ionic components is placed on the top of a column of resin and rinsed through the column with water, the ionic components will reach the bottom first since they have to displace essentially only the interstitial liquid (one bed volume). The non-ionic components, however, must displace both the interstitial liquid and the liquid in the resin phase, and will appear later in the effluent.

The resin bed consists of three parts: the solid resin network, the occluded liquid volume held within the resin beads, and the interstitial liquid volume between the resin particles. Although water is the only liquid used, the occluded liquid has a different character than the interstitial liquid because it is within a hydrocarbon matrix and contains a high concentration of ionic groups. The resin network thus serves as a boundary or semipermeable membrane between two liquid phases. Highly ionized materials, such as the strong mineral acids, are essentially excluded from the interior of the resin and pass through the column with the eluent front, emerging when the interstitial volume has been displaced. Non-ionic materials, such as most organic acids, are free to enter the interior resin phase and thus emerge some time later, after the interstitial volume plus the inner volume have been displaced.

In addition to this effect, there exist differences in the tendency of various non-ionic materials to be retarded by the resin phase. These differences are due to both polar attractions between functional groups and to van der Waals' forces between the nonelectrolytes and the hydrocarbon portion of the resin. The result is a partition of the solutes between the resin liquid and the interstitial liquid. This partition is described by the distribution coefficient $K_d = C_i/C_o$ where C_i is the concentration of the solute in the inner volume within the resin and C_o is the concentration of the solute in the void volume outside the resin. The ability to separate given solutes will depend on differences in K_d and will be affected by factors which affect K_d, such as the nature and concentration of the solute and the characteristics of the resin used.

Applications

Ion exclusion permits the separation of acids and salts from glycerin, alcohols, and from amino acids. The method can be extended to separate solutes of different ionization constants; thus, acetic acid from mineral acids or formic acid,[12] and mixtures of mono-, di-, and trichloroacetic acids. Because of the decreased electrolyte exclusion, separations become more difficult as the electrolyte concentration increases in the solution.

Since ion exclusion operates on diffusion gradients, the equilibrium existing between the concentration of a non-ionic solute in the interstitial liquid and the resin phase differs appreciably from one non-ionic solute to another. Consequently, two or more non-ionic, but water-soluble, materials may be separated from one another by washing the various components through the exchange column with water.

[12] R. M. Wheaton and W. C. Bauman, *Ind. Eng. Chem.* **45**, 228 (1953).

Through variation in size of the resin pores, it is possible to effect separations of non-ionic materials primarily on the difference in their molecule size. Methanol and glucose can be separated readily on 8% cross-linked resin. The larger glucose molecule is excluded to a greater extent from the interior of the resin beads and appears first in the effluent. Glycols, methanol, and phenol can similarly be separated by a single pass through a resin column.

A semi-continuous ion-exclusion process for the separation of glycol from NaCl has been outlined by Bauman, Wheaton, and Simpson.[13] Only the concentrated effluent fractions are taken as product cuts in every cycle.

LABORATORY WORK

Preparation of the Gel Column

Any chromatographic column can be used, but to avoid the possibility of zone remixing the column should have a minimum amount of space at the outlet. Columns 1.4 to 2.5 cm in diameter and 100 to 500 cm in height are suitable.

Coat the column interior with a solution of 1% dichlorodimethylsilane in benzene. The silane solution is heated to approximately 60°C, poured into the glass column, allowed to stand for several minutes, the solution decanted, and the benzene evaporated in a drying oven. Repeat the entire procedure to double the coating, after which the column can be used for several months without further treatment.

The dry beads should be hydrated by adding them slowly with constant stirring to a solution of the buffer that will be used for the elution. Fine-mesh beads (100–200 mesh or 200–400 mesh) give a more efficient column but a relatively low flow rate. Coarser beads (100–300 μ; 50–100 mesh) should be used on large columns and for operations such as desalting. The water uptake values in Table 12-1 show the minimum amount necessary to hydrate each gram of dry gel beads; 10 times this amount should be used to prepare a slurry suitable for pouring into a column. Gels from types 2 through 10 swell rapidly and are ready for use after 2 hr; types 20 through 50 require 3 hr; types 60 and 75, 24 hr. The larger pore size gels must be left to stand for at least three days with intermittent stirring. On a boiling-water bath the minimum swelling times are decreased to 1 hr for types 20 to 50; to 3 hr for types 60 and 75, and to 5 hr for the gels with greater porosity. A swelling time of 3 hr is suggested for the hydrophobic gels.

The smallest particles are removed by repeated sedimentation and decantation. Above the juncture of the column outlet, insert a plug of glass wool and on top of this a layer of glass beads, 0.5 mm in diameter, or a layer

[13] W. C. Bauman, R. M. Wheaton, and D. W. Simpson, in *Ion Exchange Technology*, F. C. Nachod and J. Schubert (Eds.), p. 182, Academic Press, New York, 1956.

of sand, until the curved bottom part of the column is filled. Some buffer should be present to prevent air bubbles from forming during this operation.

To form the gel bed, the swollen gel slurry is poured into the column already partly filled with buffer solution, and at the same time allowing the excess liquid to percolate through the growing gel bed. As the gel settles, more slurry is added until the desired bed volume is formed. Rinse the bed with several column volumes of buffer, using approximately the same flow rate (~6 ml/cm^2/hr) that will be used for the sample elution. Never allow the column to run dry. Protect the upper surface of the bed with a disk of filter paper and/or a 3-cm layer of highly cross-linked gel beads. Even packing of the column is checked by watching the passage through it of a band of colored substance such as blue dextran. Correct any faults.

Bacterial and fungal growth may interfere seriously with the chromatographic properties of the gels. The eluent in columns which are not in use should always contain a small amount of some bacteriostatic agent, such as sodium azide (0.02%), 1,1,1-trichloro-*t*-butanol (0.1%), or chloroform (saturation concentration at room temperature). The latter is not recommended on the looser gels as it causes the gel particles to shrink slightly. The gels may be autoclaved in the wet state at 110°C for 40 min without any changes in the gel properties.

Procedure for Column Runs

Equilibrate the column with the buffer selected using a flow rate of approximately 30 ml per hr and passing two bed void volumes.

Dissolve the sample in the equilibration buffer (0.5 to 2 ml) and carefully apply the solution to the top of the column by layering under the buffer already present without disturbing the upper surface of the gel bed. If necessary, the density of the solution can be increased by the addition of sucrose. Allow the sample to enter the bed and wash down the sides of the column carefully with a small volume of buffer. Fill the column with buffer to the desired level and attach the column to the buffer reservoir. Adjust the position of the solvent reservoir to provide the necessary hydrostatic pressure for a flow rate through the column of approximately 6 ml/cm^2/hr.

When an effluent volume somewhat smaller than the void (interstitial) volume of the column has been taken off, the rest of the effluent is collected in appropriate fractions with a fraction collector. A suggested size for fractions is 2.5 to 3.0 ml.

The fractions are analyzed by a suitable method: acids and bases by titration with dilute sodium hydroxide and hydrochloric acid, respectively; salts by titration after passage through an anion or cation exchanger; aromatic substances by their absorption in the ultraviolet.

The volume of sample which can be applied depends on the purpose of the experiment. For complete resolution of substances with similar K values, the sample volume must be small in comparison to the bed volume. Where the difference in distribution coefficients is greater (as in desalting), sample volume corresponding to 20–30% of the bed volume can be used.

Evaluation of Column Parameters

The elution volume is determined by measuring the effluent volume from the addition of the test solution to the point where the concentration gradient of the eluted substance is maximum. Estimation is often aided by extrapolating both sides of the solute peak to an apex.

The interstitial (void) volume, V_o, is determined with a colored substance whose molecular weight exceeds the exclusion limit of the gel. If the eluent is distilled water, India ink is suitable (although sometimes it leaves dark stains on the column); it is estimated at 900 mμ. Blue dextran 2000 (mol. wt. 2 million) is completely excluded from all gels in the Sephadex or Bio-Rad series. Its absorption is maximum at 280 and 620 mμ.

The internal volume, V_i, is estimated by the elution of KCl ($K_d \simeq 1$) or sucrose. The expression is: $V_e - V_o = V_i$. Potassium chloride is determined with silver nitrate; sucrose by the anthrone method.[14] The internal volume can also be calculated from the solvent uptake and the dry weight of the gel beads [Eq. (12-2)], but there may be experimental difficulties due to different degree of packing in columns. V_i, for a polymer series under test, may be taken as the difference between the elution volume at which molecular sieving ceases to be operative for small molecules and the void volume of the column.

Suggested Systems

1. Desalting a neutral amino acid. Prepare a column, 1.3 × 50 cm, of Bio-Gel P-2, 100–200 mesh, by swelling the gel in contact with distilled water. Layer the sample, 1 ml in volume and containing 10 mg each of the amino acid and NaCl, on the top of the column and elute at a flow rate of 0.5 ml/min. Collect 3-ml fractions of the effluent.

2. Separation of raffinose, maltose, and glucose (or any single mono-, di-, and tri-saccharide) as illustrated in Fig. 12-7. Elute with 0.025 *M* phosphate buffer, pH 7.0, or with distilled water. Take 20 mg of each sugar.

Problems

1. For the eight Sephadex gels (G-series) listed in Table 12-1, calculate the inner volume (V_i), the volume of the stationary phase (V_S), and the void volume (V_o) for 1 g of gel in the swollen condition.

[14] W. E. Trevelyan and J. S. Harrison, *Biochem. J.* **50,** 298 (1952).

2. For the 10 Bio-Gels (P-series) listed in Table 12-1, calculate the inner volume (V_i), the volume of the stationary phase (V_S), and the void volume (V_o) for 1 g of gel in the swollen condition.
3. For Sephadex G-15, the wet density (g/ml) is 1.19. On a prepared column the bed volume is 30 ml, and the void volume was found to be 9 ml. Calculate the inner volume.
4. For Sephadex G-100, the wet density (g/ml) is 1.04. On a prepared column the bed volume is 85 ml, and the void volume was found to be 30 ml. Calculate the inner volume.
5. From the information on the desalting and separation of raffinose, maltose, and glucose on Sephadex G-15, given in Fig. 12-7, calculate for each component these quantities: (V_e/V_o) and the partition coefficient K_{av}.
6. On a tighter gel (Sephadex G-10) column, whose bed volume was 1.6 cm × 96 cm, these elution volumes (V_e) were observed: raffinose, 84.8 ml; glucose, 102 ml; and KCl, 124 ml. Void volume was 69 ml. Calculate for each component, V_e/V_o and K_{av}.
7. During the desalting of leucine on a column of Bio-Gel P-2 (1.3 cm × 36 cm), the leucine peak was observed at 36.5 ml and the peak for NaCl at 42.5 ml. Sample size was 1.0 ml. Estimate the V_e/V_o ratio and K_{av} for each component. Calculate the separation volume (ΔV_e). Assume V_r is 0.6 weight of gel present.
8. From the information contained in Fig. 12-7, calculate the separation volume for each pair of adjacent components and compare the calculated value with the observed peak separation.
9. From the information contained in Fig. 12-5, estimate the elution volume of a substance with a molecular weight of 10,000 when the bed volume is 100 ml for these Bio-Gels: P-30, P-60, and P-100.
10. From the information contained in Fig. 12-4, estimate the elution volume of a globular protein with a molecular weight of 20,000 on each of the Sephadex gels.
11. From the information contained in Fig. 12-4, estimate the elution volume of a globular protein with a molecular weight of 40,000 on each of the Sephadex gels.
12. For a bed volume of 30 ml (i.e., 1 g of Sephadex G-200) and a 1 ml sample volume, what must be the difference in values of K_{av} if the zone broadening is assumed to be six times the sample volume? If one component is a globular protein with a molecular weight of 20,000, what would be the molecular weight of a protein that could be adequately resolved from it?
13. What improvement in separatory ability would result from increasing the bed volume of G-200 to 100 ml in Problem 12?
14. In Problems 12 and 13, what improvement would result in shifting to Sephadex G-75?
15. From the fractionation of oligomers of ethylene glycol on Sephadex G-15 (Fig. 12-10), (a) plot the elution volume vs molecular weight, (b) estimate the exclusion limit, and (c) compare the fractionation range and resolution with the information in Fig. 12-8.
16. From the fractionation of the mono-, di-, and tri-saccharides (Fig. 12-7), plot the elution volume vs the molecular weight and estimate the exclusion limit.
17. If a sample volume of 10 ml (containing compounds whose molecular weight exceeds 4000) is to be desalted using Sephadex G-25, what weight of gel will be required? When would the salt-free solution emerge from the column? Suggest suitable bed dimensions.

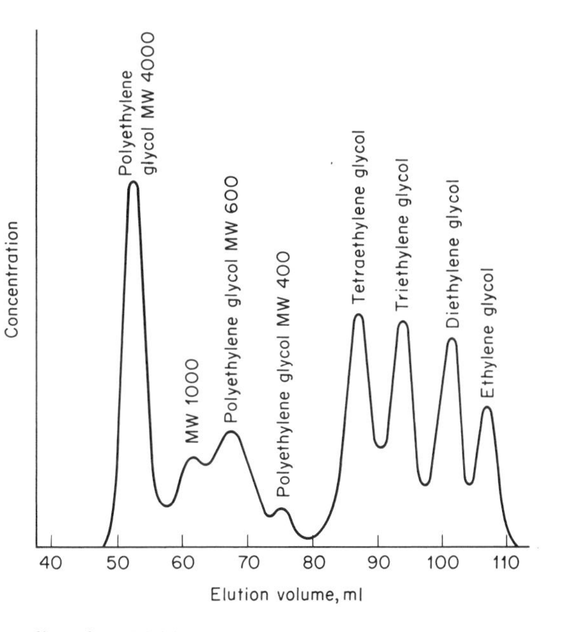

Fig. 12-10. Illustration for Problem 15. Fractionation of oligomers of ethylene glycol on Sephadex G-15. Sample volume: 0.5 ml; bed volume: 1.4 × 102 cm; void volume: 55 ml; flow rate: 6 ml/hr.

18. With glucagon (3500) and ovalbumin (45,000) present as reference markers, these elution volumes were obtained on columns of Sephadex G-75 and G-100:

Component	G-75	G-100
Glucagon	195	216
Cytochrome *c*-551	150	180
Ribonuclease	138	158
α-Lactalbumin	128	152
Ovalbumin	88	114

Estimate the molecular weights (to nearest 10%) of the proteins.

19. The elution peaks of the listed materials were obtained on a 4-ft column of polystyrene modified gel with a 500 Å permeability limit. Eluent was tetrahydrofuran.

Component	Elution volume, ml
Polystyrene S-102	
(mol. wt. 82,000)	24.7
(mol. wt. 4000)	28.5
(mol. wt. 2000)	31.8
(mol. wt. 750)	36.0
bis-Phenol-α-diglycidyl ether	40.7
o-Dichlorobenzene	45.3
Benzene	46.3
Water	44.4

Plot the logarithm of the chain length (and molecular weight) vs the elution volume. Estimate the relationship between elution volume and rejection of molecules with increase in chain length.

20. Estimate the approximate values for the maximum and minimum pore diameters in Sephadex G-75 and G-100 by assuming that the pores are circular in cross section and that the appropriate diameters are equal to the diameters of spherical protein molecules that are just excluded at those points. Protein molecules have partial specific volumes of 0.73 ml/g and are hydrated in solution to the extent of 0.3 g of water per gram of protein.
21. Devise a rapid method for analyzing butane mixtures by means of molecular sieves.
22. Devise a method for conveying a reactive peroxide (used in curing rubber) to the site of its catalytic action without destruction during the blending operations.
23. Suggest proper types of molecular sieves for accomplishing these separations: (a) carbon monoxide from argon; (b) simultaneous removal of water, *n*-paraffins, and sulfides from benzene; (c) simultaneous removal of moisture and acid gases from a surrounding atmosphere; (d) practical separation of mixtures of naphthas and petroleum distillates; (e) *n*-hexanol from cyclohexanol; (f) removal of benzene and toluene from paraffinic hydrocarbons; (g) separation of *n*-decylbenzene from 1,3,5-triethylbenzene; (h) removal of carbon dioxide, acetylene, and acetone from ethylene; and (i) removal of odors and colors from organic materials.

BIBLIOGRAPHY

Gel Permeation (Filtration)

Determann, H., *Gel Chromatography*, Springer-Verlag, New York, 1968.

Gelotte, B., and J. Porath, "Gel Filtration," in E. Heftmann (Ed.), *Chromatography*, 2nd ed., Reinhold, New York, 1967.

Gelotte, B., "Fractionation of Proteins, Peptides and Amino Acids by Gel Filtration," in A. T. James and L. J. Morris (Eds.), *New Biochemical Separations*, Van Nostrand, Princeton, N. J., 1964.

Granath, K., "Fractionation of Polysaccharides by Gel Filtration," in A. T. James and L. J. Morris (Eds.), *New Biochemical Separations*, Van Nostrand, Princeton, N. J., 1964.

Altgelt, K. H., "Theory and Mechanism of Gel Permeation Chromatography," in Vol. 7 of J. C. Giddings and R. A. Keller (Eds.), *Advances in Chromatography*, Dekker, New York, 1968.

Inorganic Molecular Sieves

Thomas, T. L., and R. L. Mays, "Separations with Molecular Sieves," in W. Berl (Ed.), *Physical Methods in Chemical Analysis*, Vol. IV, Academic Press, New York, 1961.

Ion Exclusion

Helfferich, F., *Ion Exchange*, McGraw-Hill, New York, 1962, pp. 431–433.

13 *Membrane Separation Methods*

A membrane is a barrier, usually thin, which separates two fluids. For our purposes it is formulated to be semi-permeable, that is, it permits transfer of some components and not of others. The usefulness of membrane separations lies in their ability to make separations on a molecular scale and with a preciseness of molecular order. This is enhanced by the variations possible due to the almost limitless combinations of permeants, barriers, and applied external stresses.

Transfer through membranes is by diffusion, a process of mass transfer which occurs as a movement of individual molecules. This movement may be induced by *dialysis,* a concentration gradient between the phases separated by the membrane; by *ultrafiltration,* a hydrostatic pressure gradient; or by *electrodialysis,* an electrical field applied across the membrane. Each member of the family of membrane processes is named for the major effect that occurs in the transport process.

Diffusion through membranes is fundamentally no different from diffusion in liquid systems. In theory the same laws apply. The mechanism involves transfer of a molecule to a vacancy or "hole" in the membrane. Energy of activation is required for formation of the hole. What we have is a rate process based upon a specific rate constant which measures the frequency of a molecular jump and the parameter which measures the mean distance of such a jump.

A membrane appears to possess a duality of properties. It is a form of gel which consists of two phases: one, a solid dispersed phase; the other, a continuous liquid phase. Accordingly, diffusion through the membrane is via this liquid part of the structure. On the other hand, sometimes the components of a membrane act as if they are in a state of solid solution within a membrane. There is, moreover, a duality of nature in the tendency of the membrane to dissolve and in the restraint of molecular cohesive forces which hold it intact. The net result is to imbibe liquid and swell.

A model for a membrane may be likened to a network or mosaic of linear or branched chains of atoms, crosslinked at various points and filled with molecules of solvent and of mobile solutes. The solvent which is held within the network contains molecules and holes exactly as in the free solvent. For specific separation problems, the possibility exists to tailor-make membranes.

DIALYSIS

In dialysis, separation is based simply on the relative rates of diffusion of two species through a membrane. The objective is to obtain a mass transfer which is selective for certain of the solution components. Thus, dialysis requires that the membrane separating two liquid phases permit diffusional exchange between at least some of the molecular species present while effectively preventing any convective exchange between the solutions.

Mechanism of Dialysis

A schematic diagram of equipment necessary to carry out a dialysis is shown in Fig. 13-1. One compartment contains the solution to be dialyzed and consists of two solutes of different molecular size. The other compartment contains only pure solvent. The membrane serves to physically separate the two compartments to prevent mechanical mixing and to provide a barrier which is readily permeable to the solvent and the smaller solute molecules, and impermeable to the larger solute molecules.

The nature of the pathways (or pores) through a dialyzing membrane may be likened to a "brush heap." The aggregation of micelles align themselves in a more or less random manner. Heterogeneous pathways are the result. Each individual pore is an irregular channel with cross connections and dead ends and, most likely, a length which is considerably greater than the thickness of the membrane. Because of this heteroporosity it is possible to separate solutes quantitatively only if the molecules differ greatly in size. Leakage of the larger molecules through the membrane can be prevented only if all effective pore sizes are less than the diameter of these molecules. The fractionation of a complex mixture of solutes by dialysis requires that there be available porous membranes sufficiently selective to distinguish among solutes of similar sizes and properties.

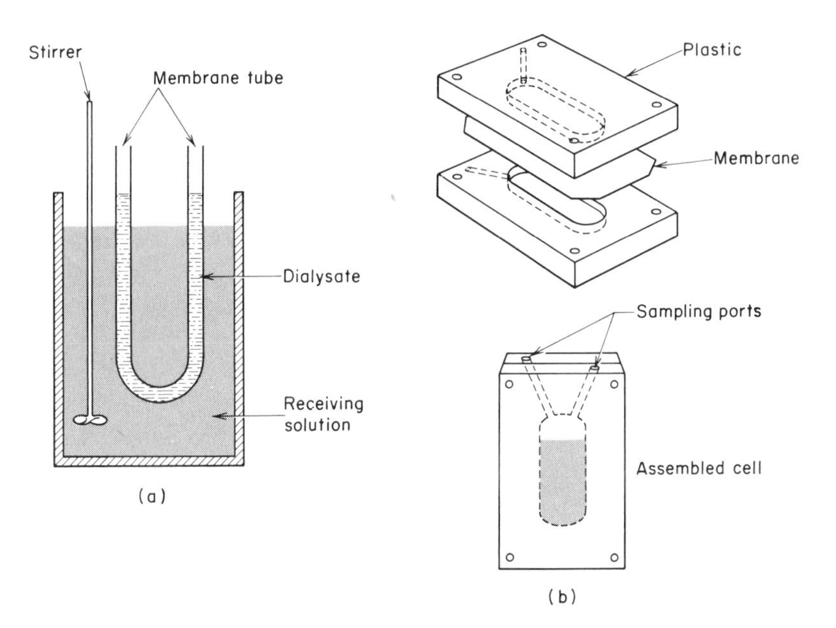

FIG. 13-1. Schematic arrangement of components in dialysis: (a) conventional cell for use with tubing; (b) sandwich-type cell for use with small volumes of sample and receiving solution (introduced by means of a hypodermic syringe).

In dialysis the driving force which impels the solute molecules through the membrane is the difference in concentration of a diffusing substance between the solutions on the two sides of the membrane. The effectiveness of a dialysis as a separation process is governed by two factors: (1) the diffusion rate of solutes through the membrane and stagnant films of fluid at the membrane interface on either side and (2) the properties of the membrane. The rate at which a substance diffuses through a membrane (and solution) is expressed by Fick's law of diffusion (which is strictly applicable only when the pore diameter of the membrane is much larger than the diameter of diffusing solute):

$$\frac{\partial Q}{\partial t} = -A_p D \frac{\partial C}{x_p} \tag{13-1}$$

Here Q is the quantity of solute diffusing in time, t, through x_p, the mean length of the pores, A_p is the area available for diffusion (i.e., the total effective cross-sectional area of pores), D is the mass transfer coefficient through the membrane and stagnant films, and ∂C is the concentration gradient. The minus sign denotes that the substance diffuses in the direction of decreasing concentration.

For the solute it is the difference between the cross-sectional size of the pore and that of the molecule which is important. A simple relationship is given by

$$A_p = A_o(1 - a/r)^2 \tag{13-2}$$

Here A_o is the cross-sectional area of the pore, r is the radius of the molecule, and a is the radius of the pore. For a molecule to pass through a pore, estimates indicate that pore diameters must be at least twice that of particle diameters. As the limiting size of the pore is approached, a small increase in the size of the diffusing molecule will be sufficient to prevent it from entering the pore. In other words, the selectivity of membrane diffusion should become very much higher when the membrane will barely allow the solute of interest to diffuse through it. For example, Size 20 Visking cellophane tubing will allow ribonuclease (mol. wt. 13,600) to pass very slowly at room temperature when the solvent is $0.01M$ acetic acid but will almost completely hold back chymotrypsinogen (mol. wt. 24,500). However, very little resolution will be noted with this membrane for a mixture of bacitracin (mol. wt. 1420) and subtilin (mol. wt. 3200). Both diffuse rapidly through the membrane.

In the dialysis of electrolytes the electrochemical behavior of the membrane will play a role in determining the effective pore diameter. Cellulose nitrate membranes, for example, have acidic groups situated in the membrane skeleton which impart a negative charge on the pore wall of the membrane which then repels anions. Since the maintenance of electrical neutrality demands an equal number of positive and negative ions on either side of the membrane, the restriction on anion permeability directly impedes the diffusion of cations. The same loss of permeability results if the membrane becomes positively charged. The smaller the opening, the greater the screening effect of fixed charges on impeding the passage of ions.

Rapid approach to equilibrium is achieved by careful attention to several pertinent design factors in dialysis equipment. (1) The membrane area per unit volume of dialysate solution should be maximized. This is best achieved through the use of long lengths of small-diameter tubing. (2) The receiving solution (diffusate) should be changed continually or periodically. (3) The membrane should be as thin as practically possible. Equation (13-1) points out that the rate is inversely proportional to the first power of membrane thickness. (4) If permissible, use of an elevated temperature is beneficial. For small solutes the rate of diffusion increases about 2%/°C, and for larger solute molecules the increase is even greater.

An inherent limitation of the dialysis method of separation centers around its driving force. Unaugmented dialysis is a "passive" process whose driving force continually diminishes. As dialysis proceeds, the concentration gradient necessarily becomes less and less but never reaches zero. By contrast, other membrane processes promote the separation by externally applied forces; pressure in ultrafiltration and electrical potential in electrodialysis. As a result, dialysis is most attractive in separating dissolved materials present in high concentration.

Osmosis must also be considered. At the beginning of a dialysis, with pure solvent on one side of a membrane and a solution of variable concentration on the other side, there is a difference in the osmotic pressure on the two sides of the membrane. This results in a flow of liquid from the diffusate toward the dialysate with consequent dilution of the solution being dialyzed. The net dilution effect is most severe when solutions of high concentration are being dialyzed.

Membrane Materials

Cellophane (regenerated cellulose) is the most commonly used material for dialysis in the laboratory. Although quite dense in the dry state, cellophane swells considerably to form a spongelike medium when placed in water. Wet cellophane is in fact a gel. It behaves as if it had pores of more or less fixed size which range from 30 to 50 Å. Cellophane carries practically no fixed charges and does not give trouble due to adsorption for most solutes when 0.01*M* acetic acid is the solvent. Any mechanical strain must be strictly avoided if the porosity is to be held constant. In fact, a stretching procedure [1] can serve as an effective method for adjusting pore size. Adjustment of pore size can also be accomplished by controlled acetylation to reduce pore size or by controlled swelling through treatment with a strong $ZnCl_2$ solution. Any pore size—the smallest, which will reject amino acids, to the largest, which will readily pass solutes of molecular weight 100,000—can be produced at will. Once a membrane has been adjusted to a desired pore size, it can be calibrated and used repeatedly until it fails mechanically.

Collodion membranes can be prepared from nitrocellulose dissolved in an ether–alcohol mixture. A thin layer of the solution is cast on a solid surface and allowed to dry completely. Then the membrane is placed in ethanol–water solutions where it will swell to an extent that is a function of time and ethanol concentration.[2] Membranes of graded porosity made from cellulose and cellulose derivatives are available commercially and cover the range 0.2 to 3.0 μ.[3]

Applications

Cellophane membranes have been used in the rayon industry to regenerate sodium hydroxide by separating it from the "steep" liquors. These membranes have a pore size of about 30 Å, sufficient to prevent or retard

[1] L. C. Craig and W. Konigsberg, *J. Phys. Chem.* **65,** 166 (1961).
[2] H. P. Gregor and K. Sollner, *J. Phys. Chem.* **50,** 53 (1946).
[3] From Schleicher & Schüll Co. under the series called Membrane Filters.

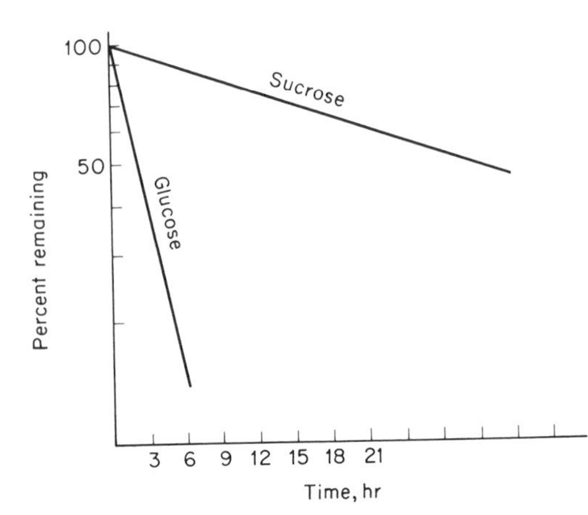

FIG. 13-2. Comparative escape rates of glucose and sucrose through a selective dialysis membrane. After L. C. Craig and A. O. Pulley, *Biochemistry* **1**, 89 (1962).

the passage of the large organic molecules while passing caustic soda. Recovery averages 90 to 93%. This amounts to a great saving in caustic soda and reduces greatly the disposal problems of the waste liquor.

Improvements in dialysis membranes have made their use possible in corrosive environments for reasonably long periods of time. Acid-resistant vinyl polymers make it possible to recover mineral acids from their salts in a variety of metallurgical processes. Examples are the recovery of copper and nickel sulfates from sulfuric acid, and iron sulfate from steel pickling liquor. In each instance all components will permeate the membrane; separations are effected by virtue of the differences in respective diffusion rates. Copper transfers at only one-tenth the rate of sulfuric acid, and nickel is slightly slower.

When the concentration of the diffusate is kept sufficiently low in relation to that of the retentate, the escape rate for a pure single solute should follow first-order kinetics. A comparison of the escape rates of glucose and sucrose is shown in Fig. 13-2. This type of chart is of considerable value in determining whether or not a given preparation behaves ideally and is homogeneous with respect to molecular size. Moreover, where the solutes give straight-line escape plots, the 50% escape time can be used for comparisons related to the sizes of the solute molecules. Conversely, where the sizes of the solute molecules are known, the ratio of the 50% escape times defines the selectivity of the membrane.

Dialysis is indispensable in the recovery and purification of materials in the chemical, food, biological, and pharmaceutical fields. In general, dialysis is indicated when salts must be removed from colloidal suspensions or when low molecular weight, but water-soluble, organic compounds are to be separated from higher molecular-weight species. Stauffer gives specific references in his excellent bibliography.[4]

[4] R. E. Stauffer: "Dialysis and Electrodialysis," in *Technique of Organic Chemistry*, A. Weissberger (Ed.), Vol. III, Interscience, New York, 1956.

ULTRAFILTRATION

Ultrafiltration, sometimes called *reverse osmosis,* is a process in which a solution is forced under pressure through a membrane with an accompanying separation of components. The driving force is the energy due to pressure difference. It is the solvent rather than the solute which moves through a membrane and this is against rather than with a concentration gradient. Any unwanted solutes and colloidal material unable to permeate the pores of the membrane are rejected (Fig. 13-3).

The usual membrane for ultrafiltration is an asymmetric cellulose diacetate type. It is constructed with a thin skin of polymer on a porous, spongy base. The pressure, in atmospheres, ranges from 135 to 300 to provide a flow rate ranging from 5 to 20 ml/hr/cm^2 on a cellulose acetate membrane with hydraulic semipermeability of 97 to 99%.

Filtration through very small pore-sized membranes (0.2 or 0.45 μ) is a means of cold sterilization for fluids, especially many heat-sensitive liquids which previously could not be sterilized at all. Draft beer in cans arose from this process; it can be sterilized without removing the esters which are the source of the "draft" flavor.

ELECTRODIALYSIS

Electrodialysis is a process in which electrolytes are transferred through a membrane as a result of the application of electrical energy. Although it resembles ordinary dialysis in its dependence upon membranes and in providing compartments and passages for the flow of solutions, the two processes differ from one another in equipment design, objectives, and membrane requirements. Energy is furnished to a system. It is possible to drive electrolytes from a dilute to a more concentrated solution. Compared with ordinary dialysis, the principal advantage of electrodialysis with non-

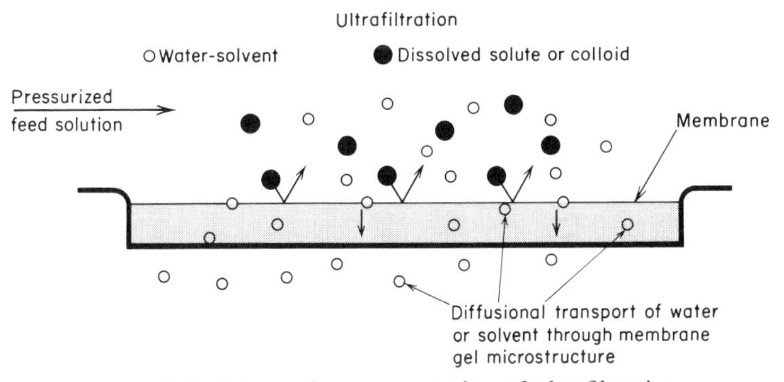

FIG. 13-3. Schematic representation of ultrafiltration.

selective membranes is that it may be used for the removal of the last traces of electrolytes from colloids. In conjunction with ion-selective membranes, an ionic material may be separated from a non-electrolyte and counter ions from co-ions.

Membranes

Compared to dialysis, a wider selection of membrane materials is possible for electrodialysis. Generally they are ion-selective membranes and, in particular, ion-exchanger membranes. The latter combine the ability to act as a separation wall between two solutions with the chemical and electrochemical properties of ion exchangers. The most important of these are the pronounced differences in permeability for counter ions, co-ions, and neutral molecules, and their high electrical conductivity. Low electrical resistance is fundamental to the economical use of electrodialysis. Good membranes have an area resistance of less than 20 Ω cm^2 in a uni-univalent electrolyte (about 0.6 *M*).

The cross-linked polystyrene-based structure is used as a basis for most of the modern types of ion-exchanger membranes.[5] Heterogeneous membranes consist of finely ground ion-exchange resin particles embedded in an inert binder which provides mechanical strength (Amberplex series, Rohm & Haas; Permaplex series, Permutit). If styrene and divinylbenzene are polymerized in the presence of an inert solvent, a homogeneous membrane results which can be made into unfractured sheet form. Subsequent sulfonation or chloromethylation followed by amination produces the ion-exchange groups. Synthetic fabrics are often embedded into these membranes to provide additional strength and ease of handling (Nepton series, Ionics, Inc.).

When in contact with electrolyte solution of low or moderate concentrations, an ion-exchanger membrane contains a large number of counter ions but relatively few co-ions. Counter ions are freely admitted to the membrane and have little difficulty in passing through it from one solution to the other. Co-ions, on the other hand, are rather efficiently excluded from the membrane. Thus, the membrane is permselective for counter ions. The permselectivity[6] gradually disappears when the solution concentra-

[5] F. Helfferich, *Ion Exchange*, McGraw-Hill, New York, 1962; Chap. 8 on "Ion Exchange Membranes," and Chap. 3, pp. 61–66.

[6] The permselectivity, *P*, of an ion-exchange membrane is a measure of the fractional increase in transport number for the mobile counter ion, compared to the ideal increase due to the presence of a perfectly selective membrane, and in relation to the transport number of the ion in the absence of an ion-exchange membrane:

$$P = \frac{t_M - t_S}{1 - t_S}$$

where t_M is the transport number for the mobile ion within the membrane and t_S is in the free solution. In general, the permselectivity of a given membrane approaches ideality, namely unity, for the permeable ion as its solution concentration decreases. Reasonable values are: over 0.95 in the 0.1*N* range, 0.90 in the 0.5*N* range, and 0.85 in the 1*N* range.

tions are increased and approach the concentration of fixed ionic groups in the membrane.

Two-Compartment Cells

The basic equipment design of an electrodialysis cell is illustrated by the two-compartment unit shown in Fig. 13-4. A cation-permeable ion-exchanger membrane divides the cell into the two compartments. An inert electrode is placed in each compartment. Schematically, the cell may be represented

$$\ominus \mathrm{Pt, H_2} \mid \mathrm{Na^+\ Sebacate^-} \mid \begin{matrix}\text{Cation-exchanger}\\ \text{membrane}\end{matrix} \mid \mathrm{Na^+\ Sebacate^-} \mid \mathrm{O_2, Pt}\oplus$$

The electric current transfers ions across the membrane; however, with a permselective membrane, there is transference only of the counter ion (Na^+). The reactions produced by the electric current are

$$\begin{matrix}\text{Cathode} & \text{Membrane} & \text{Anode}\\ 2H_2O + 2e^- \rightarrow H_2 + 2OH^- & \leftarrow\!\!\Vert\!\!-Na^+ & 2H_2O \rightarrow O_2 + 4H^+ + 4e^-\end{matrix}$$

Each faraday of electricity passed through the cell produces one equivalent of hydrogen gas (at the cathode) and oxygen gas (at the anode), which escape into the atmosphere, and one equivalent of NaOH in the cathode compartment and one equivalent of sebacic acid, which precipitates due to its slight water solubility, in the anode compartment. Of significance is the production (after a sufficient lapse of time) of pure sebacic acid uncontaminated by added salt, as would be the case if prepared from its salt solution by addition of sulfuric acid.

Unless the acid formed is undissociated or is insoluble and precipitates, hydrogen ions will also be transferred across the membrane.

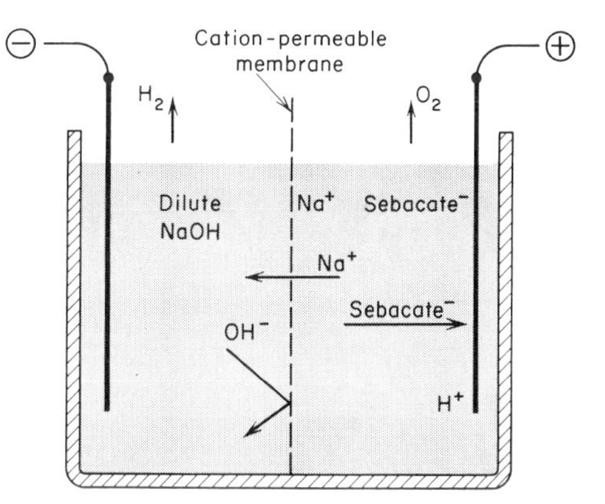

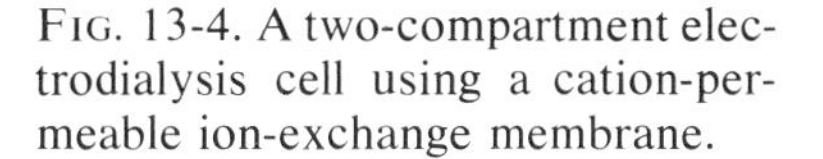
FIG. 13-4. A two-compartment electrodialysis cell using a cation-permeable ion-exchange membrane.

Multicompartment Cells

Electrodialysis can also be carried out in cells with more than two compartments and more than one membrane. Consider a three-compartment cell

	Cation-exchanger membrane		Anion-exchanger membrane	
Cathode | Na^+Cl^-	‖	Na^+Cl^-	‖	Na^+Cl^- | Anode

With the permselective cation-exchanger membrane there is transference only of sodium ions across the membrane (in the direction of the cathode). Conversely, only chloride ions are transferred through the anion-exchanger membrane (to the anode compartment). Ultimately, the net effect is the depletion of the electrolyte content of the central compartment.

Cathode compartment	Cation-exchanger membrane	Central compartment	Anion-exchanger membrane	Anode compartment
$2Cl^- \rightarrow Cl_2 + 2e^-$	$\longleftarrow\!\!\!\Vert\!\!\!-$	$Na^+ \quad Cl^-$	$-\!\!\!\Vert\!\!\!\longrightarrow$	$2H_2O \rightarrow O_2 + 4H^+ + 4e^-$

The nature of the electrodes is of minor importance since the electrode reactions have little effect on the depletion in the center compartment. Consequently, this arrangement can be utilized to isolate an electrode from a given electrolyte when undesirable reactions might otherwise occur.

In multicompartment cells with alternating cation- and anion-exchanger membranes in series, the electric current causes electrolyte depletion in every other cell, namely, in the cells with cation-exchanger membranes on the cathode side, and electrolyte accumulation in the cells in between.

Applications

Electrodialysis is applicable to processes where the concentration of electrolytes or the deionization of solutions is desired. It is being widely used for desalinization of brackish water and sea water. Compartments, several square feet in cross section and less than an inch thick, are separated by a series of alternating cation- and anion-exchange membranes. A single pair of electrodes provides the electric field. This approach is illustrated for sodium chloride in Fig. 13-5.

The method depends on the existence of membranes of sufficient ion-exchange capacity coupled with small enough effective pore size to electrostatically repel the ion of opposite charge. The maximum effective pore size for the coulombic repulsion to take place is approximately 30 Å;

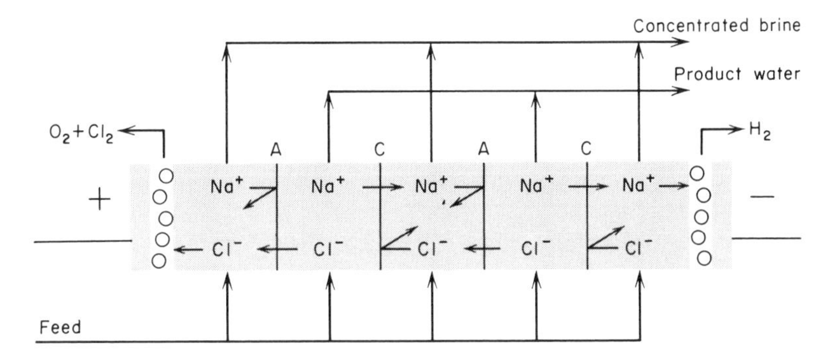

FIG. 13-5. Multicompartment electrodialysis unit for desalinization of brackish waters.

most ion-exchange membranes made commercially have effective pore sizes of 7–15 Å.

When the electric field is applied to an electrodialysis "stack" containing saline water, the cations move into or stay in the concentration compartment by being drawn to the cathode until repelled by an anion-exchange membrane. Conversely, the anions move toward the anode until stopped by a cation-exchange membrane. The result is the formation of an alternating series of dilute and concentrated compartments which are manifolded together by proper construction of equipment. The rate of hydraulic flow and the current density are regulated to yield the degree of desalinization desired. If the current density is too high for the rate of migration of the solute ions to carry the electrolytic current, the deficiency is made up by the electrolysis of water. Most of the voltage drop is normally across the dilute compartments, not the membranes. In practice desalinization is not carried out beyond less than about 300 ppm[7] because of the resistance of more dilute solutions and the competitive use of ion-exchange resin columns in the range of lower salinity.

The method is used in the dairy industry to process whey and thereby remove minerals from the whey fraction of cow's milk. In one installation four 200-membrane stacks, containing alternating cation- and anion-exchange membranes, reduce salt content from 6000 ppm to 300 ppm. About 60 gallons of whey are desalted each minute. Power requirements are high—between 200 and 400 V, with a current of 120 A.

In the photographic industry electrodialysis is used to produce photographic emulsions. Silver ions transfer through cation-permeable membranes and bromide or chloride ions through anion-permeable membranes into a gelatin stream. Silver halide precipitates, forming the emulsion.

[7] The salinity threshold is about 800 ppm for most people.

ELECTRODECANTATION

A close relative of electrodialysis is electrodecantation. Also called electrogravitational separation, electrodecantation is based on a stratification phenomenon that may take place when colloidal dispersions are subjected to an electrical field between vertical membranes permeable to the electrical current and impermeable to the colloid. The difference in density created by the transport depletion at the membrane causes a film of more dense liquid to fall on the concentrating side. Semicolloids and electrolyte also stratify in a similar manner on selectively permeable membranes. This difference in density based on concentration is augmented in the case of ionic solutes by the thermal effects of electrical resistance. With colloids the charged particles are retained and accumulated on the membranes.

The mechanism of electrodecantation is illustrated in Fig. 13-6 for the desalinization of brackish water. The ion-exchange membranes hang freely without gaskets, spacers, or manifolds—the chief advantage of electrodecantation over electrodialysis. A boundary layer of solution rich in ions and colloids builds up at the anodic membrane and sinks to the bottom, while a boundary layer of relatively pure water at the cathodic end of the cell rises to the top. This sets up a countercurrent circulation, and purified water can be withdrawn at the top of each cell. Feed enters the center of each cell. Illustrative concentrations and velocities in Fig. 13-6 are shown for 50 sheets of membrane 8 in. square separated by compartments 1/4 in. thick. The moving films are estimated to be 1/16 in. thick.

The behavior of the colloidal material retained on the membrane depends upon the migration velocity of the particles in the electrical field and the velocity of the gravitational movement of the enriched layer. If the

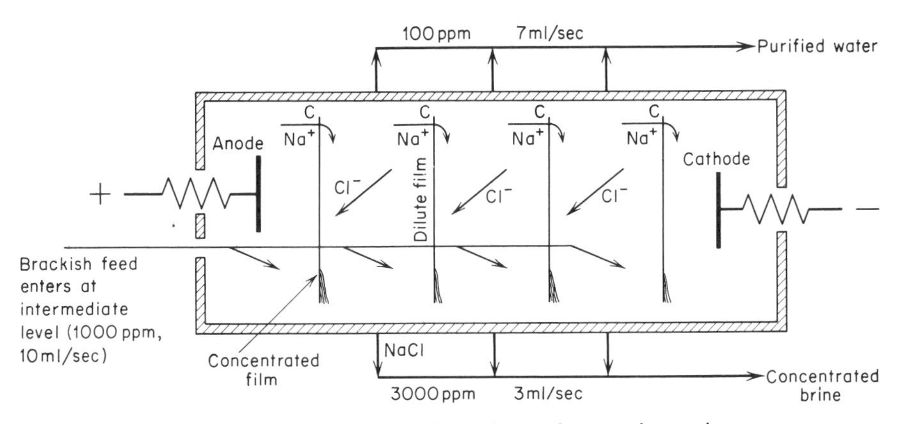

FIG. 13-6. Schematics of an electrodecantation unit.

latter is greater, stratification will take place continuously, but if the migration velocity is greater a deposit will accumulate on the membranes. If this deposit were allowed to remain in contact with the membrane, coagulation would result. This can be prevented by reversing the polarity of the electrodes periodically. At current reversal, the deposit moves away from the membrane. Being thus freed from the frictional resistance, it immediately stratifies.

Applications

Electrodecantation has been used for many years to concentrate (i.e., dewater) natural rubber latex. Any membrane can be used which discriminates between water and a colloid; for example, neutral membranes such as cellophane (cellulose xanthate) or regenerated cellulose from cellulose nitrate. In a similar manner tetrafluoroethylene (TFE) polymers are being concentrated from 4% to 55–75% solids. The Teflon TFE concentrates are used for coating cookware.

Multi-membrane electrodecantation units are available commercially (Quickfit Reeve–Angel). Each of the separation cells contains a membrane pack composed of 22 membranes which are separated by 0.8-mm spacers. Cells can be connected in series.

Electrodecantation can be used for protein separation. The method depends on the differences in isoelectric points of the various proteins in a mixture. At its isoelectric point, a protein in a buffered solution will not migrate under the influence of an electric field. Proteins in a mixture other than the isoelectric protein will migrate toward one or the other electrode until a membrane is encountered. In the thin layer next to the membrane, concentration of the unwanted proteins occurs, causing their decantation to the bottom of the cell, while the protein at its isoelectric point remains in solution. For example, the method is used to free gamma globulin from components in blood serum and operates with a buffer of pH 8.4, the isoelectric point of gamma globulin.

GAS PERMEATION THROUGH SOLIDS

When a membrane is permeated by a gas, the condensed penetrant dissolves into the surface layers, migrates under a concentration gradient via the interstices of the film, and evaporates from the opposing surface downstream.[8] The actual condensation of gas within the polymer occurs as adsorption or chemisorption. The condensation occurs within the amorphous regions of the polymer in which case true solutions may be formed. Neglect-

[8] C. E. Rogers, "Permeation," in E. Baer (Ed.), *Engineering Design for Plastics*, Reinhold, New York, 1964.

ing van der Waals' effects, the rate of transport is inversely proportional to the square root of the molecular weight.

The classical separation in this category is that of hydrogen through heated palladium, platinum, and certain other metals which permit only hydrogen and its isotopes to pass through, separating the hydrogen from other gases and from impurities. This action is due to catalysis of the hydrogen dissociation to protons and electrons which pass through the metal sheet and recombine on the other side to form molecular hydrogen. The diffusion rate through a palladium–silver alloy is superior. Hydrogen as pure as 99.999% can be produced in commercial units handling up to 20 tons per day and based on hydrocarbon–hydrogen and hydrogen–nitrogen mixtures.

High-silica glass can preferentially pass helium. This is due to voids in the glass about the same size as the helium atom (2.7 Å). As temperature increases, the voids increase in size and neon (2.8 Å) and hydrogen are passed. This approach is utilized for the separation of helium from other gases for cryogenic and nuclear applications.[9]

Helium can be recovered from natural gas through the use of Teflon FEP membranes. Teflon FEP is 25 times more permeable to helium than to nitrogen and 44 times more permeable to helium than to methane.

A silicone rubber membrane, 0.001 in. thick, will produce enriched (40% oxygen) air for hospital and aerospace uses. The permeability of a Teflon polymer membrane to oxygen dissolved in liquids is the basis of a polarographic method. The oxygen-sensing probe is an electrolytic cell with a gold cathode separated from a tubular silver anode by an epoxy casting. The anode is electrically connected to the cathode by electrolytic gel, and the entire chemical system is isolated from the environment by a thin gas-permeable membrane. The oxygen in the sample diffuses through the membrane and is reduced at the cathode with the formation of the oxidation product (silver oxide) at the silver anode. The resultant current is proportional to the amount of oxygen reduced. Operation range is from 0.2 to 50 ppm of dissolved oxygen.

A fluorocarbon membrane, which is normally hydrophobic, can hold back water until quite high pressures are reached. For example, a 0.2-μ membrane holds back 35 psi of water before it starts to flow; above this pressure the water flows at normal rates. This water-repellent property can be used to separate water from hydrocarbons. By using the correct pore-size membrane, the breakthrough pressure can be varied and controlled.

Porous membranes whose pore size is large compared to molecular size, but approaches the mean free path of the gas, are capable of making gas separations based upon diffusion. These rates are inversely proportional to

[9] K. A. McAfee, "Diffusion Separations," in 2nd ed. of *Kirk-Othmer Encyclopedia of Chemical Technology,* Reinhold, New York, 1963.

the square root of the molecular weight of the diffusing gas. The gaseous diffusion plants for the enrichment of uranium-235 are based on this principle.

Problems

1. Design a two-compartment electrodialysis cell for the preparation of pure ethylenediamine from ethylenediamine HCl.
2. Suggest an electrodialysis cell for the quantitative removal of iron, copper, lead, and cadmium salts from suspensions of SiO_2 and WO_3.
3. Suggest a method for preparing NaOH solution free from borate ion, or for determining the boron content of sodium metal.
4. Outline the method of freeing protein hydrolysates from salts by electrodialysis.
5. Design a multi-compartment apparatus for carrying out the separation of aluminum and magnesium ions. [Hint: Ammonium citrate buffer (pH 7.5) should be added to the sample compartments.]
6. Dialysis rates of selected substances are tabulated; τ is the 50% escape time. The two sections of the table contain data from different membranes.

Substance	Molecular weight	τ (hr)
Xylose	150	1.3
Arabinose	150	1.9
Glucose	180	3.5
Galactose	180	4.8
Sucrose	342	30.0
Sucrose	342	1.5
Lactose	342	2.6
Raffinose	504	5.4
Stachyose	666	15.0

   (a) Express the comparative escape rates through each dialysis membrane (two graphs) of each solute as a plot of percent remaining (logarithmic scale) vs the time (in hr). (b) Assuming that the concentration of diffusate is kept sufficiently low in relation to that of the retentate, and that a straight-line escape rate prevails, how long should the dialysis be continued with membrane 1 to lower the concentration of glucose in sucrose to 1.0%? What percent of sucrose remains in the retentate after this time interval? (c) Comment on the escape rates and molecular configuration for these pairs: xylose and arabinose; sucrose and lactose; and xylose and glucose. (d) Plot the escape rate vs the molecular weight for mono-, di-, and tri-saccharides.
7. Devise a method for the removal of citric acid from tart citrus juices by means of electrodialysis.
8. By means of electrodialysis, suggest a method for these metatheses: (a) NaOH plus $CaCl_2$ from NaCl and $Ca(OH)_2$; (b) HBr plus KCl from KBr plus HCl.

BIBLIOGRAPHY

Carr, C. W., "Dialysis," in G. Berl (Ed.), *Physical Methods in Chemical Analysis,* Vol. IV, Academic Press, New York, 1961.

Craig, L. C., "Dialysis," in *Encyclopedia of Polymer Science and Technology,* Vol. 4, Interscience, 1966.

"Dialysis," in *Encyclopedia of Chemical Technology,* 2nd ed., Vol. 7, p. 1, R. E. Kirk and D. F. Othmer (Eds.), The Interscience Encyclopedia, Inc., New York, 1965.

"Electrodialysis," in *Encyclopedia of Chemical Technology,* 2nd ed., Vol. 7, p. 846, R. E. Kirk and D. F. Othmer (Eds.), The Interscience Encyclopedia, Inc., New York, 1965.

Stauffer, R. E., "Dialysis and Electrodialysis," in *Technique of Organic Chemistry,* Vol. III, part I, A. Weissberger (Ed.), Interscience, New York, 1956.

Tuwiner, S. B., *Diffusion and Membrane Technology,* Reinhold, New York, 1962.

14 *Electrophoresis*

Electrophoresis—from the Greek "borne by electricity"—may be defined as the migration of charged species in solution under the influence of an electric field. When dispersed in an aqueous solvent, practically all substances acquire either a positive or negative charge. Principally, these arise from complete or partial ionization of the substance, from the adsorption on dispersed solids of ions or charged substances in the solution, or from ion-pair formation. Each charged species moves along the field gradient at a rate which is a function of its charge, size, and shape. Hopefully, after a suitable period of time the charged species will be separated into distinct mobility classes, forming separate zones. The position of the resultant bands or zones is observed or detected by appropriate methods. In principle, electrophoretic separation can be applied to any mixture in which the components carry a charge and have different mobilities in an electric field. It should, however, be emphasized that two components which show the same rates of migration are not necessarily identical.

The different varieties of electrophoresis are divided into three main classes: free-boundary electrophoresis, electrophoresis in a fixed medium or zone electrophoresis, and continuous electrophoresis (electrochromatography). The major portion of this chapter will be devoted to zone electrophoresis.

FREE-BOUNDARY ELECTROPHORESIS

The moving-boundary method[1] allows the charged species to migrate in a free-moving solution. The sample is fractionated in a U-tube that has been filled with unstabilized buffer. Usually the sample is carefully injected

[1] A. Tiselius, *Trans. Faraday Soc.* **33,** 524 (1937).

into the bottom of the U-tube through a capillary sidearm. The position of the moving ions forms a boundary which is detected by measuring changes in refractive index throughout the solution or by the Schlieren effect. High equipment costs and difficulties in isolating the separate zones as well as technical difficulties detract from its widespread use. It remains, however, the reference method for measuring electrophoretic mobilities. Biologically active fractions can be recovered after separation and without subjecting them to denaturing reagents or processes.

The Perkin–Elmer Model 238 Tiselius electrophoresis apparatus provides both Schlieren and interference optics. An integral thermoelectric cooling unit maintains the bath temperature at about 1°C without introducing the vibration of a mechanical refrigeration unit. The interferometric optical system makes it possible to detect components in a concentration as low as 0.05 mg per ml.

Density-gradient electrophoresis is a form of free-solution electrophoresis. In this modification,[2] fractionations are accomplished in a vertical cylinder which is filled with electrolyte containing an electrophoretically inert solute, such as dextrose, sucrose, or dextran. The density of the solution is graded so that the specific gravity diminishes toward the top of the cylinder. A sample is layered on this solution and its components forced to migrate against the density gradient. This arrangement serves to reduce convection currents, allowing more complete separation of sample components to be obtained.

ZONE ELECTROPHORESIS

Electrophoresis in a supporting medium (i.e., zone electrophoresis) avoids the complexity of the moving-boundary method. The principle is the same, however, with a gel or porous material used as a stabilizing support. The supporting medium is soaked with an electrolyte (usually a buffer) and a sample of the material to be separated is placed near the center of a horizontal or vertical piece of the supporting medium. Upon application of a controlled d-c source of potential to the ends of the supporting medium, the different migrant molecules begin to move. Each rapidly reaches a constant velocity through the stabilizing medium. As separation proceeds, the original zone breaks up into several discrete zones, hopefully one for each component in the original mixture. At the completion of a run, the electrophoretic zones remain fixed in the supporting medium where subsequently the zones may be detected as on a chromatogram. A principal function of the supporting medium is to keep convection currents from distorting the electrophoretic pattern.

[2] H. Svensson and E. Valmet, *Science Tools* **2,** 11 (1955).

Instrumentation

The essential components of an apparatus for zone electrophoresis are shown diagrammatically in Fig. 14-1. Separations may be performed either in the vertical or horizontal plane; only the latter is illustrated. The area upon which the electrophoretic separations occur, called the *bed,* can be composed of any of a number of supporting media. It is moistened with an electrolyte solution, usually a buffer. The ends of the bed are immersed in more of the electrolyte contained in two chambers designed to hold electrodes that are connected to a d-c power supply. Carbon electrodes, or electrodes composed of platinum foil bonded to plastic, are separated from the bed by diffusion barriers. These may be made of gel, filter paper, sponge, or other electrically permeable material. The purpose of the barrier is to prevent diffusion and convection processes from carrying electrolytic decomposition products into the bed. Provision is made for adjusting the electrolyte in the electrode chambers to equal levels so that siphoning action does not occur through the bed that could wash the sample away from the site of application. The entire apparatus, excluding the power supply, is enclosed in an airtight chamber of clear plastic to prevent excessive evaporation of buffer. Clearance between the bed and the cover need only be a few millimeters. The bed may be mounted on a cooling block through which tap water or a mechanically refrigerated liquid is circulated. If it is not desirable to have paper or cellulose acetate strips in contact with a base plate, a

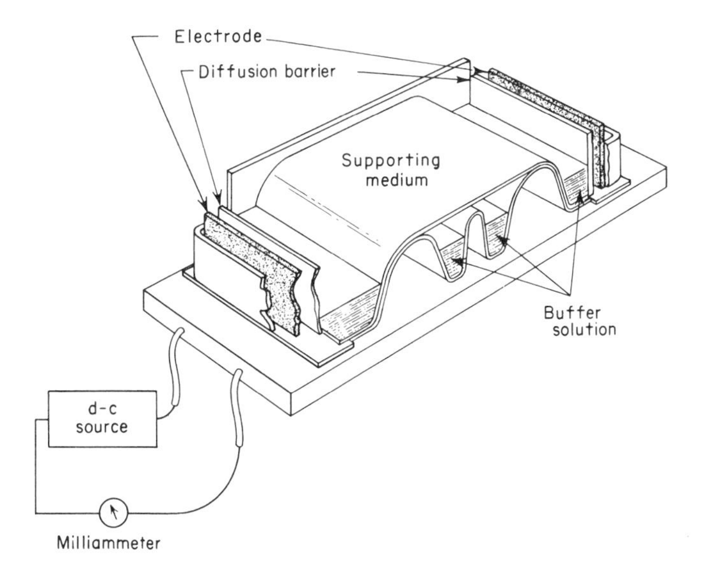

FIG. 14-1. The essential components of an apparatus for zone electrophoresis.

central bridge or rod is placed under the strip. Suspension in this manner keeps the strips from sagging and prevents buffer from accumulating at the low point which, if it occurred, would cause serious distortion of the electrophoretic pattern. When separations are made in gels or powders, the solid media are confined to the bed surface by sides which extend above the surface and by removable gates placed across the ends of the bed. A cover of plastic wrap prevents evaporation of electrolyte.

Power Supplies The power source should be capable of providing either a constant voltage or a constant current. For conventional electrophoresis the voltage gradient most likely to be used lies between 5–10 V/cm; this requires a power supply capable of developing between 50–150 V. When separation alone is desired and no exact knowledge of mobilities is needed, almost any d-c power supply, even a half-wave rectifier, will perform satisfactorily. With paper or cellulose acetate the current drain is less than 1 mA per strip, so the power requirement is very low. For many purposes a number of dry-cell batteries connected in series provide adequate voltage and will give several months of service before replacements become necessary. It should be noted that if two strips are run in parallel the current required will be double that required for one strip alone. The voltage will be the same, however, irrespective of the number of strips employed.

"High voltage" is an arbitrary term; it is usually applied to the voltage gradients lying between 10–100 V/cm. True bed voltage gradients can only be obtained by making measurements between points in the working bed. Resistances that cause voltage drops elsewhere in the apparatus contribute to the problems of heat production without enhancing migration rates.

Factors Affecting Movement of Ions

A charged particle, when dissolved or suspended in a conducting liquid medium and subjected to the influence of a uniform electrical field, attains a constant rate of migration. The positively charged species will move towards the cathode and the negatively charged species will move towards the anode. The rate of movement of both species will be determined by the motive force to which they are subjected. This will be the product of the net charge on the species and the field strength, X, defined as

$$X = E/s \tag{14-1}$$

where E is the potential difference in volts between test probes inserted into the supporting medium and spaced s centimeters apart. The motive force is opposed by the frictional force encountered by each charged species as it moves through the supporting medium, this being largely determined by the physical size and geometrical configuration of the species and the viscosity of the solution. The rate of migration of a charged species can therefore be said to depend on the field strength, the shape and size of the species, and the charge carried by the species. The characteristics of the supporting medium also exert their influence on the migrant.

Ion Mobilities The electrophoretic mobility μ of an ion is defined as the linear distance d travelled by the migrant, relative to the stabilizing structure, in time t (i.e., its velocity) in a field of unit potential gradient, or

$$\mu = (d/t)/(E/s) \tag{14-2}$$

The units are in cm/sec per V/cm or $cm^2\ volt^{-1}\ sec^{-1}$. From Eq. (14-2) it is seen that the velocity of electrophoretic migration is directly proportional to the field strength. Increasing the field strength will speed fractionation. Apart from expediency, it will also enhance the quality of resolution because the decreased running time minimizes the opportunity for diffusion of the fractions. However, the field strength permissible is determined by the amount of heat that can be removed from the supporting medium by the cooling system.

The limiting mobility μ_0 at zero ionic strength may be calculated from the limiting equivalent conductance λ_0 of the ion:

$$\mu_0 = \lambda_0/\mathscr{F} \tag{14-3}$$

$\mathscr{F}$ being the number of coulombs (96,487) in 1 faraday. Unfortunately, comparatively few equivalent conductance values of organic ions are recorded in the literature. Calculation[3] from an empirical modification of the Stokes equation yields the expression:

$$\mu_0 = (1.602 \times 10^{-12} z)/5\pi r\eta(f/f_o) \tag{14-4}$$

for an ion of valence z and van der Waals' radius r (in angstroms) in a medium whose viscosity is η. The frictional ratio (f/f_o) of an ion allows for its shape when not spherical, but prolate or oblate. For ions in water at 25°C, the mobility in $cm^2\ sec^{-1}\ volt^{-1}$ is

$$\mu_0 = (1.14 \times 10^{-3} z)/r(f/f_o) \tag{14-5}$$

Thus, the mobility of an ion is a function of its charge, size, and shape. In solutions of finite ionic strength, the ionic mobility of a monovalent organic ion is reduced by about 6–9% when the concentration of a solution is increased from 0 to $0.01M$, and reduced by 20% when it is increased to

[3] J. T. Edward in J. C. Giddings and R. A. Keller (Eds.), *Advances in Chromatography*, Dekker, New York, 1966, Vol. 2, Chap. 2.

$0.1M$. For divalent organic ions, a 36% reduction may be assumed. Relative ionic mobilities will be only very slightly affected.

Resolution Rearranging Eq. (14-2), the distance travelled by a particle is

$$d = \mu t(E/s) \tag{14-6}$$

To separate two components, it is necessary to permit migration to continue until one component has travelled at least one thickness of the volume that the sample initially occupied (the starting zone) farther than the other. However, the sharpness, and therefore the resolution, of the zones occupied by each migrant diminishes with time because of the spreading of the zones as a result of diffusion. The separation of two components is facilitated when their mobilities differ widely. After time t the difference in the distance travelled is equal to

$$d_1 - d_2 = (\mu_1 - \mu_2)(E/s)t \tag{14-7}$$

Since diffusion is time-dependent, being proportional to the square root of time, its effect can be minimized by applying the highest possible potential gradient. Remarkable resolution has been achieved when advantage is taken of the frictional properties of the gels (such as agar, starch, and polyacrylamide) to aid separation by sieving at the molecular level.

Buffer pH and Ionic Strength The pH of the electrolyte greatly affects electrophoretic mobility since the net charge carried by most species is pH-dependent. When the sample is a weak acid or base, the pH establishes its degree of ionization. The apparent or net mobility μ' of a partly ionized solute in solution is given by the expression

$$\mu' = \mu K_a/([H^+] + K_a) \tag{14-8}$$

where K_a is the ionization constant. Thus, if one measures μ for a substance under conditions where ionization is substantially complete and then μ' at a known pH at which ionization is not complete, the ionization constant of the substance may be calculated. The extent to which two incompletely ionized acids may be separated depends upon the difference in their apparent mobilities [Eq. (14-7)]. The difference is at a maximum [4] when

$$[H^+] = (K_{a,1}K_{a,2})^{1/2} \left[\frac{(\mu_1/\mu_2)^{1/2} - (K_{a,1}/K_{a,2})^{1/2}}{1 - [(\mu_1/\mu_2)^{1/2}(K_{a,1}/K_{a,2})^{1/2}]} \right] \tag{14-9}$$

When $K_{a,1} > K_{a,2}$ and except when $K_{a,1}/K_{a,2}$ lies between the values of μ_1/μ_2 and μ_2/μ_1, and if the optimum pH is used, the maximum difference

[4] R. Consden, A. H. Gordon, and A. J. P. Martin, *Biochem. J.* **40,** 33 (1946).

in apparent mobility is

$$(\mu_1' - \mu_2') = \frac{\mu_2\left[\left(\frac{\mu_1 K_{a,1}}{\mu_2 K_{a,2}}\right)^{1/2} - 1\right]^2}{(K_{a,1}/K_{a,2}) - 1} \tag{14-10}$$

Figure 14-2 shows $(\mu_1' - \mu_2')/\mu_2$ plotted against $pK_{a,2} - pK_{a,1}$ for various values of μ_1/μ_2, and the intersecting family of lines shows the pH which must be employed. It is not important that these pH values should be followed exactly; 0.1 pH unit makes little difference to $(\mu_1' - \mu_2')$. Further, in Fig. 14-2, where pH $= pK_{a,2} + \infty$ is shown, there is little advantage in going beyond pH $= pK_{a,2} + 2$. This applies also to the left of this line where the separation depends only on the difference in ion mobility. It is also possible in the case of substances in which μ_1/μ_2 is low, by working at a high pH, to separate them in the opposite direction to that shown in Fig. 14-2, using only the ion mobility difference. The sign convention adopted in the figure is suitable for the separation of anions; i.e., the acid with the greater pK_a is designated $pK_{a,2}$. For the separation of cations, the base with

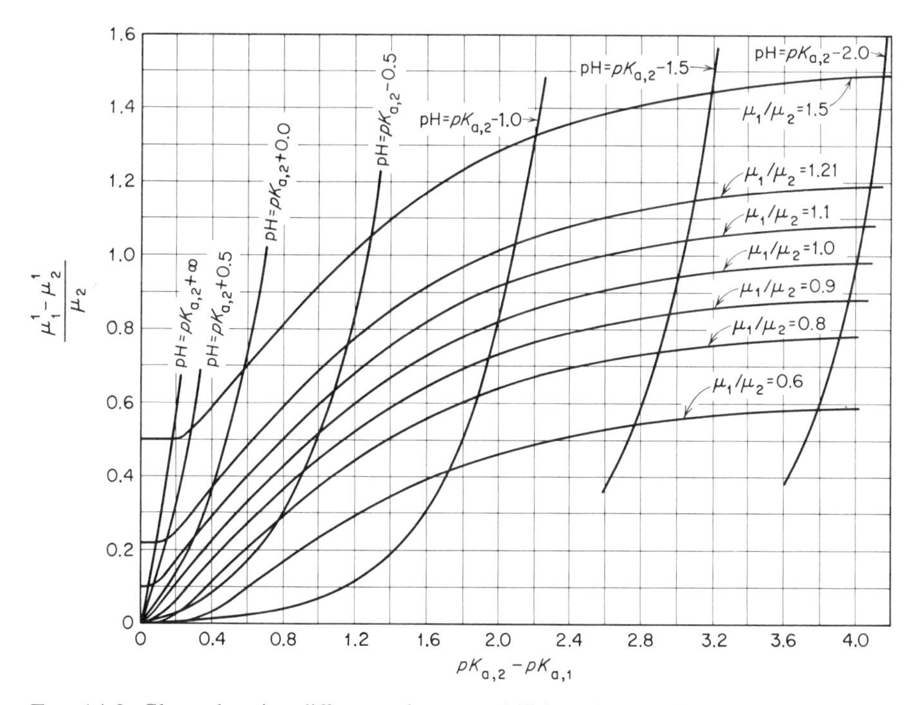

FIG. 14-2. Chart showing difference in net mobilities of two acids, for a difference in pK_a and given mobility ratio, when separating at the optimum pH, the value of which is given by the intersecting family of lines. After Consden, Gordon, and Martin, *Biochem. J.* **40,** 33 (1946).

the greater pK_a must be called solute 1. This rule holds also for ampholytes. Thus, in separating glycine and serine in alkaline solution, the ions are anions and $pK_{a,2}$ must be greater than $pK_{a,1}$, even though the separation depends on differences in the dissociation constants of the amino groups.

Ionic mobilities of most amino acids are not available. An estimate may be made, however, by assuming them to be the same as the corresponding hydroxy or fatty acid; an extensive table of such values is given in the International Critical Tables. Appendix B contains selected or estimated values for several amino acids.

Example 14-1

What is the optimum pH for the separation of the binary mixture of the amino acids alanine ($pK_a = 9.87$; $\mu_0 = 2.67 \times 10^{-4}$) and methionine ($pK_a = 9.21$; $\mu_0 = 2.06 \times 10^{-4}$)?

Substituting the appropriate terms in Eq. 14-9, the optimum $[H^+]$ is

$$[H^+] = [(6.16 \times 10^{-10})(1.35 \times 10^{-10})]^{1/2} \times \left[\frac{(2.06/2.67)^{1/2} - (6.16 \times 10^{-10}/1.35 \times 10^{-10})^{1/2}}{1 - (2.06/2.67)^{1/2}(6.16 \times 10^{-10}/1.35 \times 10^{-10})^{1/2}}\right]$$

from which pH = 9.38.

Now, in a borate buffer whose pH is 9.4, what is the apparent mobility of methionine and alanine?

Utilizing Eq. (14-8) for methionine, the apparent mobility is

$$\mu' = (2.06 \times 10^{-4})(6.16 \times 10^{-10})/[(4.0 \times 10^{-10}) + (6.16 \times 10^{-10})]$$

$$= 1.25 \times 10^{-4} \text{ cm}^2 \text{ V}^{-1} \text{ sec}^{-1}$$

and for alanine

$$\mu' = (2.67 \times 10^{-4})(1.35 \times 10^{-10})/[(4.0 \times 10^{-10}) + (1.35 \times 10^{-10})]$$

$$= 0.67 \times 10^{-4} \text{ cm}^2 \text{ V}^{-1} \text{ sec}^{-1}$$

Finally, it is possible to estimate the migration distance during an electrophoretic run that would be conducted for 5.75 hr at a voltage gradient of 180 V per 53 cm and a current of 200 mA. From Eq. (14-6), and the knowledge of the charge borne by the methionine, the distance travelled towards the anode is expected to be

$$d = (1.25 \times 10^{-4})(5.75 \times 60 \times 60)(180/53) = 8.8 \text{ cm}$$

and for alanine

$$d = (0.67 \times 10^{-4})(5.75 \times 60 \times 60)(180/53) = 4.7 \text{ cm}$$

Quite likely, a correction for electro-osmosis must be made before the calculated and experimental distances are in agreement.

For cations electrophoresis is generally run in the intermediate pH range using complexing agents such as lactic, tartaric, or citric acids to ensure solubility. Most procedures for separating anions or weakly acidic substances, such as carboxylic acids and phenols, require alkaline solutions to avoid any possible existence of two ionization states in equilibrium. Conversely, weakly ionized bases such as amines and alkaloids should be separated at low pH values.

Usually the ionic strength of a buffer is adjusted to about 0.05 to 0.1, the optimum compromise between high mobilities (in solutions of low ionic strength) and sharpness of zones. This is accomplished by adding enough neutral salt to the solution to compensate for the diminution in the ionic strength of the buffer owing to change in pH. Directions have been published for preparing buffers at different pH values but at the same ionic strength.[5] It is desirable to keep ionic strength low because conductivity, and consequently power consumption, is a function of ionic strength. This means that the use of buffers with low ionic strength helps to minimize both the production of heat and the quantity of electrode products formed. Ionic strength has important effects also upon the solubility of macromolecules; many proteins become insoluble when the ionic strength is too low or too high.

Certain organic substances have no charge at any pH, but form electrophoretically mobile complexes with substances that have a charge. For example, sugars and polyalcohols form charged complexes in the pH range between 9 and 10 with a number of anions, including borate, molybdate, and arsenite. Structures can be established by comparison of mobilities with complexes of suitable reference compounds (often *d*-glucose). Many monohydric alcohols can be converted to the corresponding xanthates by brief warming with solid sodium hydroxide and subsequent treatment with carbon disulfide. Many aldehydes and ketones migrate in bisulfite-containing electrolytes. In all cases the complexing agent must be present in the buffer in large excess so that mass action maintains the compounds in the complexed condition.

Electrolysis The transfer of anions and cations into opposite electrode chambers during electrophoresis results in a gradual divergence of the pH in the chambers. The buffer at the cathode becomes progressively more alkaline, while that at the anode becomes more acidic. This effect, even with appropriate barriers, can eventually extend itself into the electrophoretic bed as a progressive change in pH which affects the mobility of the migrants.

These undesirable effects of electrolysis can be controlled by providing a sufficient volume of buffer in the electrode chambers. When effective anticonvection barriers are present between the electrodes and the bed, at

[5] P. J. Elving, J. M. Markowitz, and I. Rosenthal, *Anal. Chem.* **28,** 1179 (1956).

least 1% of the buffering capacity can be lost through electrolysis without affecting the electrophoretic pattern to a noticeable degree.

The quantity of electricity consumed during a run determines the volume of buffer required; this is calculable by means of Faraday's law. The gram-equivalents Y of buffering substance that will be decomposed by any given current, i, during any time, t, of a run can be calculated as:

$$Y = it/26{,}802 \tag{14-11}$$

where the faraday is expressed as 26,802 mA-hr (rather than 96,487 coulombs or A-sec). Assuming that a 100-fold excess of buffer is required, the gram-equivalents of buffering substance that should be present in the electrode chambers are

$$100Y = 100it/26{,}802 = 0.00373it \tag{14-12}$$

Expressed as volume of buffer, V, of normality, N, this gives

$$V = 0.00373it/N \tag{14-13}$$

The volume of buffer required establishes the size of the electrode chambers.

Example 14-2

Let us assume that the buffer is $0.05M$ in the sodium salt of the weak acid, that the voltage gradient induces a current flow of 4 mA, and further that the time required for separation will be 3 hr. The volume of buffer needed for this operation is

$$V = \frac{0.00373\ (4\ \text{mA})\ (3\ \text{hr})}{0.05M} = 0.9\ \text{liter (or 900 ml)}$$

The run will require 900 ml of buffer divided evenly between the electrode chambers.

Electro-osmosis Another disturbing effect may be electro-osmosis. During the migration to the appropriate electrodes, hydrated ions carry with them their associated water molecules. This assemblage consists of water molecules held directly in the solvation sheath plus additional layers of water molecules attracted and held by the inner layers. Thus, it is possible for hundreds of solvent molecules to be dragged along by each migrating ion. The opposing flows caused by cations and anions seldom compensate because the ions differ in the amounts of water that they can carry with them. Inasmuch as cations usually carry more water than anions, the net flow of liquid is almost always toward the cathode. This frequently results in an apparent movement of neutral molecules towards the cathode. Because this movement varies approximately linearly with voltage gradient, it may be treated as a mobility correction to be added for anions and substracted for cations. The value of this correction may be determined by observing the migration of an uncharged substance such as acetone, dextran, glucose, urea, or deoxyribose. Electro-osmosis will, of course, vary with factors influencing the electrokinetic potential and viscosity of the solution, i.e., pH, buffer composition, ionic strength, and pretreatment of the electrophoresis bed.

In addition to affecting migration velocity, electro-osmosis causes spreading by chromatographic action. Most supporting media bind water. This bound water, along with water held mechanically in folds and cul-de-sacs of the supporting medium, forms a stationary water phase that extracts portions of sample from the fluid in electro-osmotic motion.[6]

Example 14-3

The rate of electro-osmosis can be estimated from the movement of molecules that are neutral at the pH of the buffer. The pK_a of caffeine is 0.15; hence, a net charge of zero above pH 3 is expected. The mobility of caffeine then should represent the rate of electro-osmosis. At any pH greater than 3, picrate ion ($pK_a = 0.8$) has one net charge and the mobility should be independent of the pH. These data were obtained for a migration time of 100 min at 18 V/cm on a 40-cm bed:

		Distance moved, cm [a]		
pH	Buffer	To anode (picrate) [b]	To cathode (caffeine) [c]	Net distance moved, cm
3.6	Formate	11.0	1.1	12.1
4.6	Acetate	10.9	1.8	12.7
6.8	Phosphate	9.0	2.8	11.8
9.2	Borate	8.1	4.5	12.6
9.9	Carbonate	9.6	3.7	13.3

[a] Corrected for difference in viscosity between buffer and distilled water.
[b] Referred to site of application as zero.
[c] Electro-osmotic rate.
[Taken from A. M. Crestfield and F. W. Allen, *Anal. Chem.* **27,** 422 (1955).]

The average net distance moved is 12.5 cm. From Eq. (14-6), the ionic mobility is

$$\mu = d/t(E/s) = 12.5/(100 \times 60)(18) = 1.16 \times 10^{-4}\ \text{cm}^2\ \text{V}^{-1}\ \text{sec}^{-1}$$

for the picrate ion.

Temperature The mobility of a migrant increases with increasing temperature. To intercompare results on separate runs, the temperature must be closely controlled. There exists a linear relationship for each

[6] Q. P. Peniston, H. D. Agar, and J. L. McCarthy, *Anal. Chem.* **23,** 994 (1951).

migrant when the logarithm of the mobility is plotted against the reciprocal of the absolute temperature.

The unavoidable electrical heating that accompanies electrophoresis has a number of adverse effects that can disrupt the electrophoretic pattern. Also, unsuspected sources of heat from surroundings cause distortions of patterns with the samples closest to the heat source moving more rapidly. Many of these difficulties can be circumvented by performing electrophoresis at reduced temperatures, about 4°C.

Evaporation of solvent affects the flow of liquid from the ends of the bed to the center, where salts tend to concentrate. A diminished voltage gradient arises in this region which greatly reduces the rate of electromigration. Evaporation from gels is more serious than from thin strips. Even with strips, evaporation from edges results in each zone assuming the shape of a "V" pointing in the direction of migration. When beds are thick, V-shaped patterns occur also because the absence of convection in gels results in a heated region along the longitudinal axis of the bed with cooler regions at the periphery. Since mobility increases with heat, the central portion forms a set of nested hollow cones.

Stabilizing Media

The properties of the stabilizing medium exert a marked influence on the resolution of the electrophoretic pattern. Stabilization by the solid medium is most effective when the pockets of liquid are smallest. The ideal support should approximate a series of capillary tubes tied together. Ion migration would then follow a straight path. The media should offer low resistance to current and be inert to sample components and dyes used for identification.

In stabilized electrophoresis, quantitative determination is feasible if chemical or physical methods are available. The separated zones can either be eluted and submitted to direct measurement, or the bed can be rendered transparent and the zones evaluated densitometrically. The elution method is considered to give better reproducibility whereas the direct densitometric method is speedier, expedient, and more suitable for routine analysis.

Filter-Paper Electrophoresis Paper was commonly used in electrophoresis in earlier years. Many kinds of filter paper can be used;[7] these include Schleicher & Schüll 2043A and B; Whatman 1, 2, and 3; Eaton-Dikeman 301-85, 320 and 652; and Munktells 20/50. The papers enumerated are suitable for making fractionations of samples up to 20 μl in volume. For larger volumes, an extra-thick paper, such as Whatman 3MM or 31 E.T., can be used. The additional thickness will support samples as large as

[7] B. S. Hartley, *Biochem. J.* **80,** 36 (1961).

60 μl in volume. Larger samples should be fractionated on a wide sheet of thin paper as a streak across the starting line. When the materials being used would react adversely with ordinary paper, a paper made from fibers of borosilicate glass, such as Whatman GF/B, may be used.

Advantages of filter paper include its cheapness and convenience. Paper stabilizes a relatively large volume of electrolyte solution per unit weight and can be efficiently cooled. In many respects, though, filter paper has been only a partially satisfactory support. Paper, with its fibrous or reticular structure, tends to reduce the resolution which can be obtained. Interactions between ionic sites on paper and migrating species (particularly if polar) cause tailing or blurring of the electrophoretic pattern. Electro-osmosis creates distortions in the separations because the ionic sites tend to develop a charge on the paper fibers. At the present time paper has been largely superseded by other materials. Paper, nevertheless, remains a useful medium for the separation of compounds with relatively low molecular weights.

Electrophoresis on Cellulose Acetate Microporous cellulose polyacetate is formed by chemically reacting cellulose acetate to tie up the hydroxy sites. This relatively inert cellulose ester forms strong and flexible membranes which possess a very uniform, foam-like structure. The pore size is closely controlled in the 1 μ range. As a consequence, finer resolution is achieved as compared with filter paper. The inertness of the membrane virtually eliminates adsorptive effects and trailing boundaries. Running times are roughly 20–90 min. Cellulose acetate can be cleared to glass-like transparency for densitometry by immersion in Whitemor oil 120 or simply by dipping it in a mixture of acetic acid–ethanol. Or, the bands may be cut out and dissolved in a solvent which extracts the colored species and dissolves the cellulose acetate. Cellulose acetate has a small sample capacity; 5 μl of sample constitutes heavy loading.

The disadvantages of cellulose acetate are: high electro-osmotic flow, a low water uptake, and the necessity for a low current density to prevent excessive solvent evaporation and drying out of the thin membrane. Separations must be carried out in a completely water-saturated atmosphere to ensure reproducibility of results. The chamber used for cellulose acetate electrophoresis must have electrode and buffer compartments of large surface area. A tight-fitting lid, covered on the under surface with a layer of sponge, is moistened with water prior to carrying out a run.

Electrophoresis in Gels A very important difference between electrophoresis in gels and in other media is the sieving effect that is unique to gels. For complicated mixtures, gel media have provided new tools of extremely high resolving power.

The bed is formed by casting the gel in a trough or tube that forms a bridge between the electrode chambers. Electrical contact between the chambers and the bed is achieved by wicks. One side of a paper tongue is laid on the layer to a distance of 0.5 to 1.0 cm while its other side is laid on the bottom of a compartment that contains a pad of material which is wetted with buffer and connected by a second wick to the buffer reservoir.

Excellent resolutions have been obtained using *starch gel*[8] but the preparation of beds using this medium is difficult.[9] Starch gels are formed from soluble potato-starch grains which have been ruptured by heating. Electrophoresis on starch gel can be carried out in the vertical or horizontal position, the choice between the two modes is determined by the method used to apply the sample. In the horizontal arrangement samples may be pipetted directly into a slit cut into the gel bed or applied as a slurry in starch grains or by absorption on filter paper placed in the slit. Substances of molecular weight less than 20,000 are not significantly sieved. The gel is easy to handle during staining and recovery procedures. It is rendered transparent by heating it slowly in glycerol until the gel becomes clear.

Agar is a popular medium even though it lacks the resolving power of starch gel. Of all the gels it is the easiest to use in forming a bed, partly because it is fluid in a convenient temperature range, and partly because of its relatively great mechanical strength in the solid state. This mechanical strength is also an asset in manipulating the completed electrophoregram, which with most gels is easily damaged during the process of staining and sample recovery. An agar layer can be prepared for scanning with a densitometer (or for storage) by soaking it in 5% glycerol and then allowing it to dry. The glycerol which acts as a plasticizing agent, causes the dried strip to resemble a piece of cellophane. Due to the acidic nature of agar, marked electro-osmosis takes place in most alkaline buffers.

Polyacrylamide gels are formed directly in the electrophoresis bed by polymerizing acrylamide monomers in the presence of *N,N'*-methylene-bis-acrylamide, the latter acting as a cross-linking agent. By varying the concentrations of the reagents, gels of different pore size can be formed. The electrophoresis bed may be a conventional trough or a small vertical column of gel. Application of the sample is effected in exactly the same way as described for starch gels when the polyacrylamide gel is cast in a horizontal bed. For columns the sample is applied directly on top of the gel as a suspension in starch grains or Sephadex (or Bio-Rad) beads. Electrophoresis in tubes is sometimes called *disk electrophoresis* because of the shape assumed by the fractions as they migrate in the cylindrical bed.[10] Advantage is taken of the adjustability of the pore size of the synthetic gel to produce auto-

[8] T. Johnson and O'N. Barrett, Jr., *Lab. Clin. Med.* **57,** 961 (1961).
[9] O. Smithies, *Biochem. J.* **61,** 629 (1955); *Arch. Biochem. Biophys. Suppl.* **1,** 125 (1962).
[10] See the series of articles in *Ann. N. Y. Acad. Sci.* **121,** Art. 2, pp. 305–650 (1964).

matically starting zones of the order of 10-μ thickness from initial volumes with thicknesses of the order of centimeters. High resolution is thus achieved in very brief runs. The method is capable of separating some 20 components in about 3 μl of serum in a period of actual electrophoresis of only 20 min. It is advisable to cool the gel during the course of electrophoresis since considerable heat is evolved during a run due to the high water content and low electrical resistance of the gel.

Electrophoresis in Powders Powder beds are formed by making a slurry of the powder in a buffer, pouring it onto a trough, and allowing the bed to set. Often materials used in thin-layer chromatography (*q.v.*) are employed. Since fine-grained powders can be used (the range of particle size lies between 40 and 250 μ), a homogeneous bed can be obtained that has excellent resolving power. Fractions can be excavated with a spatula, suspended in a solvent, and separated by centrifugation. Most of the techniques for manipulating these beds are borrowed from thin-layer chromatography. However, to obtain satisfactory patterns certain experimental precautions must be observed. After preparation in the normal way, the bed must be soaked with electrolyte by means of a wick applied from one end only. About 1 hr is required. Spraying is completely unsatisfactory. When mixtures of volatile solvents are employed to prepare a suspension, the thin-layer bed must be used immediately because the concentration changes with drying. When compounds with high water solubility are separated, freeze-drying the bed before the detection reagent is applied is necessary to avoid haphazard zone migration. The dry layer may be scored into strips by means of a scriber.

Preparative Zone Electrophoresis

So far the discussion has been confined to media which are capable of handling rather limited amounts of material. For preparative purposes it is necessary to increase the volume of the supporting media by using either large columns, blocks of powders, or granulated gels. A typical trough, $20 \times 10 \times 1.5$ cm, is constructed from glass or clear plastic. Troughs of greater capacity than this are not recommended since excess heating during electrophoresis then becomes a major problem. Starch blocks are prepared from a slurry of larger grains in a suitable buffer; after allowing time for the starch to settle, excess buffer is removed through paper wicks placed at both ends of the trough. Blocks can be prepared in a similar manner using slurries of finely powdered polyvinyl chloride resin, glass powder, or Sephadex beads in suitable buffers. Foam or sponge rubber in sheet form is cut into suitable dimensions, soaked in buffer, air bubbles removed by gentle squeezing under the surface of the buffer solution, excess buffer removed by blotting, and the strips assembled in the electrophoresis trough.

The apparatus shown schematically in Fig. 14-3 will accommodate samples in the range of 50 to 500 ml, containing 10 g or more of material. The separation chamber is formed by the annular space between two coaxial transparent plastic cylinders. It is open at the top, but is closed at the bottom by a porous plastic membrane. The outer column tube (11 cm) also serves as the inner wall of the cooling jacket, while inside cooling is provided for by the central tube. The separation chamber may be filled with powder to a height of about 70 cm. Supporting media include cellulose powder, granulated gels, and plastic powders. Liquid column volume varies from 25 to 50 ml per centimeter bed height. The column liquid can be kept stationary or it can be run off at a controlled rate. The latter technique is used to keep substances moving upwards rapidly within the column by counter flow or to collect those moving downwards through elution nozzles located circumferentially at the bottom of the column. After termination of the electrophoresis the compartment between the column is emptied, and the liquid containing the fractionated substances is run off at a controlled rate and collected. The advantage of column electrophoresis in large-scale preparations is that the components are subjected to the electrical field until they migrate to the end of the column.

Applications

The electrophoretic method is extraordinarily versatile. Proteins, viruses, clay suspensions, rubber emulsions, and colloidal particles may be

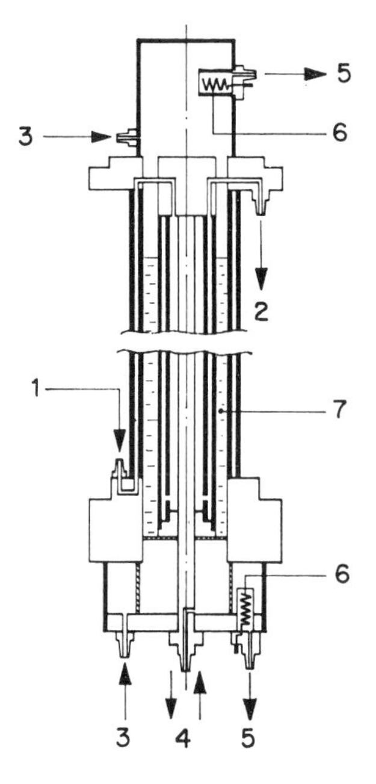

FIG. 14-3. Schematic drawing of apparatus for preparative zone electrophoresis. 1. Inlet for cooling water. 2. Outlet for cooling water. 3. Inlets for circulating electrode buffer. 4. Inlet for eluting buffer (during electrophoresis); outlet for eluate (after electrophoresis). 5. Outlets for circulating electrode buffer. 6. Platinum electrodes. 7. Separation chamber.

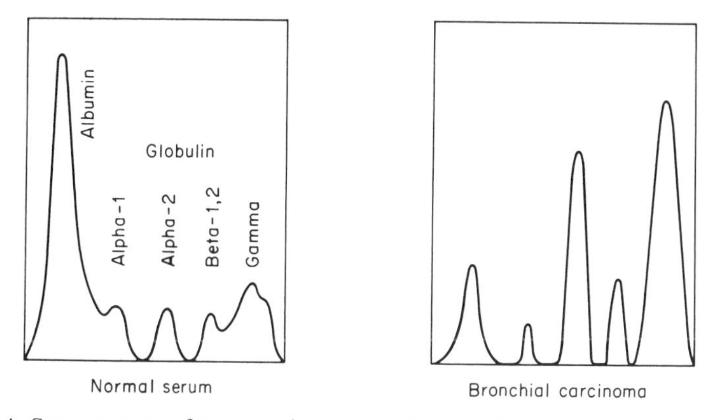

FIG. 14-4. Scan patterns for normal serum and serum from an individual with bronchial carcinoma.

separated; in fact, there is scarcely any class of compounds which has not been separated by zone electrophoresis. It has become indispensable to biochemists who have found it useful for fractionating an astonishing variety of biological materials, including body fluids and tissues. Electrophoresis is finding increasing use as a method for studying the interaction of substances. It is also being exploited for separating and identifying drugs for pharmaceutical and forensic purposes. The very high resolving power of the starch and acrylamide gels makes the technique of particular value in determining the purity of protein fractions, as in the purification of enzymes. Through the formation of complexes, such as the borates of polyhydroxy compounds, clues to the presence of particular molecular structures may be obtained. From variation of mobility with changes of pH, isoelectric points due to the presence both of acidic and basic groups may be indicated. By their effect upon the polyvalent cations, molecules with structures that form complexes may be detected.

Perhaps the greatest usefulness of zone electrophoresis has been in clinical diagnosis, for analyzing serum, urine, spinal fluid, gastric juice, and other body fluids. Many pathological conditions cause dramatic changes in serum protein fractions. For example, marked elevation of the β-globulin, or occasionally of the α_2-globulin fraction occurs in multiple myeloma. Figure 14-4 illustrates the scan patterns for normal serum and serum from an individual with bronchial carcinoma. Some illnesses can be traced to the malfunction of a particular organ by discovering through electrophoresis the presence of characteristic enzymes in the bloodstream. Still another phase of this versatile method concerns the genetic disturbances involving hemoglobin formation. The pattern of hemoglobin and hemoglobin S from patients with sickle-cell anemia is distinctive.

Separation of inorganic substances involves utilization of such properties as amphoterism, the ability to form complexes, and differences in valence state. In general, inorganic substances have larger charge-to-mass ratios than organic materials. At the same time the volumes of inorganic ions are relatively small, and thus the barrier effect of stabilizing media is less important than with macromolecules.

A combination of electrophoresis on thin layers, followed by chromatography in a second dimension, holds considerable promise. It combines the advantages of thin-layer chromatography with electrophoresis.

Combinations of zone electrophoretic methods with antigen–antibody reactions in transparent gel media are known as *immunoelectrophoresis*. Biological material is separated into fractions by electrophoresis and the fractions allowed to react with immune serum following their diffusion through the electrophoretic bed. Since this combines electrophoresis and diffusion with a highly specific detection technique, it is possible to detect many more fractions in a proteinaceous mixture by immunoelectrophoresis than by conventional electrophoresis. For details the literature should be consulted.[11]

CONTINUOUS ELECTROPHORESIS

When large volumes of sample must be handled it becomes desirable to use a continuous-flow system. The essential parts of a continuous electrophoresis apparatus are shown diagrammatically in Fig. 14-5. The heart of the system is an electrophoresis cell in which separations are performed by simultaneous flow of an electrolyte and electrical migration transverse to the solvent flow. In this method the solution to be examined is introduced continuously through a wick (or pen) at the top of the supporting medium. This medium, held vertically, may be a sheet of soft, thick industrial filter paper, a layer of starch, a bed of resin particles or a bed of any particulate material, supported between two plates of glass or clear plastic. The eluting electrolyte flows downward by gravity through the supporting medium. The electrical field is applied at right angles to this flow by means of a wick arrangement whereby the lower corners of the paper dip into the electrode compartment, or by means of a pair of flat platinum electrodes that clamp onto the vertical edges of the sheet. This form of electrophoresis is often called *curtain electrophoresis* because separations are accomplished in a sheet or curtain of buffer.

The sample stream, about 0.12 to 0.50 mm in width, is subjected to chromatographic adsorption and partition forces (and possible sieving action) in the descending electrolyte and to a lateral d-c field produced by the flanking electrodes. This splits the sample into identifiable fractions. The pathway of each individual fraction describes the vector of these two forces,

[11] P. Grabar and P. Burtin, *Immuno-Electrophoretic Analysis*, Elsevier, Amsterdam, 1964; M. D. Poulik, *Can. J. Med. Sci.* **30**, 417 (1953).

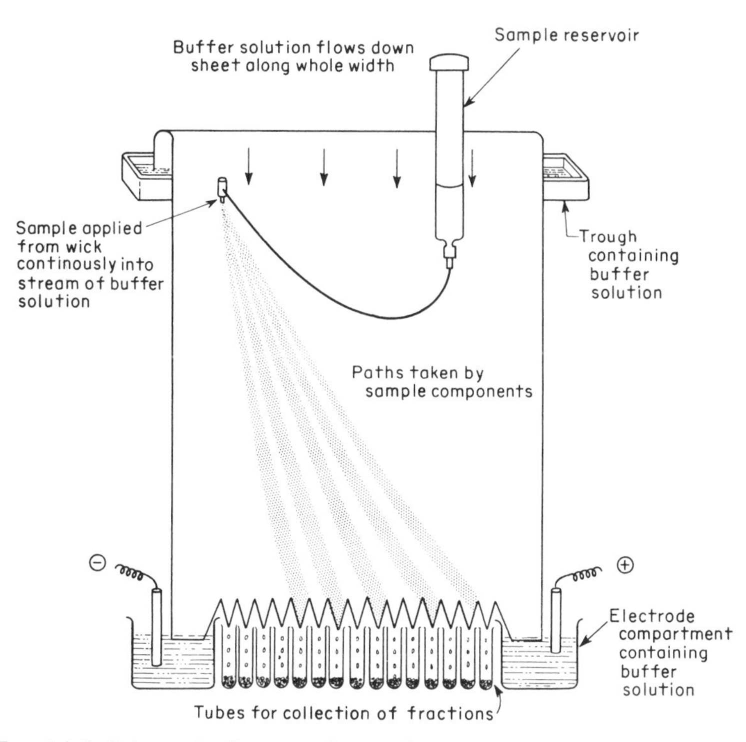

FIG. 14-5. Schematic diagram of a continuous electrophoresis apparatus.

causing the electrophoretic pattern to assume the form of a fan, with fractions emerging from the bed at different points along the bottom edge of the supporting medium.

By proper selection of buffer pH (and positioning of the sample wick), compounds can be fractionated into groups depending upon their charge. Neutral species move straight down the supporting medium, whereas positively charged species migrate toward the negative electrode and conversely the negatively charged components migrate toward the positive electrode. Pronounced differences in electrophoretic mobility and zeta potential lead to distinct migration paths even among species of the same charge type. In the presence of complex-forming reagents, many cations exhibit anionic properties, and with different reagents the order of separations can be changed. If certain substances follow the same pathway but at different rates, they would be separable by one-way chromatography or electrophoresis alone. The lack of separation is due to the counterbalanced effects of differential migration and chromatographic displacement.

It is possible to use a free solution by allowing a thin film of buffer to flow uniformly between two plastic plates. By making the film velocity high enough (1 cm per sec), the sample can be drawn out into a fine flowing line

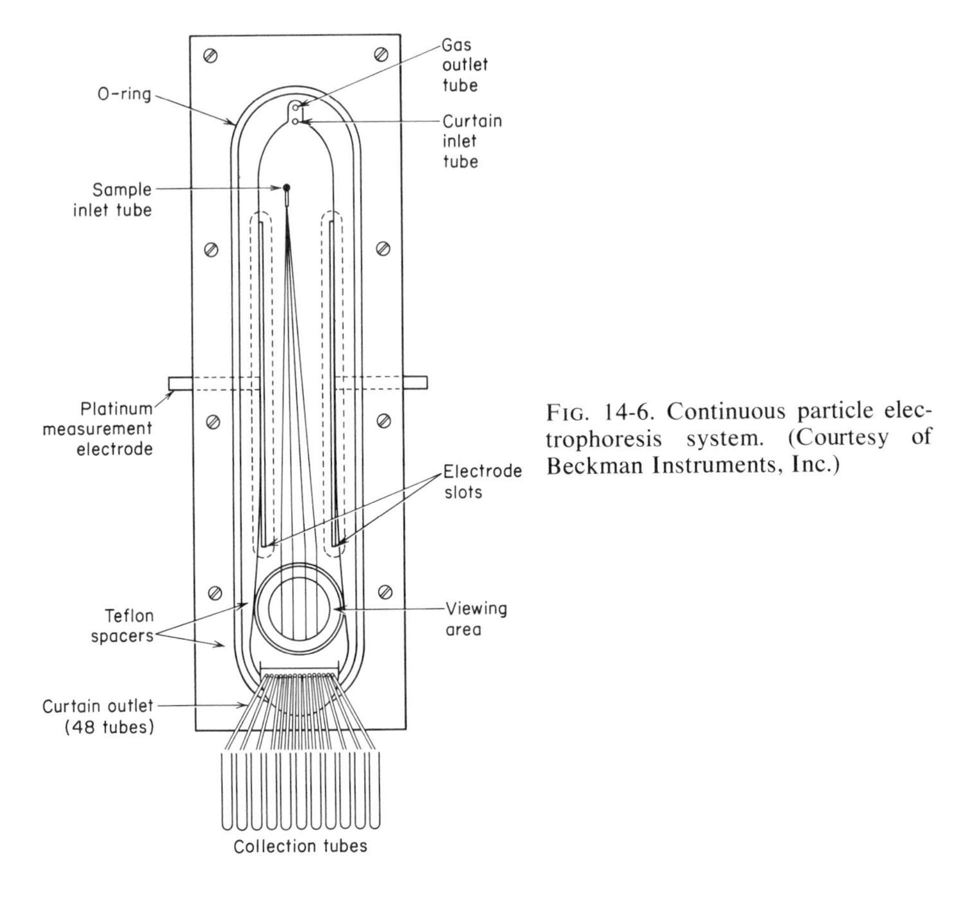

FIG. 14-6. Continuous particle electrophoresis system. (Courtesy of Beckman Instruments, Inc.)

as small as 0.12 mm wide. Lateral migration need last only long enough to separate the components by an amount equal to their band width. The fast-flowing and narrow curtain (25 mm wide) enables an electrophoretic band pattern to develop 24 to 30 sec after start of sample flow. The continuous particle electrophoresis system is shown in Fig. 14-6.

LABORATORY WORK

A number of experiments on cellulose polyacetate are described in *Techniques and Apparatus for Electrophoresis,* published by the Gelman Instrument Company, Ann Arbor, Michigan, 1968. A similar set of experiments in paper electrophoresis, using the Shandon Unikit, is described in J. G. Feinberg and I. Smith, *Paper and Thin Layer Chromatography and Electrophoresis,* Shandon Scientific Co., Ltd., London, 1965. The latter work is also to be found in *Advances in Chromatography,* Vol. 1, pp. 61–92, edited by J. C. Giddings and R. A. Keller, Dekker, New York, 1965. In a different manner, J. R. Sargent, *Methods in Zone Electrophoresis,* The British Drug Houses Ltd., Poole, Dorset, England, approaches the applications of zone electrophoresis from the viewpoint of the media used rather than the classes of compound separated. A fairly large bibliography is included with each supporting media.

Sample Application

To obtain the finest separations, the sample must be applied to the supporting medium in a straight, narrow, distinct band. This can be done with an applicator consisting of two evenly spaced, parallel stainless steel wires tensioned across a yoke the width of the strip. The applicator is loaded with a micropipet or capillary pipet; the sample is deposited by pressing the wires firmly against the supporting medium for a few seconds.

Cellulose Acetate

Hold the strip at both ends by means of curved forceps and bring the strip as closely as possible over the surface of the buffer, contained in a shallow tray. Release both ends of the strip at the same time. The strip will then float on the surface of the buffer which will soak in almost immediately. (The buffer must not be allowed to flow on to the top surface of the strip before the under surface is wetted, nor can the strip be simply immersed in the buffer.) After the strip is thoroughly wetted (no opaque areas), immerse it in the buffer until ready to conduct a run. Blot the strips lightly between sheets of filter paper to remove excess buffer. The strips are then immediately placed in position in the chamber. The current is turned on and allowed to run for about 15 min before application of the sample to allow establishment of equilibrium conditions in the chamber.

With the current switched on, the sample is applied to the strip in a straight continuous line starting 0.5 cm from the side of the strip. Prior to soaking the strips the starting line is marked with a ball-point pen at the side of the strip or a small spot of marker dye (apalon yellow, amaranth red, and brilliant blue is one mixture) is placed at the point of application.

At the end of the time allocated for a separation, turn the power off and remove the cellulose acetate strips without allowing buffer to come into contact with the electrophoretic bands. Place the strips in the stain immediately after electrophoresis. For proteins, stain the strips in Ponceau S (0.5% in 5% trichloroacetic acid solution) for at least 5 min. Remove the excess and background stain by washing 1 min in each of 4 trays of 5% acetic acid.

At this point, the strips can be quantitated by elution. Blot the strip between two absorbent pads. Cut the various bands and place them in a series of marked tubes; also cut a clear area to serve as a blank. Add exactly 2.0

ml of 0.1N NaOH to each tube and shake at intervals until all the color has been eluted from each segment. Ponceau S forms a purple eluent in NaOH whose color can be measured at 525 mμ against the blank. Alternatively, individual bands may be dissolved in a solvent which extracts the dye and dissolves the cellulose acetate. Reagents which dissolve cellulose acetate include: glacial acetic acid, acetone, pyridine, methylene chloride, butyl acetate, ethyl acetate, and methyl acetate.

Optical scanning is possible after clearing. All moisture must be removed either by drying at 37°–45°C for 20–30 min or by dipping twice into methanol. To clear the strip, dip the dehydrated strip into a solution of 10% acetic acid in methanol for 30–60 sec until completely wet. Remove the strip and place on a clean flat plate of glass slightly shorter than the length of the strip, and fold the ends under the plate. Bubbles should be removed immediately by rolling the strip onto the plate. Dry the strip in a 60°C oven for 15 min. Although translucent while wet, the strip becomes transparent as it dries.

EXPERIMENT 14-1 *Amino Acid Electrophoresis*

Amino Acid Standards Amino acid standards are prepared by making 0.05M solutions, in a 10% isopropyl alcohol solution. If necessary, add 1–2 drops of concentrated HCl to the solution as an aid to solubility. Prepare a standard solution of reference amino acids by adding equal volumes of the separate amino acid solutions to a small test tube and mixing thoroughly. Lysine, histidine, alanine, glycine, serine, glutamine, and tryptophan are useful for urinary and serum amino acid samples.

Buffer Solution Mix 31.2 ml formic acid with 59.2 ml glacial acetic acid and make up to 1000 ml with distilled water; pH 2.0. Store in a refrigerator. The buffer is stable for 1 month.

Ninhydrin Reagent Dissolve 0.1 g in 3 ml of ethyl alcohol and dilute to 50 ml with diethyl ether. Stable at refrigerator temperatures for 1 week. Just before use, pipet 0.5 ml of collidine into 49.5 ml of ninhydrin reagent. Mix well and use immediately.

Procedure Apply the mixture to one strip, the test solutions of single amino acids to additional strips. The base line should be at the anode side. Use 0.5–1.0 μl sample. An electrophoresis run of 30 min at 50 V/cm will separate the more common amino acids on cellulose acetate. On filter-paper strips, apply 100 V/cm for 4 hr.

After development of the electrophoretic pattern, the strip is dried as described elsewhere. Filter-paper strips are sprayed with ninhydrin reagent and dried at 110°C for 10 min. Cellulose acetate strips are dipped rapidly, right-side up, in the ninhydrin reagent. The strips are then placed on glass plates and incubated at 45°C for 60 min.

Trace the outlines of the zones and measure the distance the amino acids have travelled. On cellulose acetate the bands may be quantitated by cutting out each amino acid band and dissolving it in 0.5 ml of a 50% $CHCl_3$–50% acetone mixture. Absorbance may be read at 560 mμ.

After the various plasma proteins have had adequate opportunity to migrate with the electric current, the electrophoresis is stopped. The filter-paper strip is removed and heat dried to coagulate the proteins and thus fix them. After fixation, proteins are stained with a dye such as bromophenol blue. Since each of the serum proteins differs in its isoelectric point and its molecular weight, each travels along the filter paper at a different speed. The further the pH of the buffer solution is from the isoelectric point of a particular protein, the more that protein dissociates, and thus the better it is able to travel toward the positive or negative pole of the filter-paper strip. Likewise, the lower its molecular weight, the further the protein can travel.

EXPERIMENT 14-2 *Separation of Lower Fatty Acids on Paper*

Spot individual samples of the first six acids close to the cathode on Whatman No. 3MM paper. Moisten the paper with 0.1*M* ammonium carbonate solution adjusted, if necessary, to pH 8.9. Use this same solution in the electrode compartments. Carry out the electrophoretic separation at 10 V/cm for 60 min. Dry the paper at room temperature and spray with 0.025% bromophenol blue solution in acetone–water (9 : 1).

To illustrate the relationship between time of electrophoresis and distance of travel of a solute zone, prepare three individual strips of paper. Spot each strip with formic acid and develop at 10 V/cm in a 0.1*M* ammonium carbonate buffer at pH 8.9. Remove one strip after 20 min, the second after 40 min, and the third after 60 min development time.

From the experimental data obtained, estimate the ionic mobilities of each solute. For formic acid, correlate the distance travelled with the time of electrophoresis.

EXPERIMENT 14-3 *Separation of Indicators*

On Whatman No. 3MM paper spot samples of bromocresol green, bromophenol blue, methyl red, and methyl violet. Carry out the electrophoresis 20 V/cm for 2 hr in a buffer that is 0.1*M* in acetic acid and 0.1*M* in sodium acetate (pH 4.5). Dry the paper and mark the location of each zone. Correlate the direction and distance of travel with the type of ion predominant at the pH of development.

The experiment can be repeated in 0.6*M* formic acid and in a buffer composed of 0.1*M* aqueous ammonia and 0.1*M* ammonium acetate (pH

9.0). From the three sets of data, the ionization constant of the indicators may be estimated.

Problems

1. Suggest a reason why the electrically neutral α-globulin fraction of serum moves towards the cathode during electrophoresis. Suggest a method for correcting for this apparent movement towards the cathode.
2. What is the optimum pH for the separation of these binary mixtures of amino acids? (a) Lysine and histidine; (b) glycine and serine; (c) glycine and glycylglycine; (d) glutamic acid and aspartic acid. See Appendix B for necessary values of ionic mobility.
3. What separations would be anticipated if the electrophoresis of amino acids were to be carried out in a pyridine acetate buffer, pH 5.3?
4. Calculate the limiting mobilities at zero ionic strength for these ions. In parentheses are given the van der Waals' radius, in angstroms, and the ratio (f/f_o). (a) $CH_3CH_2NH_3^+$ (2.30, 1.015); (b) $CH_3(CH_2)_{11}NH_3^+$ (3.75, 1.21); (c) $CH_3CH_2)_2CO_2^-$ (3.12, 1.08); (d) $CH_3CH(CO_2^-)_2$ (3.50, 1.02).
5. What volume of 0.2N buffer solution would be adequate for an overnight run of 15 hr at a current of 3.5 mA?
6. For these mixtures, predict the direction and relative migration rates: (a) lysine, arginine, and histidine at pH 7.5; (b) aspartic and glutamic acids at pH 3.0; (c) alanine, valine, proline, and tryptophan at pH 1.7; (d) glycine, isoleucine, phenylalanine, and hydroxyproline at pH 1.7.
7. Calculate the apparent mobilities of glycine and serine in a phosphate buffer adjusted to pH 9.2.
8. In a citrate buffer adjusted to pH 3.8, predict the electrophoretic pattern of a mixture composed of methylamine, ethylamine, and n-propylamine. The mobilities at zero ionic strength are 5.99, 4.85, and 4.30×10^{-4} cm^2 V^{-1} sec^{-1}, respectively.
9. The separation of the amines in Problem 8 was conducted at 5 V/cm for 2.5 hr. Where would one expect the zones to be located relative to the origin?
10. What would be the minimum time required to separate the centers of the zones (Problem 9) by a distance of 1.0 cm for (a) ethylamine and propylamine, and (b) methylamine and ethylamine?
11. On a slab of silica gel impregnated with a carbonate buffer adjusted to pH 6.6, the electrophoretic pattern of lysine and histidine was run for 7 hr at 100 V on a bed 53 cm in length. Where are the zones located relative to the origin?
12. Using a formate buffer (pH 3.5) these ribonucleotides were run on paper for 60 min at 30 V/cm. The distances travelled towards the anode are given in millimeters. Calculate the apparent mobilities for each ribonucleotide. Adenosine 3′-phosphate, 10 mm; guanosine 3′-phosphate, 20 mm; cytidine 3′-phosphate, 4 mm; and uridine 3′-phosphate, 25 mm.
13. In a borate buffer (pH 9.2) the adenosine 3′- and 5′-phosphate species are located 45 and 57 mm, respectively, from the origin in the direction of the anode. Suggest an explanation for the ability of the borate system to resolve the A^3 and A^5 isomers.
14. In an acetate buffer (pH 4.6), electrophoresis was conducted for 3.5 hr at 6.73 V/cm. From the origin, vanillin moved 49.5 mm towards the cathode whereas in the direction of the anode ferrulic acid moved 10.5 mm and vanillic acid

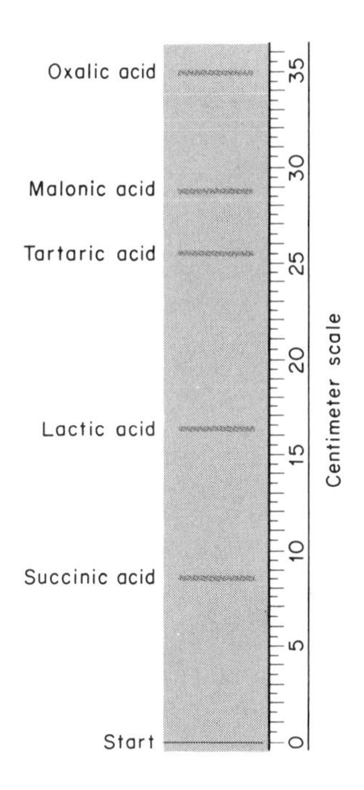

FIG. 14-7. Illustration for Problem 19. Electrophoretic pattern of some organic acids.

moved 28.0 mm. During the electrophoretic run acetone was observed to move 50.9 mm towards the cathode. With this information, calculate the apparent mobilities and correct the apparent mobility values for electro-osmosis.

15. In Problem 14, the acetate buffer was $0.12M$ in each component. What volume of buffer solution would be required for a current of 37.5 mA?
16. Derive an expression for the apparent mobility of monoaminomonocarboxylic acids in a buffer whose pH is less than the pH at the isoelectric point.
17. Estimate the two acid dissociation constants of glycine from these apparent mobilities ($\times 10^{-4}$) observed at the pH values in parentheses: 2.94 (1.0); 2.83 (2.0); 0.65 (3.0); 0.07 (4.0); 0.14 (8.5); 0.88 (9.2); 2.36 (10.0); 2.94 (11.0); and 3.07 (12.5). Graph the results and estimate the two acid dissociation constants of glycine.
18. These experimental mobility values were obtained for bovine β-lipoprotein: 2.75×10^{-5} at pH 2.67; 0.3×10^{-5} at pH 4.9; and -2.3×10^{-5} at pH 7.0. Estimate the isoelectric point of the bovine serum beta lipoprotein.
19. The electrophoretic pattern obtained with several water-soluble, non-volatile organic acids is shown in Fig. 14-7. The systems were run in a pyridine–acetic acid buffer (pH 4.1) at 40.3 V/cm and 10°C. Under the same conditions, the chloride ion travelled 73.0 cm. The mobility of the chloride ion at 8°C is approximately 2.9×10^{-4} cm^2 V^{-1} sec^{-1}. Calculate the apparent mobilities of each acid anion. From the known acid dissociation constants in the literature, estimate the ionic mobility of each species.
20. From the electrophoretic run of natural amino acids shown in Fig. 14-8, estimate the apparent mobilities of each species and by intercomparison with values available for some of the amino acids, estimate the ionic mobilities.

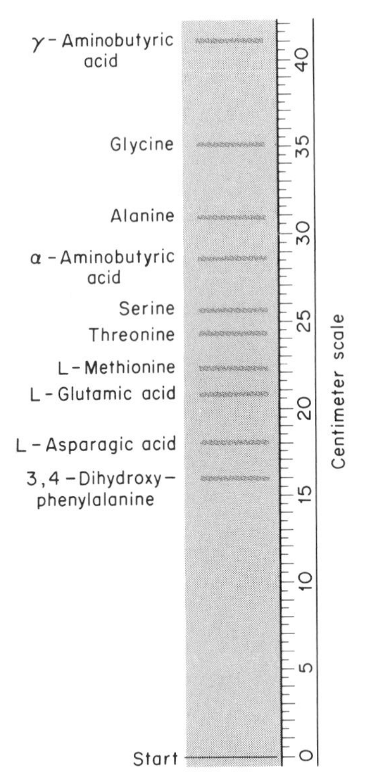

FIG. 14-8. Illustration for Problem 20. Electrophoretic pattern of natural amino acids.

21. A convenient desalting technique is available through the use of volatile buffers with filter-paper electrophoresis. To illustrate the method, locate the position of these substances, operating at 25 V/cm for 15 min with a sheet of filter paper saturated with pyridine–acetic acid (pH 6): K^+, Na^+, SO_4^{-2}, Cl^-, $H_2PO_4^-$, aspartic acid, and ornithine.
22. In a buffer (pH 7) electrophoresis was conducted for 3.5 hr at 6.73 V/cm. From the point of sample application, these phenolic substances were located as follows; Acetovanillone, 43 mm, and guaiacol 92 mm, towards the cathode; vanillin, 13 mm, and vanillic acid, 97 mm, towards the anode. At the same time acetone was observed to move 50.9 mm towards the cathode. Calculate the apparent mobilities and correct these values for electro-osmosis.
23. Compare the apparent mobilities obtained for vanillin and vanillic acid in Problem 14 and in Problem 22. Suggest a reason for the increased mobility at pH 7.
24. In the course of an electrophoretic run, the bands often broaden to twice the width of the starting zone at the point of sample application. In both Problems 14 and 22 the starting zone was 10 mm in width. Comment on the separability of the solute zones. What length of time would be necessary to obtain complete resolution of ferrulic and vanillic acids?

BIBLIOGRAPHY

Bier, M., *Electrophoresis,* Academic Press, New York, 1959.

Michl, H., "Techniques of Electrophoresis," in E. Heftmann (Ed.), *Chromatography,* 2nd ed., Reinhold, New York, 1967.

Strickland, R. D., "Electrophoresis," in F. J. Welcher (Ed.), *Standard Methods of Chemical Analysis,* 6th ed., Vol. 3, Part A, Van Nostrand, Princeton, N. J., 1966.

Wieme, R. J., "Theory of Electrophoresis," in E. Heftmann (Ed.), *Chromatography,* 2nd ed., Reinhold, New York, 1967.

15 *Miscellaneous Methods*

In this final chapter we shall examine briefly some selected separation methods. Although limited in scope, these methods provide useful means for accomplishing unique separations.

DISTILLATION AND EVAPORATION

Analytical distillation, as a strictly analytical technique for mixtures of volatile organic compounds, is completely outclassed and outdated by gas–liquid chromatography. Fractional distillation still functions uniquely to assess distillation for ultimate use in a commercial process and in establishing operating conditions, expected temperature levels, azeotrope formations, and so on.[1] Its surviving, useful offspring is called small-scale distillation. Leslie[2] has discussed this aspect of organic analytical distillation.

Separating elements through the vapor phase is less convenient than separating them by solvent extraction or precipitation, but it is usually clean and quantitative. The lighter elements are those most often separated by vaporization, especially the non-metals and pseudometals. The method can be extended to a number of metals by using higher temperatures or reduced pressures. Vaporization is also used to remove unwanted elements before determining other constituents.

Organic Elementary Analysis

Automated equipment is commercially available for the determination of carbon, hydrogen, and nitrogen in organic compounds. In a typical unit (Fig. 15-1), a sample is weighed on a microbalance, placed in a conventional

[1] F. E. Williams, *Anal. Chem.* **40,** 62R (1968).
[2] R. T. Leslie, *Ann. N. Y. Acad. Sci.* **137,** 19 (1966).

platinum (aluminum or ceramic) boat, and covered with catalysts or oxygen donors if required. The sample boat is placed in a quartz ladle and inserted into the inlet, which is then closed. The combustion system is flushed and filled with oxygen, or any mixture of helium and oxygen. The sample ladle is next moved into the furnace using a magnetic manipulator. The sample is burned at 900°C. The combustion products are scrubbed, if necessary, to remove halogen and sulfur-containing gases. Standard materials are MgO for fluoride, Ag° wool for other halogens, and silver vanadate or tungstate for sulfur oxides. The combustion gases then pass through the reduction tube (650°C) where oxides of nitrogen are reduced to molecular nitrogen by copper (and where excess oxygen is removed). The carbon dioxide, water vapor, and nitrogen are then flushed by helium gas flow into a "mixing volume" where they are mixed at a precise pressure and temperature. This uniform mixture is analyzed by passage through a series of high-precision thermal-conductivity detectors operated in pairs. Between the first and second pairs of detectors is an adsorption tube containing magnesium perchlorate for the removal of water, and between the second and third pair of detectors is an absorption tube of soda asbestos (plus a short section of magnesium perchlorate) for the removal of carbon dioxide and water formed by adsorption of carbon dioxide. The residual material, consisting only of nitrogen and helium, is measured against a pure helium reference.

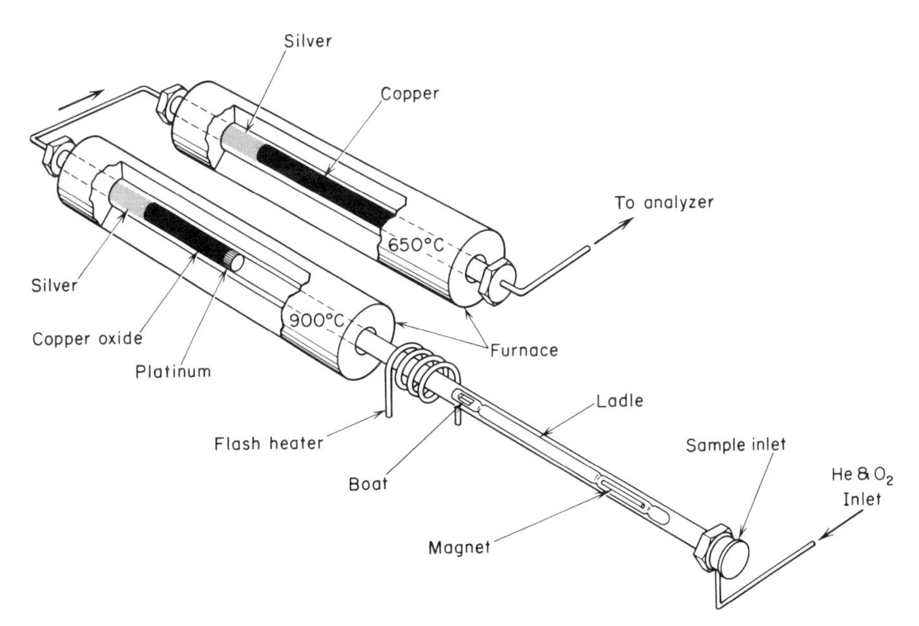

FIG. 15-1. Schematic view of the combustion and reduction train of a CHN analyzer. (Courtesy of Perkin–Elmer Corporation).

Oxygen, in the Unterzaucher method, is converted to carbon monoxide by passing the vaporized compound over carbon granules heated to 1200°C in a stream of dry nitrogen or helium. The carbon monoxide reacts with iodine pentoxide to yield iodine vapor, which in turn is adsorbed and determined by titration. Reaction of oxygen in organic and inorganic compounds with BF_3 yields gaseous oxygen.

In the Kjeldahl method nitrogen in organic compounds is converted to ammonium sulfate by digesting with hot concentrated sulfuric acid and appropriate catalysts; then ammonia gas, liberated by adding sodium hydroxide, is distilled and absorbed in boric acid, then determined by titration. Nitrogen in nitrates can be reduced to ammonia in alkaline solution with Devarda alloy (Cu 50%, Al 45%, Zn 5%).

Distillation from Solutions

Distillation from solutions involves compounds in which covalent bonds prevail (an analogy to extraction by nonpolar solvents). Volatile inorganic substances form typical molecular lattices in which there exists an intimate association of a small number of atoms. This association is preserved on volatilization. Only small cohesive forces exist between individual molecules. Forces holding the individual atoms together within the molecule are much stronger. Thus, application of small amounts of energy to these systems easily disrupts the weak molecular bonds between these molecules and those of the solvent. Examples include the non-saltlike hydrides (H_2O, H_2S, NH_3), non-saltlike halides ($GeCl_4$, $AsCl_3$, $SeOCl_2$), a few oxides (CO_2, OsO_4), some of the metal carbonyls, and a few special cases among the carbon compounds.

As little as 0.1 μg/ml of boron in aqueous solutions can be separated by distillation as the ester methyl borate $B(OCH_3)_3$ (b.p. 68.5°C) from methanol solutions at 75–80°C in all-silica glassware.[3]

Fluoride ion, an impediment to many analytical determinations, is easily removed as HF by evaporation with sulfuric or perchloric acid. Separation is accomplished by distillation as H_2SiF_6. The sample is placed in a distilling flask with glass beads, sulfuric or perchloric acid is added, and the mixture distilled at 135–140°C.[4]

As their chlorides, the elements germanium, arsenic, antimony, and tin may be easily separated from other elements and from each other. $GeCl_4$ may be distilled from a 3–4*N* HCl solution in an atmosphere of chlorine. Arsenic will be in the pentavalent state which is not volatile. Tin is kept in solution as the stable hexachlorostannate ion, $SnCl_6^{-2}$.

Arsenic in the trivalent oxidation state (hydrazine added as reductant)

[3] C. L. Luke, *Anal. Chem.* **30,** 1405 (1958); A. R. Eberle and M. W. Lerner, *Anal. Chem.* **32,** 146 (1960); M. Freegarde and J. Cartwright, *Analyst* **87,** 214 (1962).

[4] H. H. Willard and O. B. Winter, *Ind. Eng. Chem., Anal. Ed.* **5,** 7 (1933).

can be distilled quantitatively as $AsCl_3$ from 6*N* HCl at 110–112°C. It may be carried over in a stream of carbon dioxide or by adding hydrochloric acid from a dropping funnel into the distilling flask. After arsenic has distilled, phosphoric acid is added to the distilling flask to combine with the tin (and raise the boiling point) and the antimony is distilled as $SbCl_3$ at a temperature of 160°C. Finally, at 140°C (in the presence of bismuth; otherwise 165°C) tin is distilled over by dropping in a mixture of HCl–HBr (3 : 1). The larger size of the bromine atom precludes arranging more than four around the rather small tin atom, and thus molecular $SnBr_4$, rather than ionic $SnBr_6^{-2}$ (analogous to $SnCl_6^{-2}$) is formed.

Simultaneous removal of arsenic, antimony, and tin is readily accomplished by dropping in a mixture of HCl–HBr (3 : 1) to the sample contained in a wide-mouth test tube (with reducing agent) and heated to 140°C.

Chromium can be removed as chromyl chloride CrO_2Cl_2 by evaporating the sample to fumes with perchloric acid, which converts the chromium to the hexavalent state, then concentrated hydrochloric acid is dropped into the boiling solution.

Osmium is easily separated from the other platinum metals. The sample is treated in the distilling flask with 8*N* HNO_3 or boiling $HClO_4$ and a slow stream of air passed through the boiling solution carries away the OsO_4 (b.p. 129°C). The osmium tetroxide is absorbed in 6*N* HCl saturated with sulfur dioxide to immediately reduce osmium to the nonvolatile lower valence states.

Vacuum Fusion

Gases in metals are removed by melting in vacuum; the individual gases are then determined by different methods. This method, which is called vacuum fusion, involves melting the sample in a graphite or platinum crucible by means of high-frequency heating. The gases are collected and then determined by various gas analytical techniques. Commercial units are available.

Evaporation of Metals at High Temperatures

Direct distillation of a more volatile metal from a sample leaves the less volatile components in the residue. By distilling at low pressures the temperatures needed are reduced, and differences in vapor pressure are magnified. When the matrix is volatile, this approach affords an appreciable degree of concentration, usually about 100 times. This procedure has been applied for the concentration of impurities in zinc and cadmium. The reversed operation – distillation of minor components – has been used for the separation of zinc and sodium traces in aluminum of high purity, and to extract traces of lead from meteorites. The vaporized metal is condensed in quartz wool or deposited as a metallic mirror on the cold portion of a quartz ignition tube. A stream of inert gas aids the separation.

Mercury may be volatilized as the metal by ignition of its compounds or ores alone at a red heat. It is collected on gold or silver foil, or under water.

ZONE REFINING

Zone refining was first developed by Pfann (1952) for the purification of materials used as semi-conductors—specifically purifying germanium for transistors—where purity standards are fantastically high (less than 1 part in 10^9). The method is a most useful technique for preparing ultrapure metals and compounds for use as melting-point, microanalytical, mass, ultraviolet, and infrared standards; in fact, standards for any precise physico–chemical measurement where purities greater than 99.9 mole % are required. Zone refining is usually of particular value and most effective when applied to samples that are already of 99 mole % or greater purity. The starting material for zone refining has therefore often undergone preliminary purification by gas–liquid chromatography, fractional distillation, or some other chemical separation method.

Principle

Zone melting depends upon the difference in solubility of an impurity in the liquid and in the solid main component. In a eutectic-type mixture, if thermodynamic equilibrium is reached at the interface of a cooling solid in contact with its melt, the impurities will remain in the melt and the major component will freeze out in a purer state. Thermodynamic equilibrium means a temperature and velocity differential between the solid and liquid freezing interface. This often ranges from 0.1 to 0.01°C/mm for the solidifying interface.

In a zone refiner a heater moves along a solid rod of material 1–55 ml in volume (Fig. 15-2). A small molten zone forms and travels slowly along the column of solid. Melting occurs in front of the zone, solidification from an air blast at its rear. The zone refiner can be set for a single pass or as many multiple passes as necessary. The freezing solid is the purified zone when impurities lower its melting point. If impurities raise the melting point, they go into the solid phase, purifying the liquid. Assuming that the impurities are more soluble in the liquid than in the solid, they are carried forward with the molten zone so that the recrystallized material is purified at the end where the zone starts its traverse. Impurities are concentrated at the opposite end. Where desired in trace analysis, impurities can be concentrated until there is an adquate sample for precise quantitative analysis.

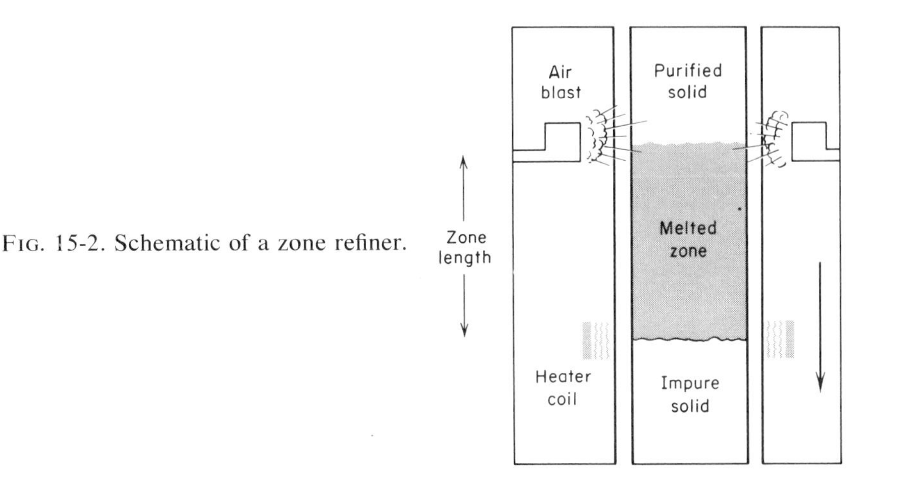

FIG. 15-2. Schematic of a zone refiner.

When the following conditions are satisfied, a single passage of a zone leads to the equation for impurity concentration C (Eq. 15-1).

(1) Complete mixing of the liquid is achieved at all stages.
(2) No diffusion takes place in the solid.
(3) The effective partition coefficient, $K_d = C_{\text{solid}}/C_{\text{liquid}}$, is constant.
(4) The concentration of impurity, C_0, is initially uniform.
(5) A constant cross section and constant length of molten zone, L, is used.

$$C = C_0[1 - (1 - K_d)e^{-K_d(d/L)}] \qquad (15\text{-}1)$$

where d is the distance along the column of material. Equation (15-1) is not applicable to the last zone length.

Detailed calculations show that there is an advantage in using a relatively long molten zone in the early passes to bring about a rapid movement of impurity and a short zone in the later passes to bring about a greater degree of purification. Ratios of column length to zone length of 10 or greater in the later stages generally permit a good final purification to be achieved. The number of passages of the zone necessary to produce a concentration distribution approximate to the ultimate one (i.e., $n \rightarrow \infty$) depends upon the value of K_d and on the ratio of the column length to zone length (Fig. 15-3). Where K_d is 0.2, and if the ratio of column length to zone length is 10, ten passes will bring about a relative impurity concentration of only 2×10^{-5} at one end and a 10-fold increase at the other. If nothing is known about the system, it is usually reasonable to try between 10 and 20 passes.

The thermal conductivities of the solid and liquid phases limit the column diameter to a maximum of 5 cm and determine the method used to create molten zones. Zone speeds are predicated upon the speed of crystal growth and the rate of diffusion of impurities from any local pockets of high concentration. The growth rate must be slow enough and the movement of

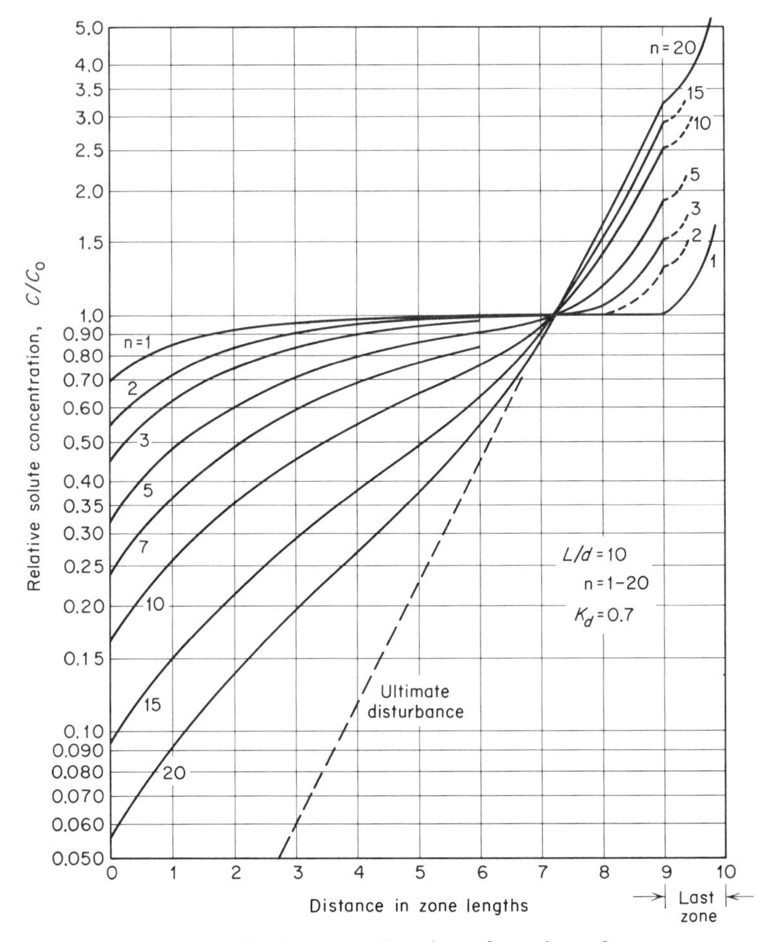

FIG. 15-3. Concentration of solute as a function of number of zone passages when ratio of column length to zone length is 10 and $K_d = 0.7$.

zone similarly slow such that the impurities have sufficient time to be transported away from the advancing solid face and into the bulk of the molten zone by diffusion, convection, or stirring.

The mode of column operation, whether a "floating-zone" technique in a horizontal tube or liquid zones in a vertical column, is dependent upon the density difference between solid and liquid and the surface tension of the liquid. The high surface tension of liquid metals enables them to be treated in horizontal tubes but the low surface tension of organic compounds necessitates treatment in vertical tubes because the low surface tension of the melt allows liquid from the molten zone to seep back under the resolidified material. When a vertical column is used the heater must be started at the top if the solid expands on melting and at the bottom if it contracts.

The efficacy of a zone-refining apparatus for purifying a particular material may be studied by using a suitable tracer—for metals a radioactive tracer, and a dye or fluorescent material for organic substances. Zone refining is a more convenient purification process than fractional freezing because the effect of multiple zone passes is accumulative and it is not necessary to remove the impure fractions after each zone pass. Automation of the apparatus is easily achieved. No reagents are added to contaminate the sample. The method is applicable for nearly any organic or inorganic chemical with a melting point between 50° and 300°C. Although the separation is not rapid, the process requires little attention, so that the cost in man-hours is small.

The disadvantage of the method is a different partition coefficient between the crystalline and liquid phases of various elements and compounds. Some secondary processes, as volatility or oxidation of particular impurities, may influence the course of the concentration process. Large volume changes often occur when handling organic compounds.

Zone Leveling

Zone leveling is a term applied to zone-melting processes undertaken with the object of achieving as uniform a concentration as possible over as great a length of the column of material as possible. If K_d is near unity, a single passage of a zone through a column of material of fairly uniform composition will yield a central portion of even more uniform composition. A method of wide applicability is to pass the heater repeatedly at constant speed in alternate directions. This procedure, which brings about zone leveling in all parts of the column except the part to freeze last, can be used when the value of K_d is not known, although it is then difficult to produce a rod of selected composition.

If K_d is known, it is possible to arrange that the solute concentration in the first zone length has the value C_0/K_d, where C_0 is the average initial concentration of the dilute component in the remainder of the column. If a molten zone is now moved along the column, the amounts of impurity entering and leaving the zone are equal, so that the concentration in the column after passage of the zone is uniform and equal to C_0, except for the last one.

Zone leveling is useful for preparing reference samples of uniform composition.

INCLUSION COMPOUNDS

An inclusion compound consists primarily of two molecular species, one of which provides space in its structure in which molecules of the other are accommodated. The "host" species forms the essential structure in which the "guest" is spatially included or contained. If the guest molecules to be separated have the proper geometry and size to form an inclusion compound with a suitable host material, and there are no other molecules in the mixture that may also form an inclusion compound, a pure product is readily obtainable. Although the interaction energy between the host and guest is weak, it is necessary that the interaction energy be sufficiently high (in excess of 5 kcal/mole of guest) for the equilibrium to lie well on the side of the inclusion compound. Differences in bonding of various guest molecules within the host may be used in refining the separation process. In a mixture the inclusion compound of the more strongly bonded molecules, by van der Waals' bonds, will predominate.

Three general types of molecular combinations are formed. The *clathrate*[5] or cage structure is one in which molecules of the guest fit into molecular cages formed by the host molecules. Decomposition of a clathrate can occur only if the host structure is broken by such means as melting, sublimation, or dissolution. In *channel* structures the host molecules form a cylindrical channel in which the guest constituent is enclosed. Only the cross-sectional dimension of a molecule is critical for it to be a guest, and this is set by the smallest constriction that occurs in the channel space. Release of guest molecules from an open end of a long channel is possible without first breaking the host structure. The third category is the *layer* type. Here, molecules of the host are arranged in layers between which guest molecules may be accommodated. Layer structures are fairly flexible as compared to the clathrate and channel structures. These latter types are quite inflexible and are, therefore, much more selective in regard to size and shape of guest molecules than are layer host compounds. There is no precise line dividing types of inclusion compounds since what is a closed cavity for a large molecule may be part of a continuous channel for a smaller one.

Clathrates

Host molecules completely surround the guest molecules in a clathrate. The hydroquinone cage structure will illustrate this type of inclusion compound. The hydroxyl groups of hydroquinone are linked through their oxygens with other hydroquinone molecules to form a plane hexagon stabilized by hydrogen bonds. Each of the six hydroxyls forming the hexagon originates from a different molecule (Fig. 15-4). The hydroquinone molecules are located alternately above and below the plane of the hexagon. When one cage structure is formed, simultaneously a second identical cage structure will interpenetrate it and be displaced vertically halfway between the top and the bottom of the first hexagon. In this way, hydroquinone forms a clathrate with itself. At the same time the double interpenetrated framework will contain a cavity in which smaller molecules may be trapped. This cavity

[5] From the Latin *clathratus*—enclosed by the crossbars of a grating.

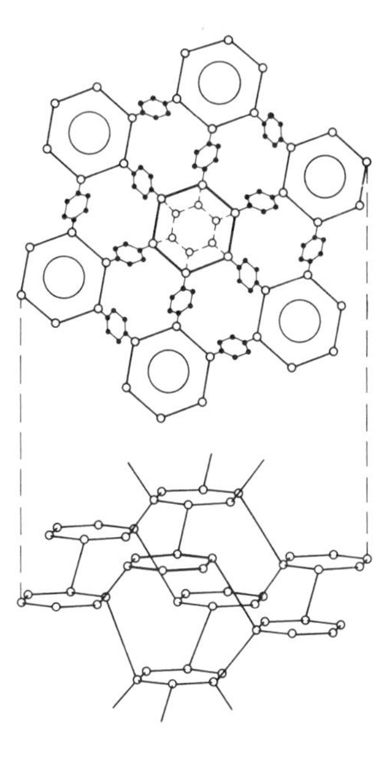

FIG. 15-4. Clathrate inclusion compound. *Above* Projection of a portion of the network of hydrogen-bonded hydroquinone molecules. Filled circles denote carbon atoms; small open circles, oxygen atoms; and larger open circles, the guest. Each regular hexagon denotes six hydrogen bonds between oxygen atoms. *Below* Perspective drawing corresponding to above. The hexagons denote the hydrogen bonds; the longer lines connecting different hexagons denote the O—O axis of the hydroquinone molecule.

is roughly spherical and is about 4Å in diameter. The hydroquinone cage has an amazing flexibility which enables it to elongate in order to entrain long molecules when necessary. This distortion is the result of both a slight extension and a puckering of the hexagon of hydrogen-bonded oxygens.

The host structure imposes a barrier to easy diffusion of guest molecules into or out of the host. Clathrates are formed by crystallizing the structure from a solution of unassociated host and guest molecules using a solvent whose molecules cannot be accommodated in the host. Guest gases require the use of pressure vessels. Release of the guest from a clathrate usually requires decomposition of the host structure.

Clathrates may be used to store inert gases. The hydroquinone–argon clathrate, $3C_6H_4(OH)_2 \cdot Ar$, in a theoretical volume of 260 ml contains 1 mole (25.1 liters at 1 atm and 15°C) of argon. Dissolution of the clathrate gives a controlled release of the gaseous argon. Benzene has been essentially freed of thiophene by the formation of the clathrate of benzene-monoammine nickel(II) cyanide. Clathration species are highly selective for a number of isomers. Tetra-(4-methylpyridino) nickel dithiocyanate will separate anthracene from phenanthrene, naphthalene from diphenyl, and is selective for the *para* forms of xylene and cymene.

Channel Inclusion Compounds

When urea is allowed to crystallize from a solvent, a rather loose tetragonal crystalline lattice is obtained. However, if urea is crystallized in the presence of an unbranched hydrocarbon, the solid obtained consists of spiraling chains of urea molecules with each molecule forming a hexagonal structure with a channel 5 Å in diameter along its central axis. Oxygen atoms are located on the parallel edges. Each oxygen is hydrogen bonded to four nitrogen atoms and each nitrogen to two oxygen atoms (Fig. 15-5). Because the channels extend throughout the structure, there is no restriction on the length of the guest molecule. The distribution of the chains along the channels is completely random and there is no regularity in the molar ratio of host to guest. However, for the latter there is an increase somewhat proportional to the length of the chain. For molecules with cross section too small, the introduction of functional groups of adequate size leads to inclusion. In general, the stability of urea inclusion compounds varies inversely with the vapor pressure of the included compound. Although this is not the sole criterion for stability, it does determine the rate of diffusion from the channels of the host structure.

The presence of a sulfur atom instead of an oxygen in thiourea accounts for an increase in the size of the thiourea molecule and of the crystalline structure as a whole. This leads to a larger central channel of 8 Å in diameter. Branched and cyclic hydrocarbon chains become enclosed because their cross section will be of sufficient size to warrant retention. The narrower linear chains are able to slip out.

Decomposition of urea and thiourea inclusion compounds can be accomplished by addition of water to dissolve the host molecule, leaving the included compound(s) as an oil or a solid. Conversely, the inclusion com-

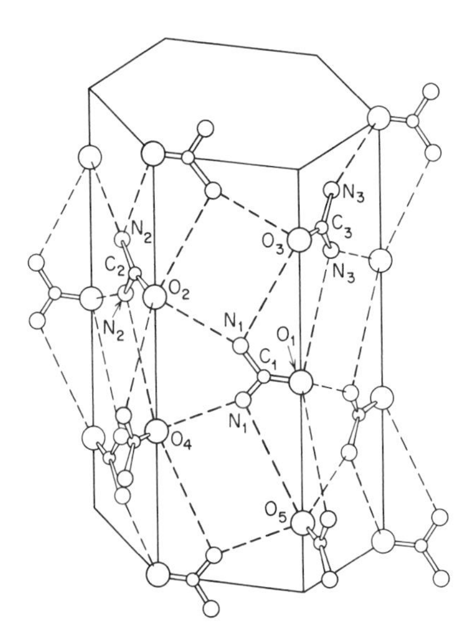

Fig. 15-5. Perspective of a urea inclusion compound. Dotted lines denote hydrogen bonding. Hexagonal cylinder delineates the channel extending through the structure where the guest resides. [From A. E. Smith, *Acta Cryst.* **5**, 224 (1952).]

pound can be heated with a solvent in which urea is insoluble. The included compound(s) will be extracted from the host.

Molecular sieves are host materials whose macromolecular crystalline structure is permanent. They are discussed in Chapter 12.

Separations can be based on differences in degree of unsaturation, chain length, and branching. The urea channel is about 6 Å at its widest part and about 5 Å at its narrowest. Straight-chain hydrocarbons have a cross section of about 4.1 Å and form inclusion compounds readily. Hydrocarbons with a single methyl branch require a channel diameter of about 5.5 Å. When the straight-chain portion of such compounds is relatively short, as in 3-methylheptane, the compound itself does not form an inclusion compound but forms one readily when it is mixed with a straight-chain compound which appears to drag it into the host structure. When the straight-chain part of a singly branched compound is long, no difficulty is experienced in obtaining the inclusion compound. Compounds with a double branch on one carbon atom require a channel diameter of about 6 Å in all directions and no inclusion compound with urea has yet been obtained with such materials.

In separations based on differences in chain length, advantage is taken of the fact that the longer chain compounds form inclusion compounds preferentially. Therefore, if insufficient host material is employed to combine with all the components in a mixture, the longer chain components will combine and crystallize. The components to be separated should differ in chain length by at least four carbon atoms and preferably by six. As a long-chain component becomes more unsaturated, it shows greater deviation from the normal straight-chain structure. Therefore, at a given chain length, saturated components of a mixture would be expected to form inclusion compounds preferentially to mono-unsaturated, mono-unsaturated to di-unsaturated, and so on.

Layer Compounds

Graphite has a layer structure that consists of one plane of hexagonal arrays of carbon atoms, each atom being located at the corner of a hexagon and associated with three such hexagons. Other host materials for layer compounds are micas, vermiculites and montmorillonites. The interlayer distance often depends in part on the size of the guest molecules. Layer inclusion compounds are formed by diffusion of the guest molecules into the host solid. Removal of the guest and regeneration of the host for further use are accomplished by leeching or by thermal means.

A large number of metallic chlorides have been explored using graphite. The decisive factor seems to be the electronic configuration of the guests because they are the salts of multivalent metals in their higher state of valency, while those in their lower states do not become enclosed.

FOAM SEPARATION

Foam separation is a technique utilizing stable foams to separate components in solution. It is based upon the tendency of surface-active solutes to collect at the gas–liquid interface. Foaming is simply a practical means of producing and collecting large quantities of gas–liquid interface. Unlike froth flotation which involves removal of solids in suspension, foam separation is used with true solutions or colloids. The bubbles or droplets move through the bulk liquid which is stirred by their passage. It is possible to separate nonsurface-active ionic species from aqueous media by addition of an oppositely charged surfactant. The surfactant serves the dual purpose of an extracting agent (formation of a complex with the solute) and as foaming agent. By suitable alteration of the charge on the solutes it becomes possible to selectively remove various species. In a sense foam separation is a chromatographic process in which the separation depends on the distribution of the solutes between the bulk liquid phase and the interfacial film between this phase and bubbles or droplets formed in it.

A foam separation unit in its simplest form consists of a column atop a vessel containing the solution to be foamed. Air or other suitable gas is dispersed in the liquid reservoir through a fritted glass disk or other suitable sparger. A receiver at the top collects the foam produced. The collected foam is collapsed and the liquid which results differs in concentration from the original solution. Batch, continuous, and multistage systems have been used. In a continuous operation, feed is steadily introduced into the system while foam is withdrawn. The similarity to distillation columns is readily seen.

The ability of a foaming agent to produce froth results from Gibbs' adsorption at the gas–liquid interface. For an ideal aqueous solution containing a single solute at constant temperature, the adsorption isotherm that relates the surface excess of a solute to its bulk concentration at equilibrium may be expressed by the equation

$$\Gamma = \frac{-C}{\mathscr{R}T}\frac{d\gamma}{dC} \qquad (15\text{-}2)$$

where Γ is the surface excess (or depletion), C is the concentration of solute, and γ is the static surface tension. The surface excess depends upon the slope of the surface tension versus concentration plot.

In foam separation the factor Γ/C indicates the degree of separation attainable. In order to remove metal ions from solution, the added foaming agent must also be a complexing agent for the ion to be removed. For example, the removal of radioactive strontium and cesium from nuclear process streams was accomplished using polyaminopolycarboxylic acids for stron-

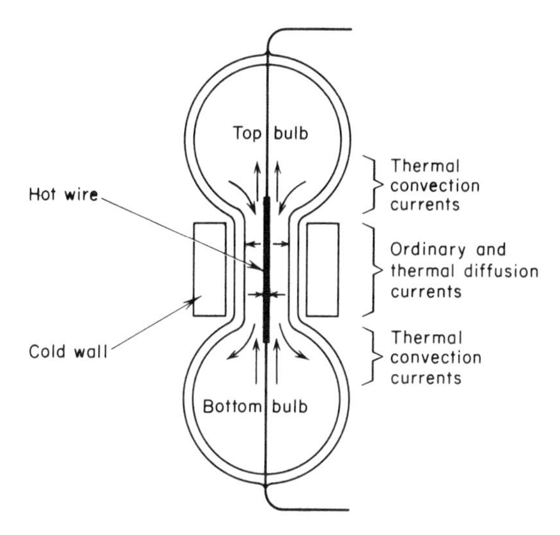

FIG. 15-6. Schematic of a thermal diffusion unit. In a cooled thermogravitational column with a hot coaxial wire, thermal diffusion moves material to be separated toward the walls, convection carries it upward.

tium, and the same foaming-complexing agent in combination with tetraphenyl boron for cesium.[6]

THERMAL DIFFUSION

Application of a temperature gradient on a homogeneous system establishes a concentration gradient. A partial unmixing takes place. The effect, known as thermal diffusion, offers a separation method that in the liquid phase depends upon shape differences between molecules more than any other property. As shown in Fig. 15-6, in a cooled thermogravitational column with a hot coaxial wire, thermal diffusion moves material to be separated toward the walls, convection carries it upwards to the top bulb. The convective flow induced by the thermal gradient produces a cascading (or refluxing) effect which multiplies the separation otherwise obtainable.

Equipment applicable to gases or solutions can be assembled from concentric tubes with external electrical heating. Some equipment, mainly for liquid separation, is constructed of flat plates. Vertical columns can be operated in a batch or continuous flow manner.

Thermal diffusion has been successfully used to separate gaseous isotopes such as He^3, $C^{13}H_4$, and the isotopes of chlorine. It has also been used to effect separations in the liquid phase of isomers and close boiling liquids. In most cases the heavier component concentrates at the cold wall. Mixtures of hydrocarbons are easily separated by liquid-phase thermal diffusion be-

[6] E. Schonfeld, R. Sanford, G. Mazzella, D. Ghosh, and S. Mook, *U.S. Atomic Energy Comm. Rep. NYO* 9577 (1960).

cause there are few cases of interaction between hydrocarbon molecules. Each molecule of such a mixture acts independently under the influence of a temperature gradient. Fractionation of tall oil has proved to be especially susceptible to this technique.

Problems

1. To what structural factor can the volatility of the trihalides of arsenic and antimony be attributed?
2. Suggest a method for removing SnO_2 and weighing the residual impurities that accompany the precipitation of tin as SnO_2.
3. Outline the distillation of OsO_4 and RuO_4 from each other. [Hint: Find a selective pair of oxidizing agents.]
4. Suggest an explanation for these observations: Methyl borate fails to distill from aqueous solutions containing fluoride ions in excess unless a large excess of anhydrous aluminum chloride in methanol is added before the distillation.
5. Devise a method for the separation of neon from a mixture with argon, krypton, and xenon.
6. Suggest a method for the separation of primary alcohols from secondary alcohols in the C_6–C_8 range.
7. Outline a method for the separation of fatty acids, free or as their ethyl esters, from natural oils.
8. State the reasons for the separation of these mixtures through formation of inclusion compounds: (a) Benzene and cyclohexane from 1-phenyloctadecane and 1-cyclohexyleicosane. (b) Methyl 2-methyloctanoate from 3-methylbutyl octanoate. (c) Failure of 2,2,4-trimethylpentane to form an inclusion compound with urea. (d) Why do not methyl and ethylbutanols and hexanols form inclusion compounds whereas 3-methylbutyl hexanoate and 2-ethylhexyl octanoate do?
9. Discuss the theoretical basis for the successful pyrohydrolysis of many metal fluorides: $MF_2 + H_2O \rightleftharpoons MO + 2HF$.

BIBLIOGRAPHY

Distillation

Francis, H. J., Jr., "Automated Elemental Microanalysis," *Anal. Chem.* **36** (No. 7), 31A (June 1964).

"High Precision CHN by Automated Instrument," *Lab. Management* (December 1965), p. 23.

Leslie, R. T., *Ann. N. Y. Acad. Sci.* **137,** 19 (1966).

Meinke, W. W., and B. F. Scribner (Eds.), *Trace Characterization, Chemical and Physical,* NBS Monograph 100, U.S. Government Printing Office, Washington, 1967.

Sloman, H. A., and C. A. Harvey, *J. Inst. Metals* **80,** 391 (1951/52). Vacuum fusion.

The Determination of Gases in Metals, The Iron and Steel Institute, London, 1960.

Walton, H. F., *Principles and Methods of Chemical Analysis,* 2nd ed., Prentice-Hall, Englewood Cliffs, N. J., 1964.

Zone Refining

Herington, E. F. G., "Zone Melting with Some Comments on Its Analytical Possibilities," *Analyst* **84,** 680 (1959).
Herington, E. F. G., *Zone Melting of Organic Compounds,* Blackwell, Oxford, 1963.
Herington, E. F. G., *Ann. N. Y. Acad. Sci.,* **137,** 63 (1966).
Pfann, W. G., *Zone Melting,* Wiley, New York, 1958.
Wilcox, W. R., R. Friedenberg, and N. Back, *Chem. Rev.* **64,** 187 (1964).

Inclusion Compounds

Barón, M., "Analytical Applications of Inclusion Compounds," in W. G. Berl (Ed.), *Physical Methods in Chemical Analysis,* Vol. IV, Academic Press, New York, 1961.
Hagan, M. M., *Clathrate Inclusion Compounds,* Reinhold, New York, 1962.
Hagan, M. H., *J. Chem. Educ.* **40,** 643 (1963).
Mandelcorn, L., *Non-Stoichiometric Compounds,* Academic Press, New York, 1964.
Mandelcorn, L., *Ann. N. Y. Acad. Sci.* **137,** 72 (1966).

Foam Separation

Cassidy, H. G., *Fundamentals of Chromatography,* Vol. X in A. Weissberger (Ed.), *Technique of Organic Chemistry,* Interscience, New York, 1957, p. 327.
Schoen, H. M., *Ann. N. Y. Acad. Sci.* **137,** 148 (1966).
Shedlovsky, L., *Ann. N. Y. Acad. Sci.* **49,** 279 (1947).

Thermal Diffusion

Dickel, G., "Separation of Gases and Liquids by Thermal Diffusion," in W. G. Berl (Ed.), *Physical Methods in Chemical Analysis,* Vol. IV, Academic Press, New York, 1961.
Jones, A. L., and G. R. Brown, "Diffusion Methods," in I. M. Kolthoff and P. J. Elving (Eds.), *Treatise on Analytical Chemistry,* Part I, Vol. 2, Interscience, New York, 1961.
Prabhudesai, R. K., and J. E. Powers, *Ann. N. Y. Acad. Sci.* **137,** 83 (1966).

APPENDICES

APPENDIX A *Acid Dissociation Constants at 25°C and $\mu = 0.1$*

Acid	pK_a
Acetic acid	4.65
Acetylacetone	8.9
Ammonium ion	9.37
Anilinium ion	4.7
Arsenic acid, H_3L	2.1
H_2L^-	6.7
HL^{-2}	11.2
Arsenious acid, H_3L	9.1
H_2L^-	12.2
Benzoic acid	4.10
Boric acid	9.1
Carbonic acid, H_2L	6.3
HL^-	10.1
Chloroacetic acid	2.7
Chromic acid, HL^-	6.2
Citric acid, H_3L	3.0
H_2L^-	4.4
HL^{-2}	6.1
Dichloroacetic acid	1.1
Ethylenediammonium ion, en·$2H^+$	7.30
en·H^+	10.11
Ethylenediaminetetraacetic acid (EDTA), H_4L	2.0
H_3L^-	2.67
H_2L^{-2}	6.16
HL^{-3}	10.26

APPENDIX A (*continued*)

Acid	pK_a
Formic acid	3.65
Hydrazine (2nd step)	8.1
Hydrocyanic acid	9.14
Hydrofluoric acid	3.05
Hydroxylammonium ion	6.2
Lactic acid	3.76
Nitrous acid	3.2
Oxalic acid, H_2L	1.1
HL^-	4.0
Phenol	9.8
Phosphoric acid, H_3L	2.0
H_2L^-	6.9
HL^{-2}	11.7
Phthalic acid, H_2L	2.8
HL^-	5.1
Picric acid	2.3
Pyridinium ion	5.21
Pyrophosphoric acid, H_4L	1.0
H_3L^-	2.5
H_2L^{-2}	6.1
HL^{-3}	8.5
Salicylic acid, H_2L	2.90
HL^-	13.1
Succinic acid, H_2L	4.13
HL^-	5.38
Sulfamic acid	1.0
Sulfosalicylic acid, H_2L	2.6
Sulfuric acid, HL^-	1.8
Sulfurous acid, H_2L	1.76
HL^-	6.8
Tartaric acid, H_2L	2.90
HL^-	4.09
Trichloroacetic acid	0.5
Triethanolammonium ion	7.76

APPENDIX B *Apparent pK_a Values, pI_e Values, and Ionic Mobility Values of Selected Amino Acids at 25°C*

		pK_1(COOH)	pK_2(NH_3^+)	pK_3	pI_e	μ_o(×10^{-4})
α-Alanine	CH_3—CH(NH_2)—COOH	2.34	9.69		6.0	2.67
Arginine	HN=C(NH_2)—NH—(CH_2)$_3$—CH(NH_2)—COOH	2.18	9.09	13.2 (Guanidyl)	10.9	3.82
Asparagine	H_2N(CO)—CH_2—CH(NH_2)—COOH	2.02	8.80		5.4	
Aspartic acid	HOOC—CH_2—CH(NH_2)—COOH	1.88, 3.65	9.60		2.8	3.20
Cysteine	HS—CH_2—CH(NH_2)—COOH	1.9	8.7(SH)	10.4(NH_3^+)	5.0	
Cystine	[—S—CH_2—CH(NH_2)—COOH]$_2$	<1, 2.0(COOH)	8.02	10.3	5.0	
Glutamic acid	HOOC—CH_2—CH_2—CH(NH_2)—COOH	2.3, 4.25	9.67		3.2	2.90
Glutamine	H_2N(CO)—CH_2—CH_2—CH(NH_2)—COOH	2.17	9.13		5.7	
Glycine	H_2N—CH_2—COOH	2.44	9.66		6.0	3.07
Histidine	HC=C—CH_2—CH(NH_2)—COOH (ring: N, NH, C–H)	1.8	6.00 (Imidazole)	8.97(NH_3^+)	7.6	
Hydroxyproline	HO—CH—CH_2 H_2C CH—COOH N–H (ring)	1.92	9.73		5.8	
Isoleucine	CH_3—CH_2—CH(CH_3)—CH(NH_2)—COOH	2.36	9.68		6.0	2.06
Leucine	(CH_3)$_2$CH—CH_2—CH(NH_2)—COOH	2.36	9.60		6.0	2.06
Lysine	H_2N—CH_2—(CH_2)$_3$—CH(NH_2)—COOH	2.20	8.90(α)	10.28(εNH_3^+)	9.7	
Methionine	CH_3—S—CH_2—CH(NH_2)—COOH	2.28	9.2		5.7	2.06
Ornithine	H_2N—CH_2—(CH_2)$_2$—CH(NH_2)—COOH	1.94	8.65(α)	10.76(δNH_3^+)	9.7	
Phenylalanine	C_6H_5—(CH_2)—CH(NH_2)—COOH	1.83	9.13		5.5	

APPENDIX B ***(continued)***

		pK_1(COOH)	pK_2(NH_3^+)	pK_3	pI_e	$\mu_o(\times 10^{-4})$
Proline	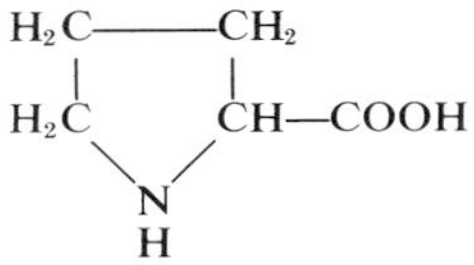 H_2C—CH_2 H_2C CH—COOH N H	1.99	10.60		6.3	
Serine	HO—CH_2—CH(NH_2)—COOH	2.21	9.15		5.7	2.64
Threonine	CH_3—CH(OH)—CH(NH_2)—COOH	2.15	9.12		5.6	
Tryptophan	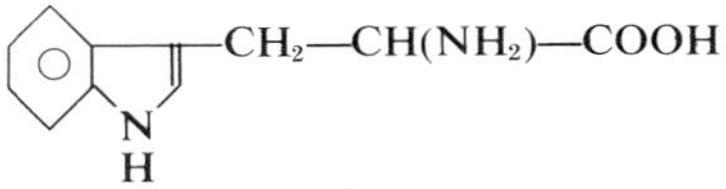 —CH_2—CH(NH_2)—COOH N H	2.38	9.39		5.9	
Tyrosine	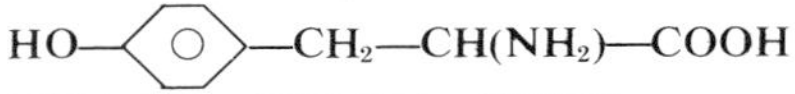 HO—⟨○⟩—CH_2—CH(NH_2)—COOH	2.20	9.11(NH_3^+)	10.07(OH)	5.7	
Valine	$(CH_3)_2$CH—CH(NH_2)—COOH	2.32	9.62		6.0	2.18

APPENDIX C *Logarithmic Values of $\alpha L(H)$*

pH	0	1	2	3	4	5	6	7	8	9	10	11	12
Acetylacetonate	8.8	7.8	6.8	5.8	4.8	3.8	2.8	1.8	0.9	0.2			
Citrate	13.5	10.5	7.5	4.8	2.7	1.2	0.25	0.05					
Ethylenediamine-tetraacetate	21.4	17.4	13.7	10.8	8.6	6.6	4.8	3.4	2.3	1.4	0.5	0.1	
Oxalate	5.1	3.35	2.05	1.05	0.3	0.05							
Phosphate	20.7	17.7	15.0	12.7	10.6	8.6	6.7	5.0	3.7	2.7	1.7	0.8	0.2
Sulfosalicylate	14.2	12.2	10.3	8.7	7.6	6.6	5.6	4.6	3.6	2.6	1.6	0.7	0.1
Tartrate	7.0	5.0	3.1	1.4	0.4	0.05							

APPENDIX D ***Formation Constants of Some Metal Complexes at 25°C***

	log k_1	log k_2	log k_3	log k_4	log k_5	log k_6
AMMONIA						
Cadmium	2.65	2.10	1.44	0.93	−0.32	−1.66
Cobalt(II)	2.11	1.63	1.05	0.76	0.18	−0.62
Cobalt(III)	7.3	6.7	6.1	5.6	5.1	4.4
Copper(I)	5.93	4.93				
Copper(II)	4.31	3.67	3.04	2.30	−0.46	
Nickel(II)	2.80	2.24	1.73	1.19	0.75	0.03
Silver(I)	3.24	3.81				
Zinc	2.37	2.44	2.50	2.15		
CHLORIDE						
Bismuth(III)	2.4	1.1	1.9	0.7		
Cadmium	1.6	0.5	−0.6	−0.6		
Copper(II)	0.1	−0.6				
Indium	1.4	0.8	1.0			
Iron(II)	0.4					
Iron(III)	0.6	0.1	−0.8			
Lead(II)	1.2	−0.6	1.8			
Manganese(II)	0.6	0.2	0.2			
Silver(I)	2.9	1.8	0.3	0.9		
Tin(II)	1.15	0.55				
Zinc	−0.2	−0.4	0.75			
CITRATE (as ML)						
Aluminum	log $k_1k_2k_3 = 20.0$					
Cadmium	log $k_1k_2 = 11.3$					
Cobalt(II)	log $k_1k_2 = 12.5$					
Copper(II)	log $k_1k_2 = 18$					
Iron(II)	log $k_1k_2 = 15.5$					
Iron(III)	log $k_1k_2k_3 = 25.0$					
Nickel	log $k_1k_2 = 14.3$					
Zinc	log $k_1k_2 = 11.4$					
CYANIDE						
Cadmium	5.48	5.12	4.63	3.55		
Copper(I)	log $k_1k_2 = 24$		4.59	1.70		
Nickel	log $k_1k_2k_3k_4 = 31.3$					
Silver	log $k_1k_2 = 21.1$		0.6			
Zinc	log $k_1k_2k_3k_4 = 16.7$					

	log k_1	log k_2	log k_3	log k_4	log k_5	log k_6
ETHYLENEDIAMINETETRAACETIC ACID (EDTA) (as ML)						
Aluminum	16.1					
Barium	7.8					
Bismuth	22.8					
Calcium	10.7					
Cadmium	16.5					
Cerium(III)	16.0					
Cobalt(II)	16.3					
Cobalt(III)	36.0					
Copper(II)	18.8					
Iron(II)	14.3					
Iron(III)	25.1					
Magnesium	8.7					
Lead	18.0					
Mercury(II)	21.8					
Nickel	18.6					
Silver	7.3					
Zinc	16.5					
FLUORIDE						
Aluminum	6.10	5.05	3.85	2.7	1.7	0.3
Beryllium	5.1	3.7	3.8			
Iron(III)	5.2	4.0	2.7			
Thorium	7.7	5.8	4.5			
Titanium (as TiO^{+2})	5.4	4.4	3.9	4.3		
Zirconium	8.8	7.3	5.8			

APPENDIX E ***Conversion Table for U.S. Standard Screen Series***

Mesh	Particle radius (cm)	Mesh	Particle radius (cm)	Mesh	Particle radius (cm)
5	0.20	50	0.015	140	0.0053
10	0.10	60	0.013	170	0.0044
16	0.060	70	0.011	200	0.0037
20	0.042	80	0.0089	230	0.0031
30	0.030	100	0.0075	270	0.0027
40	0.021	120	0.0063	325	0.0022

Answers to Problems

Chapter 2

1. At pH 8, $\alpha_1 = 0.067$; at pH 9, $\alpha_1 = 0.42$; at pH 10, $\alpha_1 = 0.88$.
2. See G. Goldstein, *U.S. AEC Report ORNL* 3620, p. 45.
3. See G. Goldstein, *ibid.*, p. 47.
4. $[Zn^{+2}] = 5 \times 10^{-8}M$.
5. $[Cd^{+2}] = 1.7 \times 10^{-22}M$.
6. $\beta^{0}_{\mathrm{Cd}} = 4 \times 10^{-4}$.
7. Conditional formation constants are 2×10^{10} (pH 4), 1×10^{14} (pH 6), 3×10^{16} (pH 8), 4×10^{17} (pH 9).
8. $K_f' = K_f \alpha_4 \beta_o = 1.8 \times 10^8$; smaller.
9. At pH 8, $[Cd^{+2}] = 8.3 \times 10^{-16}M$; at pH 9, $[Cd^{+2}] = 5.4 \times 10^{-18}M$, and at pH 10, $[Cd^{+2}] = 2.8 \times 10^{-20}M$.
10. The cyanide, sulfide, and amine ligands bind the proton.
11. Combine boron with mannitol to form a nonvolatile compound.
12. Sodium carbonate transforms the uranium(VI) into a soluble carbonate complex while the majority of the traces are isolated in the form of hydroxides or carbonates.

Chapter 3

1. Weight remaining in aqueous phase after each extraction: (a) 0.167 g, 0.0278 g, and 0.0046 g; (b) 0.0625 g.
2. (a) 7; (b) 10.
3. (a) 0.019 mg; (b) 0.04 μg.
4. (a) 99.0%; (b) 90.9%; (c) 50%; (d) 99.99%; (e) 99.2%; (f) 75%.
5. System I has single solute, $K_d = 1.0$. System II has equal amounts of two solutes: $K_d = 0.5$ and 2.0.
6. A dimer exists.
7. $K_f = 750$ (ignoring activity corrections); $K_d = 86.9$ (calculated) and 82.6 (from plot of $1/D$ vs $[I^-]$). See also *J. Am. Chem. Soc.* **74,** 2748 (1952).
8. Log $K_d = 0.30$, $pK_a = 6.0$. See *Anal. Chem.* **37,** 1137 (1965).
9. $K_d = 2.85$, $K_a = 1.09$. See *Anal. Chem.* **35,** 988 (1963).
10. $K_d = 3.24$, $K_a = 1.1 \times 10^{-9}$.

11. $K_d = 20.0$, $K_2 = 793$.
13. $K_d = 27.6$, $K_a = 0.12$. See *Acta Chem. Scand.* **11,** 1771 (1957).
14. $K_d = 2.5 \pm 0.3$; $K_2 = (6.8 \pm 2) \times 10^4$.
15. See *J. Inorg. Nucl. Chem.* **4,** 354 (1957).
16. For HgI_3^-, $k_3 = 5.3 \times 10^3$. For HgI_4^{-2}, $k_{34} = 6.9 \times 10^5$. These stepwise formation constants do not include a correction for the activity coefficients.
17. System A: $[ML_2]_o/[M^{+2}]$ or $[ML_2(HL)_2]_o/[M(HL)_2^{+2}]$. System B: $[ML_2]_o/[ML_3^-]$. System C: $[ML_2 \cdot HL]_o/[M^{+2}]$ or $[ML_2(HL)_2]_o/[M(HL)^{+2}]$. System D: $[ML_2 \cdot HL]_o/[ML_2]$. System E: $[ML_2 \cdot HL]_o/[ML_3^-]$. System F: $[ML_2(HL)_2]_o/[M^{+2}]$.
18. System A: 1,1,1; system B: −2,−2,−2; system C: 1,1,2; system D: 3,3,6; system E: 2,3,3.
19. Log $K_f = 17.2$ for NiL_2; $K_d = 350$. See also *Acta Chem. Scand.* **13,** 50 (1959).
20. (b) The pH ranges are: indium, 1.2 to 2.8; thorium, 3.2 to 4.5; cadmium, 4.8 to 7.3; silver, 5.8 to 10.8. (c) Possible to separate indium from thorium; barely feasible to separate thorium from cadmium; not possible to separate cadmium from silver.
21. (a) For Ga, %E values are 0.20, 0.89, 1.48, 54.8, 96.7, 95.0. For In, %E values are 15.3, 85.2, 98.6, 99.9, 99.4, 93.5. (b) At 3.0M HBr, 1.48% Ga contamination. No. 97% In and 1.8% Ga. (c) 1.5–2.0M HBr. (d) 2.5–3.0M HBr.
22. Log $K_d = 2.77$ for ZnL_2; log $K_f = 3.91$. See *Anal. Chem.* **35,** 1163 (1963).
23. Third power; species extracted is InL_3.
24. Log $K_e = 3.26 \pm 0.12$. See also *Acta Chem. Scand.* **11,** 1277 (1957).
25. (a–b) See graphs in *Ann. N. Y. Acad. Sci.* **53,** 1015 (1951); (c) twelve tubes in both 24 and 100 transfers.
26. Tubes 10 and 20 contain the solute maxima; tubes 12–17 contain both solutes.
27. This is analogous to column partition methods. Separation factor is 2. For equal amounts of each solute, the fractional impurity is about 3×10^{-4}.
28. (a) Tube 12; 15 tubes occupied; 63% of total tubes. (b) Tube 24, 21 tubes, 44%. (c) Tube 49, 30 tubes, 30%. (d) Tube 99, 40 tubes, 21%. Note that percent occupied decreases.
29. (a) 50, (b) 100, (c) 500, (d) 1000.
30. (a) Maximum of three components; (b) at least 10 components if the partition coefficients are properly spaced apart in magnitude.
31. (a) 3741, (b) 1096, (c) 512, (d) 326, (e) 230, (f) 181, (g) 146.

Chapter 4

1. It goes to zero as R approaches 1, and thus vanishes, unlike N, as resolution itself is destroyed by the approach of R toward unity.
2. If the quantity $1 - R$ is near unity, and $Rs = 1$, then $L/d_p \geqslant 32(K_d/\Delta K_d)^2$ or $L/d_p \geqslant 32(R/\Delta R)^2$.
3. No effect when $\nu \ll \nu_{opt}$; H increases with d_p at ν_{opt}; H_{min} increases with d_p at high flow rates.
4. $H/H_{min} = 1/2[(\nu/\nu_{opt}) + (\nu_{opt}/\nu)]$.
5. $Rs/Rs_{max} = 1\sqrt{0.5[(\nu/\nu_{opt}) + (\nu_{opt}/\nu)]}$; extremes of ν/ν_{opt} are 0.5 to 2.
6. The column dead time, the time required for passage of inert, nonadsorbing molecules.
7. $\nu = 1.0 \times 10^{-3}$ cm/sec.
8. $\bar{\nu} = 0.45$, 4.5, 45, and 225.
9. The coupled expression when $\bar{\nu} = 0.45$ and 4.5, and the high-flow expression for the remainder of the flow rates.

10. H = 0.064 cm for ν = 0.001 cm/sec; 0.016 cm at 0.01 cm/sec; 0.017 cm at 0.1 cm/sec; and 0.036 cm at 0.5 cm/sec.
11. D_M = 0.25 cm²/sec in helium, 0.073 cm²/sec in nitrogen, and 0.059 cm²/sec in argon.
12. H = 0.0026 cm for ν = 0.01 cm/sec; and 0.0027 cm at 0.1 cm/sec.
13. (b) H_{min} = 0.155 cm, ν_{opt} = 2.74 cm/sec. (c) From 1.37 to 5.48 cm/sec.
14. (a) Changing a condition such as type of stationary phase to increase Δz through increases in α and k_2; not by increasing N 25-fold which would be cumbersome except with capillary columns. (b) Most logical procedure would be to increase N by a suitable change in conditions such as increasing the column length 1.56-fold.
15. $\Delta T = 0.693\mathscr{R}T^2/\Delta H^o$ where T is the geometric mean of the two temperatures and $\Delta T = T_2 - T_1$. See also *J. Chem. Educ.* **39,** 570 (1962).
16. 30°; range usually falls within 20–40°.
17. $\sqrt{N/16}$.
18. N = 7050.
19. R_1 = 0.500; R_2 = 0.476.
20. (b) A = 0.045 cm, B = 0.475 cm²/sec, E = 0.073 sec. (c) H_{min} = 0.415 cm at ν_{opt} = 2.55 cm/sec. (d) Slopes (in sequence as data was given): 5.5, 2.1, 0.865, 0.68, 0.61, 0.56, 0.53, 0.51 . . . with limit 0.50 at $\nu \to \infty$; $\nu/\nu_{opt} \not> 2.2$ recommended.
21. At 89%, 0.5 to 2; at 77%, 0.33 to 3; at 69%, 0.25 to 4; and at 62%, 0.2 to 5. Actual linear velocity at 89% ranges from 1.28 to 5.1 cm/sec; at 77%, from 0.82 to 7.65 cm/sec.
22. (a–c)

	V'_R			N			
Column	ϕ-NO_2	meta	para	ϕ-NO_2	meta	para	Rs
1	87	142	157	1200	1170	1110	0.87
2	643	1120	1260	1680	1370	1780	1.18
3	760	1363	1548	1690	1700	2000	1.35
4	600	1173	1336	1370	1630	2000	1.37
5	1380	2440	2800	2037	2780	3070	1.95

(d) On column 1, Rν = 2.13 cm/sec for ϕ-NO_2, and 1.3 and 1.2 cm/sec for *m*- and *p*-chloronitrobenzenes. See also *Anal. Chem.* **33,** 30 (1961).
23. See *Anal. Chem.* **37,** 1015 (1965).
24. See *Anal. Chem.* **37,** 1017 (1965).

Chapter 5

1. 10^{21} exchange groups; 10^{19} exchange sites.
2. V_b = 10 ml, V_r = 3.33 ml, V_o = 6.67 ml.
3. 2.2 mEq/ml of bed.
4. 0.83 mEq/ml of bed.
5. (a) K_d(A) = 7, K_d(B) = 21. (b) For α = 2.75, N = 40.
6. Solute A: V_{max} = 53.4 ml and 8 V_o's. Solute B: V_{max} = 147 ml and 22 V_o's.
7. N = 28 for solute A and 25 for solute B. Lengthening the column to 12 cm will provide 30 plates.
8. pK_a = 4.9.
9. H (in cm): 0.00545, 0.0105, 0.0173, 0.0232, 0.0299, 0.0379. For the lowest velocity, N = 19,800. See also *Anal. Chem.* **32,** 1782 (1960).

10. $H = 0.0362$ cm.
11. Since $\alpha = 1.03$, $N = 40{,}000$, and $L = 10.0$ meters.
12. $N = 23{,}000$ and $L = 57.5$ cm.
13. $H = 0.25$ cm, $C_{HCl} = 0.435M$, $(D_v)_K = 6.3$, $(D_v)_{Na} = 4.3$, $\nu = 0.011$ cm/sec (0.73 ml/min), $N = 310$, $L = 78$ cm, $V_b = 86$ ml, $(V_{max})_{Na} = 408$ ml, $(V_{max})_K = 575$ ml, $W_{Na}(4\sigma) = 93$ ml, $W_K(4\sigma) = 130$ ml; make cut at 480 ml; separation time is 15.4 hr.
14. $H = 0.025$ cm, D_v values unchanged from problem 13, $\nu = 0.11$ cm/sec or 7.3 ml/min (a pump or over-pressure probably required), $N = 310$, $L = 7.8$ cm, $V_b = 8.6$ ml, $(V_{max})_{Na} = 41$ ml, $(V_{max})_K = 57.5$ ml; make cut in interval from 46–51 ml; $t = 0.154$ hr or about 9.3 min to reach 67 ml where potassium is 99.9% eluted.
15. Only changes: $C_{HCl} = 0.53M$, $(D_v)_{Na} = 3.64$, $(D_v)_K = 6.91$, $\alpha = 1.8$, $N = 100$, $L = 25$ cm, $V_b = 27.5$ ml, $(V_{max})_{Na} = 110$ ml, $(V_{max})_K = 200$ ml; make cut at 148 ml; $t = 6$ hr.
17. (a) $N = 174$(Tb), 149(Gd), and 167(Eu). (b) $H = 0.013$ cm (from particle size) and 0.015 cm (from average plate number). (c) $Rs = 1.6$(Tb–Gd) and 0.9(Gd–Eu); $N_{req} = 140$(Tb-Gd) and 450(Gd–Eu). (d) $K_d = 54$(Tb), 90(Gd), and 120(Eu).
18. (a) $N = 438(C_3)$ and $465(C_4)$; $K_d = 4.0(C_3)$ and $4.95(C_4)$. (b) $N_{req} = 700$ and $L = 110$ cm. (c) Make cut at <397 ml for C_3 and take C_4 above 407 ml.
19. (a) At 18°C: $N = 49$(Y), 46(La), 92(Ce), and 144(Pr). At 50°C: $N = 41$(Y), 106(La), 110(Ce), and 164(Pr). (b) At 18°C: $Rs = 0.76$(Y–La), 1.77(La–Ce), and 1.04(Ce–Pr). At 50°C: $Rs = 2.7$(Y–La), 1.78(La–ce), and 1.27(Ce–Pr). (c) At 18°C: $N_{req} = 190$(Y–La), 50(La–Ce), and 250(Ce–Pr).
20. For lithium, elution commences at 122 ml and ceases at 158 ml with V_{max} at 140 ml. For sodium the values are 217, 250, and 283 ml. For potassium the values are $N = 500$, 312, 360, and 408 ml. For magnesium the values are 430, 476, and 522 ml.
21. (a) $(D_v)_{Mg} = 3.8$ and $(D_v)_K = 3.9$. (b) $(D_v)_{Mg} = 0.038$ and $(D_v)_K = 0.39$; $\alpha = 2$.
23. With $0.1M$ $NaNO_3$ the separation is excellent. $(D_v)_{Ox} = 5.0$ and $(D_v)_{Br} = 15$ as compared with values of 20 and 30, respectively, when eluting with $0.05M$ $NaNO_3$. Note: N = 122 theoretically but only about 50 experimentally, indicating perhaps too high an eluent velocity.
24. (a) $V_b = 6.2$ ml, $V_r = 1.7$ ml, and $V_o = 4.5$ ml. (b) For tantalum $N = 300, 285$, 240, 190, . . . For niobium, $N = 246$, 178, 109, 118, and 178. (d) $Rs = 1.65$, 1.6, 0.43, 1.4, and 2.5, respectively. (e) $20M$ HF is recommended since resolution exceeds 6σ for reasonable elution times. $12M$ HF is useful if elution of niobium is desired ahead of tantalum.
25. See K. A. Kraus and F. Nelson in *ASTM Spec. Publ.* **195,** 32 (1956).
26. Anionic chloro-complexes are forming with iron(III) and mercury(II) at all HCl concentrations, and with manganese and zinc at [HCl] > $1M$. Nickel forms no chlorocomplexes.
27. In succession, elute Hg^{+2} with $0.5M$ HCl, cut at 12 ml (6 V_o's); elute Cd^{+2} with $1M$ HCl, cut at 10 ml (5 V_o's); elute Zn^{+2} with $2M$ HCl, cut at 8 ml (4 V_o's). Mn, Fe, and Ni would elute together if elution is continued with $2M$ HCl (9–20 ml) without separation into individual bands.
28. See *Anal. Chem.* **34,** 1425 (1962).
29. With $1.5M$ HCl as eluent, nickel is eluted completely at 307 ml (within 0.1%) and aluminum commences at 370 ml. Increase eluent strength to $2.0M$ after the cut to hasten the elution of aluminum. For $1.5M$ HCl, elution of nickel is complete at 243 ml but aluminum starts at 232 ml, a slight overlap.

30. (f) See *Chem. Eng. News* **32** (No. 19) 1898 (1954).
31. (h) 9*M* HCl solution passed through Dowex 1. (i) 1–2*M* HCl solution passed through Dowex 1.

Chapter 7

1. (a) See *J. Chem. Soc.* (1960) 3973. (b) See *Trans. Faraday Soc.* **61,** 2024 (1965).
2. Adsorption increases in order listed.
3. Intramolecular hydrogen bonding is possible with the l-isomer.
4. Chelate formation with surface aluminum atoms; see *J. Chem. Soc.* (1954) 4360 and *ibid.* (1959) 535.
5. Steric hindrance of strongly adsorbed sample group.
6. Presumably the acidity of these compounds is increased when the N—H bond is strained.
7. See *J. Org. Chem.* **22,** 739 (1957); *ibid.* **24,** 1458 (1959).
8. $K_d = 8.35$.
9. (a) $K_d = 73$ ml, (b) $K_d = 16.2$ ml, (c) $K_d = 21.4$ ml.
10. The solute parameter $S^o = 5.17$. Values of K_d for the systems: I, 20.4; II, 35.4, III, 10.5; IV, 1.00; V, 446.
11. (a) $K_d = 40$. (b) $K_d = 7.4$. (c) $K_d = 5.1$. f = 2.3 for acenaphthalene-pentane system.
12. $f = 2.00$.
13. 858 meter^2/g.
14. $V_a = 0.090$ ml/g; $K_d/V_a = 29$ for naphthalene; $\phi = 0.715$ (interpolated from data in Problem 11); $f = 2.3$; and $K_d = 3.9$ for acenaphthalene.
15. Approximately 1 mg of solute per gram of adsorbent.
16. See *Anal. Chem.* **33,** 1527 (1961).
17. (a) For 0.5% water-deactivated gel: $V_a = 0.049$ cm^3/g and $\phi = 0.90$. For 1% gel: $V_a = 0.044$ cm^3/g and $\phi = 0.84$. For 2% gel: $V_a = 0.034$ cm^3/g and $\phi = 0.75$. For 3% gel: $V_a = 0.024$ cm^3/g and $\phi = 0.69$. For 4% gel: $V_a = 0.014$ cm^3/g and $\phi = 0.63$.
18. $(K_d)_{CCl_4}/(K_d)_{benzene} = 23$.
19. Both ends anchored.

Chapter 9

1. (a) Little influence. (b) Alters partition in favor of water. (c) Little difference. (d) Displaces molecules from phenol to water.
2. The imino group of proline is a stronger proton acceptor than the amino group of valine, and phenol is a stronger proton donor than butanol.
3. (a) Limits of separability: propionic acid through *n*-heptanoic acid.
4. (a) The longer-chain fatty acids, being more hydrophobic, will have lower R_f values. (b) Decreased R_f values, allowing extension to longer-chain homologs. (c) Increased R_f values.
5. (b–c) Optimum pH (and pK_a): <5 for phenobarbital (8.3); <6 for free base of perphenazine (3.0, of amine salt); 4.2 for sulfadiazine ($pK_1 \simeq 2$ and $pK_2 \simeq 6.5$).

Chapter 11

1. For benzene: $V_R' = 982$ ml, $V_N = 739$ ml, $V_g = 247$ ml/g, $K_d = 287$.
   For cyclohexene: $V_R' = 856$ ml, $V_N = 637$ ml, $V_g = 214$ ml/g, $K_d = 240$.
   For cyclohexane: $V_R = 642$ ml, $V_N = 475$ ml, $V_g = 159$ ml/g, $K_c = 176$.

2. $N = 525$, $H = 0.175$ cm.
3. For cyclohexane/cyclohexene: $\alpha = 1.33$; at 4σ, $N = 200$ and $L = 35$ cm; at 6σ, $N = 400$ and $L = 70$ cm. For cyclohexene/benzene: $\alpha = 1.15$; at 4σ, $N = 1450$ and $L = 252$ cm; at 6σ, $N = 2500$ and $L = 490$ cm.
4. (a–b)

Column	Flow rate (liters/hr)	Linear velocity (cm/sec)	Plate number	Plate height (cm)
31% substrate	1	1.65	1060	0.340
	2	3.10	1155	0.310
	4	6.17	890	0.405
	6	8.68	720	0.500
	10	13.46	515	0.705
23% substrate	1	1.43	1075	0.335
	2	2.77	1480	0.245
	3	4.03	1664	0.210
	4	5.20	1650	0.217
	5	6.24	1580	0.228
	10	10.60	1190	0.302
13% substrate	2	2.49	1460	0.246
	4	4.78	2135	0.170
	6	7.15	2260	0.160
	10	9.32	2150	0.167

(d) For 31% substrate: $A = 0.09$, $B = 0.26$, $E = 0.042$. (e) For 31% substrate, $H_{opt} = 0.29$ cm at $\nu = 2.65$ cm/sec. For 23% substrate, values are 0.218 cm and 5.8 cm/sec. For 13% substrate, values are 0.160 cm and 7.25 cm/sec.

5. On 31% column: $K_d = 15.1 \pm 0.5$, $k = 5.75$.
   On 23% column: $K_d = 16.2 \pm 0.2$, $k = 3.60$.
   On 13% column: $K_d = 18.8 \pm 0.3$, $k = 2.67$.
6. (a) The larger mesh size increases the A term to about 0.3 cm from about 0.07 cm for the 30/50 mesh. (b) Plate heights are less with nitrogen as carrier gas: With hydrogen, $H_{opt} = 0.34$ cm (C_4) and 0.25 cm (C_3), and $\nu_{opt} = 1.58$ cm/sec (C_4) and 1.33 cm/sec (C_3). With nitrogen, $H_{opt} = 0.12$ cm (C_4) and 0.13 cm (C_3), and $\nu_{opt} = 1.60$ cm/sec (C_4) and 1.84 cm/sec (C_3).
7. (a) See Fig. 11-22. (b) $H_{min} = 0.345$ cm at $\nu = 2.45$ cm/sec.
8.

Column		1	2	3	4
K_d	1-MeN	5660	3100	540	154
	2-MeN	5035	2760	485	140
k	1-MeN	36.5	61.2	4.86	66.5
	2-MeN	32.4	54.5	4.40	60.6
N	1-MeN	27,500	16,200	12,650	13,200
	2-MeN	19,400	15,500	12,900	15,000
Phase ratio		155	50.6	11.1	2.30
Resolution		4.5	3.7	3.06	3.3

9. Approximately 14,000 plates.
10. In all cases, $Rs = 1.5$. For case 1, N is 4360 to 6240 (limit). For case 2, $N = 39{,}200$. For case 3, $N = 7740$.
11. (b) An isothermal column temperature in the interval 144–162°C (or in the low-temperature region around 50°C). About 8500 plates required since $\alpha = 1.045$ for the least satisfactory separation. (c) $V_g = 45$ ml/g for all the n-alkanes. (d) $\Delta H = 6$ kcal/mole.

12. $\alpha = 1.17$ for $\eta = 0.01$ and $N = 800$. (a) Column temperature must be 139°C or lower. (b) Temperature may not exceed 240°C. (c) Above 152°C. (d) Isothermal column temperature should not exceed 220°C (excluding ethylbenzene) or 157°C (including ethylbenzene).
13. A, *n*-hexane; B, *n*-butylbenzene; C, toluene; D, *n*-octane; E, *n*-pentane; F, ethylbenzene; G, *n*-heptane; H, *n*-propylbenzene.
14. V'_R (in ml) on Convoil-20 and tricresylphosphate: A, 177 and 71.5; B, 3157 and 4420; C, 670 and 714; D, 862 and 306; E, 77 and 34; F, 1348 and 1326; G, 400 and 150; H, 2618 and 2386.
15. *n*-Alkanes through C_6. *n*-Alkanes through C_9.
16. (c) For *n*-butyl acetate, $V'_R = 190$ ml (Carbowax) and 320 ml (Nujol). For *n*-amyl alcohol, $V'_R = 800$ and 210 ml.
17. See *Anal. Chem.* **34,** 477 (1962).
18. (a) Assuming $N = 900\text{–}1200$, α must exceed 1.15–1.17 for separations within 1%. The paraffin wax column, operating at all temperatures, effected excellent separation and resolution; the higher temperatures would shorten the time of each run. (b) On a TCP column, the pair CCl_3CH_3/CCl_4 would cause difficulty unless a capillary column with $N > 15{,}000$ were available. The temperature should be kept below 97°C to avoid coalescence of many peaks on Carbowax 4000. The compounds $CHCl_2CH_3/CCl_3CH_3/CCl_4$ would be difficult to separate without a column possessing about 20,000 plates.
19. (a) $\Delta H = 12.5$ kcal/mole. (b) $t'_R = 4.95$ min at 150°C, 1.01 min at 200°C, and 0.52 min at 225°C.
20. (c) For $\eta = 1\%$ at 138° (and 175°), N_{req} are the following: 80 (100) for 1,5-, 1,6- and 1,7- from each other; 2400 (1800) for DETA/1,6-; 2400 (4700) for TEDA/1,5-; 45 (70) for DETA/TEDA. (d) $\alpha = 1.17$ ($\eta = 1\%$), TEDA separated from 1,5- and 1,4- between 80°–130°C; DETA from 1,7- and 1,6- at 220°C.
21. (a) Peak 4 is ethyl adipate; peak 5 is ethyl azeleate. (b) For peak 6, $N = 3550$ and $H = 0.86$ mm.
22. A family of parallel straight lines, one for each homologous series, is obtained.
23. A, 14:0; B, 16:1; C, 12:0; D, 14:1; E, 16:2; F, 22:0; G, 16:3; H, 17:0; I, 16:4; J, 24:0; K, 10:0.
24. A (0.73, 0.65); B (3.8, 1.9) these are estimated from a line drawn parallel to X:3 family and spaced apart the average distance of each homologous family; C (6.3, 3.6); D (6.0, 13.4); E (6.6, 11).
25. Beginning with the first strong peak following the air peak, the components are caprylate, caprate, laurate, myristate, palmitoleate, stearate, oleate, linoleate and linolenate.
26. (a) $N = 307$ (isopentane), 432 (pentene-1) and 260 (2-methyl butene-2). (b) Since $\alpha = 1.24$, $N = 650$. (c) No difficulty since $\alpha = 1.3$ and 300 plates should suffice. $Rs = 1.5$.
27. (b) At 25°C, $N = 1620$, 1920, 1990, and 2300. At 16°C, $N = 1720$, 2040, 2120, and 2400. (c) At 25°C, $Rs = 3.13$, 1.62, and 2.65. At 16°C, $Rs = 3.33$, 1.75, and 2.79.
28. (b) $N = 330$, 485, 1020, 1350, and 1680 in order of appearance.
29. The terminal diols (peaks 2, 4, and 6) fall on a linear graph of log t_R vs carbon number. Peaks 3 and 5 are the 1,2-C_3 and 1,3-C_4 diols, respectively.
30. Peaks 2, 4, 6, and 9 are *n*-C_1 to C_4 formate esters; peaks 3, 5, 8, and 12 are the *n*-C_1 to C_4 acetates; peaks 7 and 11 are the iso-C_3 and iso-C_4 acetates; and peak 10 is *sec*-butyl acetate.
31. Peaks 1, 2, 4, 8, 11, and 14 are *n*-C_1 to C_6 alcohols; peaks 3, 6, 9, and 11 are the 2-OH alcohols; peaks 7, 10 and 13 are 2-ME-1-OH alcohols; peak 5 is *t*-

butanol and peak 12 is 4-methyl-2-pentanol, both isolated members of their families.

32. (left to right): 1311, 690, 659, 731, 663, and 446.
33. (left to right): 152, 654, 10, 49, 57, 1023, and 640.
34. C-14, 2.4%; C-16, 4.4%; C-18, 0.8%; C-20, 6.5%; C-22, 4.0%; and methyl ricinoleate (hydroxy oleate), 81.7%.
35. Ethyl benzene, 1470; *p*-xylene, 975; *m*-xylene, 1175; *o*-xylene, 760.
36. As R increases through the series: 0, 0.1, 0.25, 0.5, and 1, the plate number must be increased in the ratio: $1:1.2:1.8:4:\infty$. Above $R = 0.5$, resolution deteriorates rapidly.
37. When $k = 0.25$, $t_R = 2.50$ units. The fraction traveled by the band in unit time is 0.40 of the column. When $k = 4.0$, $t_R = 10$ units and the distance traveled per unit time is 0.10 of the column. In the first case, the gas/liquid concentration ratio is 4:1 at each unit of column length; in the second case, the ratio is 1:4.
38. Retention times are 2.3, 3.045, and 4.375 units.
39. See *J. Gas Chromatog.* **4,** 163 (1966).

Chapter 12

1. For G-10, $V_i = 1.0$ ml, $V_s = 1.6$ ml, $V_o = 0.9$ ml.
   For G-50, $V_i = 5$ ml, $V_s = 5.6$ ml, $V_o = 4.4$ ml.
   For G-200, $V_i = 20$ ml, $V_s = 20.6$ ml, $V_o = 9.4$ ml.
2. For P-200, $V_i = 13.5$ ml, $V_s = 14.1$ ml, $V_o = 33$ ml.
3. $V_i = 15$ ml.
5. For raffinose: $V_e/V_o = 1.70$ and $K_{av} = 0.30$. For maltose: $V_e/V_o = 1.87$ and $K_{av} = 0.37$. For glucose: $V_e/V_o = 2.08$ and $K_{av} = 0.46$. For KCl: $V_e/V_o = 2.36$ and $K_{av} = 0.58$.
6. For raffinose: $V_e/V_o = 1.23$ and $K_{av} = 0.092$. For KCl: $V_e/V_o = 1.80$ and $K_{av} = 0.32$.
7. For leucine: $V_e/V_o = 1.18$ and $K_{av} = 0.13$. For NaCl: $V_e/V_o = 1.37$ and $K_{av} = 0.27$. From column dimensions, $V_t = 73.5$ ml; 19.4 g gel was present. $V_i = 31$ ml, $V_s = 11.6$ ml, and $V_o = 31$ ml. $\Delta V_e = 6.0$ ml (observed).
8. For raffinose-maltose: $\Delta V_e = 8.0$ ml (experimental) and 7.5 ml (calculated). For maltose-glucose, values are 9.5 and 9.7 ml. For glucose KCl, both values are 13 ml.
9. On P-30, $V_e = 63$ ml; on P-60, $V_e = 68$ ml.
10. On G-75: $K_{av} = 0.33$ and $V_e = 6.87$ ml per gram of dry gel ($V_r = 13$ ml).
    On G-100: $K_{av} = 0.50$ and $V_e = 11.5$ ml/g dry gel ($V_r = 17$ ml).
    On G-200: $K_{av} = 0.70$ and $V_e = 24.5$ ml/g dry gel ($V_r = 30$ ml).
12. Since ΔV_e is assumed to be 6 ml, K_{av} must be 0.29. From Fig. 12-4, ΔK_{av} of component with molecular weight of 20,000 is 0.70, the second component must possess a molecular weight in excess of 80,000.
13. $\Delta K_{av} = 0.088$; components with molecular weights 12,000 or less and 28,000 or more could be separated.
14. For $V_b = 30$ ml, $\Delta K_{av} = 0.32$ and a molecular weight of 6,600 could be separated from 20,000. For $V_b = 100$ ml, $\Delta K_{av} = 0.1$ and molecular weights 15,000 or less and 30,000 or more could be separated.
15. The wider fractionation range (up to a molecular weight of 1500) is achieved at the cost of decreased resolution.
17. Bed volume should be at least 4 times sample volume or 40 ml, equivalent to 8 g of dry gel. $V_o + V_i = 35$ ml, the elution volume of the salt, and $V_s = 24$ ml. Assuming $\Delta V_e \simeq V_s$ (since $K_d = 1$ for the salt and $K_d = 0$ for the other com-

pounds), the peak separation is 24 ml and the salt-free solution should emerge around 11 ml.

18. See *Biochem. J.* **91,** 222 (1964).
19. See *J. Polymer Sci.* (Part C) (1965) 233.
20. See *Biochem. J.* **91,** 222 (1964) and *J. Chromatog.* 14, 317 (1964).
21. See *Petroleum Refiner* **35,** (7) 175 (1956).
22. See Thomas and Mays in *Physical Methods in Chemical Analysis,* W. G. Berl (Ed.), Vol. IV, Academic Press, New York, 1961, p. 93.
23. (a) Type 4A, (b) type 5A, (f) type 13X, (h) either types 5A or 13X.

Chapter 13

1. Place the amine-HCl in the cathode compartment and separated from a dilute HCl solution in the anode compartment by an anion-exchange membrane.
2. See V. A. Zarinskii *et al., Zh. Anal. Khim.* **12,** 577 (1957).
3. Place the borate ions (in NaOH solution) in the anode compartment and separated from a dilute NaOH solution in the cathode compartment by a cation-exchange membrane. See M. J. Owers, *U.K. AERE Rept. 3010* (1959); *Anal. Abstr.* **8,** 2719 (1961).
4. Adjust the pH of the solution to the isoelectric point, add 0.2*M* NaOH to the anode compartment and 0.1*M* sulfuric acid to the cathode compartment of a three-compartment cell. Place the solution in the central compartment with a cation-exchange membrane isolating it from the cathode compartment and an anion-exchange membrane isolating it from the anode compartment.
5. See E. Blasius and G. Lange, *Z. Anal. Chem.* **160,** 169 (1958); *Chem. Tech.* (Berlin) **10,** 524 (1958).
6. See L. C. Craig and A. O. Pulley, *Biochemistry* **1,** 89 (1962).
7. See *Encyclopedia of Chemical Technology,* R. E. Kirk and D. F. Othmer (Eds.), Wiley–Interscience, New York, 1965, Vol. 7, p. 861.

Chapter 14

2. A: pH = 7.6. B: pH = 9.3. C: pH = 8.8. D: Separation depends only on mobility difference as pK_a values are virtually the same.
3. This results in the separation of the amino acids into an unresolved neutral group, a basic group in which lysine, histidine, and arginine are separated, and an acidic group in which glutamic, aspartic, and cysteic acids are separated.
4. A: 4.88, B: 2.48, C: 3.38, D: 6.36 (in units of 10^{-4} cm^2 V^{-1} sec^{-1}).
5. Almost 1 liter.
6. A: Lysine and arginine move toward the cathode; histidine remains stationary. B: Aspartic acid moves faster toward the anode than glutamic acid.
7. Glycine: 0.95×10^{-4} cm^2 V^{-1} sec^{-1}. Serine: 1.43×10^{-4} cm^2 V^{-1} sec^{-1}.
8. All exist as cations; the mobility decreases in the order enumerated.
9. Towards the anode: *n*-propylamine, 14 cm; ethylamine, 17 cm; methylamine, 20 cm.
10. A: 26.3 min; B: 53.6 min.
11. Lysine moved 22 cm towards the cathode; histidine moved only 7.8 cm in the same direction.
12. Adenosine 3′-phosphate: 0.925×10^{-5} cm^2 V^{-1} sec^{-1}.
13. See *Anal. Chem.* **27,** 424 (1955).
14. Corrected values of mobility (in units of 10^{-5} cm^2 V^{-1} sec^{-1}): vanillin, 0.16; ferrulic acid, 7.24; vanillic acid, 9.3.

15. 4.075 liters.
18. $pI_e = 5.2$.
21. Towards the cathode are located ornithine, Na^+ and K^+, and towards the anode are located aspartic acid, phosphate, sulfate, and chloride ions (in order of increasing distance from the origin). The desalted area lies between the positions of Na^+ and $H_2PO_4^-$ on either side of the starting zone.
22. In units of 10^{-5} cm^2 V^{-1} sec^{-1}, the corrected apparent mobilities are guaiacol, −4.9; acetovanillone, +0.95; vanillin, 7.6; and vanillic acid, 17.6.
23. See *Anal. Chem.* **23,** 994 (1951).

Chapter 15

1. See W. Hückel, *Structural Chemistry of Inorganic Compounds,* Vol. II, Elsevier, Amsterdam, 1951, p. 477.
2. Cover precipitate with layer of ammonium iodide and heat gently; SnI_4 is volatile.
3. See *J. Am. Chem. Soc.* **57,** 2565 (1935).
4. Fluoride forms BF_4^-, but AlF_6^{-3} is more stable and releases the boron.
5. See *J. Chem. Soc.* (1950) 298, 300, 408.
6. See *Ind. Eng. Chem.* **47,** 219 (1955).
7. See *J. Nutrition* **53,** 461 (1954).
8. See *Ind. Eng. Chem.* **47,** 219 (1955).
9. See *Anal. Chem.* **26,** 342 (1954).

Subject Index

Common Logarithms

N	0	1	2	3	4	5	6	7	8	9
10	0000	0043	0086	0128	0170	0212	0253	0294	0334	0374
11	0414	0453	0492	0531	0569	0607	0645	0682	0719	0755
12	0792	0828	0864	0899	0934	0969	1004	1038	1072	1106
13	1139	1173	1206	1239	1271	1303	1335	1367	1399	1430
14	1461	1492	1523	1553	1584	1614	1644	1673	1703	1732
15	1761	1790	1818	1847	1875	1903	1931	1959	1987	2014
16	2041	2068	2095	2122	2148	2175	2201	2227	2253	2279
17	2304	2330	2355	2380	2405	2430	2455	2480	2504	2529
18	2553	2577	2601	2625	2648	2672	2695	2718	2742	2765
19	2788	2810	2833	2856	2878	2900	2923	2945	2967	2989
20	3010	3032	3054	3075	3096	3118	3139	3160	3181	3201
21	3222	3243	3263	3284	3304	3324	3345	3365	3385	3404
22	3424	3444	3463	3483	3502	3522	3541	3560	3579	3598
23	3617	3636	3655	3674	3692	3711	3729	3747	3766	3784
24	3802	3820	3838	3856	3874	3892	3909	3927	3945	3962
25	3979	3997	4014	4031	4048	4065	4082	4099	4116	4133
26	4150	4166	4183	4200	4216	4232	4249	4265	4281	4298
27	4314	4330	4346	4362	4378	4393	4409	4425	4440	4456
28	4472	4487	4502	4518	4533	4548	4564	4579	4594	4609
29	4624	4639	4654	4669	4683	4698	4713	4728	4742	4757
30	4771	4786	4800	4814	4829	4843	4857	4871	4886	4900
31	4914	4928	4942	4955	4969	4983	4997	5011	5024	5038
32	5051	5065	5079	5092	5105	5119	5132	5145	5159	5172
33	5185	5198	5211	5224	5237	5250	5263	5276	5289	5302
34	5315	5328	5340	5353	5366	5378	5391	5403	5416	5428
35	5441	5453	5465	5478	5490	5502	5514	5527	5539	5551
36	5563	5575	5587	5599	5611	5623	5635	5647	5658	5670
37	5682	5694	5705	5717	5729	5740	5752	5763	5775	5786
38	5798	5809	5821	5832	5843	5855	5866	5877	5888	5899
39	5911	5922	5933	5944	5955	5966	5977	5988	5999	6010
40	6021	6031	6042	6053	6064	6075	6085	6096	6107	6117
41	6128	6138	6149	6160	6170	6180	6191	6201	6212	6222
42	6232	6243	6253	6263	6274	6284	6294	6304	6314	6325
43	6335	6345	6355	6365	6375	6385	6395	6405	6415	6425
44	6435	6444	6454	6464	6474	6484	6493	6503	6513	6522
45	6532	6542	6551	6561	6571	6580	6590	6599	6609	6618
46	6628	6637	6646	6656	6665	6675	6684	6693	6702	6712
47	6721	6730	6739	6749	6758	6767	6776	6785	6794	6803
48	6812	6821	6830	6839	6848	6857	6866	6875	6884	6893
49	6902	6911	6920	6928	6937	6946	6955	6964	6972	6981
50	6990	6998	7007	7016	7024	7033	7042	7050	7059	7067
51	7076	7084	7093	7101	7110	7118	7126	7135	7143	7152
52	7160	7168	7177	7185	7193	7202	7210	7218	7226	7235
53	7243	7251	7259	7267	7275	7284	7292	7300	7308	7316
54	7324	7332	7340	7348	7356	7364	7372	7380	7388	7396
N	0	1	2	3	4	5	6	7	8	9